MW00351985

THE GEORGE FISHER BAKER
NON-RESIDENT LECTURESHIP
IN CHEMISTRY AT
CORNELL UNIVERSITY

BORANES IN ORGANIC CHEMISTRY

By Herbert C. Brown

BORANES IN ORGANIC CHEMISTRY

by HERBERT C. BROWN

R. B. Wetherill Research Professor
Purdue University

Cornell University Press | ITHACA AND LONDON

Copyright © 1972 by Cornell University

All rights reserved. Except for brief quotations in a review, this book, or parts thereof, must not be reproduced in any form without permission in writing from the publisher. For information address Cornell University Press, 124 Roberts Place, Ithaca, New York 14850.

First published 1972 by Cornell University Press.
Published in the United Kingdom by Cornell University Press Ltd.,
2–4 Brook Street, London W1Y 1AA.

International Standard Book Number 0–8014–0681–1
Library of Congress Catalog Card Number 79–165516

COMPOSED BY KINGSPORT PRESS, INC.
PRINTED IN THE UNITED STATES OF AMERICA
BY VAIL-BALLOU PRESS, INC.
Librarians: Library of Congress cataloging information
appears on the last page of the book.

H
546.67
B878b
163058

TO MY WIFE—

who made the writing of
my previous book a pleasure
and the writing of the
present one a necessity

PREFACE

Early Baker Lectures played a major role in the initial phases of my professional career. (Two such instances are discussed in Chapter II.) Consequently, it was with special pleasure that I accepted the invitation extended by Professor Robert A. Plane to serve as the George Fisher Baker Non-Resident Lecturer at Cornell University in the fall of 1969. My wife and I spent an exceedingly pleasant eight weeks in Ithaca in October and November of that year.

At the time I also promised to prepare these lectures for publication in book form. This promise was made in good faith. Unfortunately, many recent Baker Lecturers have found it difficult to live up to such commitments, and I also found it difficult to set about the task of putting my lectures into publishable form.

My wife, however, proved to have a more demanding conscience than my own, and she made it clear that this was an obligation I could not avoid. I cleared my desk of all other commitments and turned my full attention to this book. (I hereby offer my apologies to all those whose letters, manuscripts, research proposals, and the like received deferred attention and action.)

As it turned out, the delay may have been beneficial, for it led to a significant alteration in my original plans. The Baker Lecturer is usually invited to emphasize the contributions of his own group to research topics that have interested him. Originally I had intended to follow this practice by presenting the usual formal reviews of such topics. Certain recent developments, however, convinced me that it might be more helpful to adopt another approach.

For approximately twenty years we in science have received generous and ever-increasing financial support. New students with Ph.D.'s have been eagerly sought after. Recently, this situation underwent an abrupt change. Research funds became more difficult to obtain and good research positions became quite scarce. The shock this change produced in the younger men, inexperienced in facing such problems, brought home the realization that in recent years our system of chemical education has failed to provide students with any background in the past. They knew nothing about how earlier generations faced and surmounted problems of the kind they suddenly encountered.

I decided to try a new approach. I would undertake to provide the reader with more than an exposition of scientific accomplishments. I would try to give him some feeling for the circumstances under which research was carried out, the manner in which programs were initiated, and the factors that influenced their development.

Such a book might help the younger members of our profession gain perspective, sharing vicariously the educational experience to which we were subjected. Perhaps the historical background will help them in meeting and overcoming their current problems. At the same time older members of the profession may find it interesting to relive some of their own early trials and difficulties. It will be of interest to me to learn the reaction of readers to this experiment in chemical communication.

I wish to express my deep appreciation to the staff at Cornell University who extended the invitation and whose hospitality made our stay in Ithaca most memorable. I also want to express my indebtedness to N. Ravindran, who made the drawings for the diagrams, to Annette Wortman, who typed the manuscript, and to Dr. Ei-ichi Negishi and Dr. Chang Jung Kim, who provided generous assistance with the preparation of the manuscript for the publisher, the proofreading, and the preparation of the index.

HERBERT C. BROWN

Department of Chemistry
Purdue University

CONTENTS

Contents

PART TWO: STERIC STRAINS

Contents

PART FIVE: HYDROBORATION

PART SIX: ORGANOBORANES

Contents

PROLOGUE

I. Autobiographical Introduction

1. This Book—An Experiment in Research Education

The holder of the George Fisher Baker Non-Resident Lectureship in Chemistry at Cornell University is invited to present his own contributions to a particular area of chemistry in a series of lectures and then to publish those lectures in book form for more general dissemination. In the thirties and forties the Baker Lectures regularly appeared in book form and made a major contribution to the advance of chemistry. More recently, however, publication of the lectures has become a relatively rare occurrence. Possibly some Lecturers, appalled by the present massive surge of publications, have felt reluctant to add to the flood.

The present author delivered the Baker Lectures at Cornell University in October and November of 1969. At that time I discussed various topics that my students and I had explored over many years, topics such as "Steric Strains," "The Nonclassical Ion Problem," "Selective Reductions," "Hydroboration," and the "Organoboranes as Synthetic Intermediates"; but I also had doubts about adding to the flood by reproducing these lectures in book form. Such lectures could only be relatively brief reviews of areas of investigation. Perhaps more detailed monographs on each of the subjects would be more useful and desirable.

At Cornell I noted that the students appeared particularly interested in personal accounts of the influences that led to the initiation of certain research programs and the unexpected experimental results that led to the initiation of other research programs. These are the details that one usually excludes from printed accounts of one's investigations. Perhaps an account of my investigations over the past thirty-five years that did not exclude these personal touches would be of value to present-day students in showing how one can open up major new research areas by recognizing the significance of minor anomalies in the experimental results.

This feeling was fortified by an experience at the M. S. Kharasch Memorial Symposium, held in September 1970 at the National Meeting of the American Chemical Society at Chicago. On the spur of the moment I took a few minutes at the start of my lecture to describe the conditions under which academic chemical research operated when I was carrying on my doctoral research with Professor H. I. Schlesinger (1936–1938) and my postdoctoral research with Professor M. S. Kharasch (1938–1939). After the lecture many individuals told me how much they had enjoyed and appreciated this glimpse into the past.

Perhaps in the effort to bring our students to a level of competence in the leading edge of current chemical progress, we have been denying them significant contact with the past. Perhaps if they had more experience with the difficulties under which we operated in the past— difficulties which did not prevent the discovery and exploration of major new areas—they would feel more optimistic about the future possibilities for chemistry.

These experiences and considerations persuaded me to undertake an experiment. I would write up the Baker Lectures in an informal personal vein, attempting to convey the thoughts and considerations that led my associates and me to undertake each research program and to direct it along the particular course that it took. Accordingly, a principal objective of this book was made the tracing of my research interests and activities over the past thirty-five years. Perhaps I could perform an especial service to the younger generation of chemists by transmitting my experiences in encountering new phenomena and ideas and in developing them into major new research areas.

The Baker Lecturer is usually asked to emphasize his own contributions in his presentation. As it happens, the special objective I set myself has led me to emphasize even more strongly than I otherwise would the contributions of my own research group—a kind of scientific autobiography, if you will. There has been no attempt to provide balanced reviews. It is my hope that as scientific biography this book will prove of greater value to the young developing chemist, and of greater interest to the mature chemist, than just another book of balanced reviews on a number of current topics.

2. Experiences in the 1930's

I entered Crane Junior College in Chicago in February 1933. After one semester, because of the Depression, it was closed. In this crisis Dr. Nicholas D. Cheronis, who had been an instructor at Crane, made available to several students, including myself, his own private laboratory (the Synthetical Laboratories), operated in a small building to the rear of his home. There we kept ourselves occupied doing experiments, sometimes carrying out syntheses that the Laboratories required, but more generally doing such experiments as interested us individually.

I signed up for a correspondence course in Qualitative Analysis at the University of Chicago Home Study Division and performed the required chemical analyses at the Laboratories. On one occasion I was exasperated because a given unknown (supposed to be phosphate free) failed to behave properly. I called up the instructor listed for the course, one Julius Steiglitz, and discussed my problem with him. He deduced that my unknown must be contaminated by phosphate and promised to see that a more suitable unknown would be sent out to me. (I later discovered that Professor Steiglitz was the Chairman of the Department at the University of Chicago, and he later persuaded me to undertake study for the Ph.D. degree.)

I also took Fales' *Quantitative Analysis* and worked my way through it independently. Later, when I reached the University of Chicago, I showed my laboratory notebooks to Professor W. Conway Pierce and he saw to it that I received full credit for the two undergraduate Quantitative Analysis courses.

In September 1934 Wright Junior College opened its doors, and I entered, graduating with the first class in June 1935.

I was persuaded to try for a competitive scholarship at the University of Chicago during the spring of 1935. All my course work had been in chemistry, physics, mathematics, English, and German. This was before the days of a "general education," a "new" concept emphasized strongly at the time at the University of Chicago. I was astounded at the examination to discover that very little of it dealt with the scientific subjects I had studied. Instead, the examination dealt primarily with history, literature, philosophy, music, art, and the like. I could only grit my teeth and do what I could. To my amazement I was awarded a scholarship that made it possible for me to attend the University of Chicago.

This was a revolutionary period at the University of Chicago, where the President, Dr. Maynard L. Hutchins, was attempting to bring about major alterations in the undergraduate program. One of his pet hobbies was to encourage students to pass through the undergraduate curriculum as rapidly as possible. The tuition at that time was $100 per quarter. The normal load was three courses per quarter. Yet one could take as many courses as one wished without additional payment. A great bargain! Accordingly, I signed up for as many as ten courses per quarter in order to complete both my junior and senior years in three quarters, by June 1936. This I succeeded in doing.

I did not apply for admission to Graduate School or for an assistantship. My ambitions ran only to obtaining a B.S. degree in order to obtain a job as a chemist and begin making a living. (I had a girl friend—one of my classmates, Sarah Baylen, now my wife—and in those days one did not get married without a livelihood.)

Professor Julius Steiglitz altered those plans. I was taking his magnificent set of courses in Advanced Organic Chemistry. He had the custom of lecturing for five to ten minutes, then stopping to ask a question and to call for a volunteer. I was a brash young man, and invariably I volunteered. Apparently, my eagerness impressed him. One day he called me into his office to inquire why I was not applying for graduate study. When I explained my financial situation and plans, he urged me to undertake study toward the Ph.D. degree and offered to make available an assistantship, even though it was past the deadline for applications. I accepted his offer.

For those graduate assistants who are unhappy about present-day stipends, I should say that the stipend was $400 for three quarters. Out of this I had to pay $300 for tuition and the costs of my research!

As a graduation gift, Sarah Baylen had presented me with a copy of Alfred Stock's *Hydrides of Boron and Silicon*, published by Cornell University Press in 1933. (As I write this, I note that the price is marked as $2.00. How times have changed!) This book interested me in the boron hydrides, and I decided to undertake research with Professor H. I. Schlesinger in this area. I started with a study of the reaction of diborane with organic compounds containing a carbonyl group (Chapter II). Thus, the Baker Lectures of 1932 influenced me to undertake studies in the borane area, largely the subject of the present volume. On another occasion another set of Baker Lectures was responsible for one of the most pleasant experiences of my life. This will be discussed in Chapter III.

I received my Ph.D. degree in 1938. The economic situation was still quite poor. An academic position was too much to hope for. (As far as I know, not a single man from the Department of Chemistry at the University of Chicago obtained an academic position during my years there, 1935 to 1938.) Even industrial positions were rare. I heard of an opening at the Sherwin-Williams Paint Company and applied—unsuccessfully. I then applied for a position in the patent department of Universal Oil Products, again without success.

My every effort to avoid my future in an academic career was foiled.

At this point Professor M. S. Kharasch offered me a postdoctoral appointment in his group at an annual stipend of $1,600, and the die was cast. Originally I was supposed to work with him on the isolation of an active principle from pituitary glands, a far cry from my doctoral studies. Fortunately for me, he encountered difficulties in his negotiations with one of the packing houses for a gift of $2,000 worth of glands. He suggested that while these difficulties were being resolved I should explore the possibility of using sulfuryl chloride to achieve the chlorosulfonation of paraffinic hydrocarbons. These studies proved so fruitful (Chapter III) that I was not diverted from my studies even after the glands arrived.

3. Initial Stages of Career

In 1939 an academic opening developed at Armour Institute of Technology (now Illinois Institute of Technology). I applied. In due course I received a very nice letter from the Chairman, Professor Benjamin Freund. He informed me that I had received very high recommendations and had an exceptionally good record. Indeed, he had been forced to the conclusion that I was too good for the Institute. Apparently, Professor Freund did not feel the same way about the late Saul Winstein, since he received the appointment.

In 1951 I spent the summer at U.C.L.A. as Visiting Professor.

Prologue

In my final lecture I told this little story to my audience, which included Professor Saul Winstein. It brought the house down.

At the time, Armour Institute did not have a doctoral program. It may indicate something of the competition for academic positions at that time that Saul Winstein, with his outstanding record and background (M.S. from U.C.L.A., Ph.D. from California Institute of Technology, National Research Council Postdoctorate Fellow at Harvard University), should have been willing to accept appointment to this position of Instructor. Of course, he did not remain there long, but shortly returned to his alma mater, U.C.L.A., where he remained until his untimely death in 1969.

In recognition of his duties as Executive Secretary of the Department of Chemistry at the University of Chicago, Professor H. I. Schlesinger was provided with a research assistant, carrying the rank of Instructor. The man occupying this post did the usual academic teaching, helped Professor Schlesinger in directing his research students, and carried on research on Professor Schlesinger's problems. He did have the opportunity to attract and direct the research of M.S. candidates and in this way could carry on some independent research. This post had been held by Anton B. Burg. In 1939 he left to become Assistant Professor at the University of Southern California, and Professor Schlesinger offered the post to me at a salary of $2,400. I accepted.

Just prior to this time, my laboratory partner, R. T. Sanderson, had synthesized aluminum borohydride, $Al(BH_4)_3$,[1] and Burg had synthesized beryllium borohydride, $Be(BH_4)_2$.[2] Accordingly, it appeared appropriate to undertake the synthesis of the first alkali metal borohydride, lithium borohydride, $LiBH_4$. The reaction of diborane with ethyllithium provided the desired compound, and its chemistry was explored.[3]

We were well on our way through our next project, an exploration of the synthesis of gallium borohydrides, when a development occurred that appeared to rule out research in this area for some time.

[1] H. I. Schlesinger, R. T. Sanderson, and A. B. Burg, *J. Amer. Chem. Soc.*, **62**, 3421 (1940).

[2] A. B. Burg and H. I. Schlesinger, *J. Amer. Chem. Soc.*, **62**, 3425 (1940).

[3] H. I. Schlesinger and H. C. Brown, *J. Amer. Chem. Soc.*, **62**, 3429 (1940).

In 1940 Professor Schlesinger was advised that it was important for the national defense to discover volatile compounds of uranium other than uranium hexafluoride. He was asked to assemble a group to investigate the possibility, and he in turn asked me to give up my normal research activities and to assist him with the organization and direction of the group. Initially we undertook to prepare inner complexes of uranium containing 1,3-dicarbonyl chelating agents. The parent compound, the acetonylacetonate, exhibited a volatility too low to meet the specifications. We then prepared the trifluoromethyl chelating agent, $CH_3COCH_2COCF_3$. Its uranium derivative was more favorable.[4] We were not far from preparing the bistrifluoromethyl derivative, $CF_3COCH_2COCF_3$, when the blow fell. We were informed that the molecular weight of the compound was an important consideration. The compound $U(CF_3COCH-COCF_3)_4$ would have a molecular weight of 1066, far too high to be useful. What was wanted was a compound of uranium whose molecular weight, if possible, did not exceed 238!

In this extremity it was decided that I should undertake to prepare uranium (IV) borohydride, $U(BH_4)_4$. After all, aluminum borohydride was the most volatile aluminum compound known, and beryllium borohydride the most volatile beryllium compound known. Even though it was an enormous extrapolation, perhaps uranium borohydride could exist and exhibit the desired volatility.

Accordingly, I treated uranium (IV) fluoride with aluminum borohydride in a sealed evacuated tube. On gentle warming, green crystals readily sublimed out of the reaction mass and condensed onto the cooler surfaces of the tube.[5] As it happened, Professor H. C. Urey, then director of the S.A.M. Laboratories at Columbia University, under whose management our project came, was visiting on that day at the University of Chicago. He was greatly interested in this development. He asked me to isolate the product and to analyze it as soon as possible. Indeed, when he called from the airport several hours later, I was able to tell him that the product was in-

[4] H. I. Schlesinger, H. C. Brown, J. J. Katz, S. Archer, and R. A. Lad, *J. Amer. Chem. Soc.*, **75,** 2446 (1953).

[5] H. I. Schlesinger and H. C. Brown, *J. Amer. Chem. Soc.*, **75,** 219 (1953).

9

deed the desired uranium (IV) borohydride, volatile, with a molecular weight even lower than that of uranium hexafluoride.

Shortly afterward we were asked to expand greatly our research effort and to learn how to make the product in quantity. This required learning how to make diborane and alkali metal borohydrides readily and in quantity (Chapter IV).

Consequently, instead of being forced from our usual area of investigation by the exigencies of the war research, we were brought back to it on a greatly expanded scale of operation. We had to assemble suddenly some two dozen individuals for research on these problems. Although we recruited three of Professor Schlesinger's former students from their industrial positions, the great majority were young people with no prior training in high-vacuum technique. It was necessary to assemble a large quantity of equipment, to plan the training, and to organize a systematic attack on the problems. Professor Schlesinger was heavily burdened by his duties as Executive Secretary at a time when the routine of the Department was badly disrupted by various war activities. Consequently, many of these duties fell upon me, and life became quite hectic.

Moreover, there were still many courses to give. Many of the professors were on leave to engage in war research. When all else failed, Professor Schlesinger would fall back on me to give a needed course. Indeed, during the four years (sixteen quarters) that I was on the staff, I rarely had the pleasure of giving a course for a second time. I ran the gamut from General Chemistry through both courses in Qualitative Analysis, both courses in Quantitative Analysis, the courses in Organic Chemistry, and the courses in Physical Chemistry. During one hectic quarter I recall giving three different courses. There was an unexpected bonus. Since the professors were too busy to supervise the research of graduate students, I was permitted to take on several very competent Ph.D. candidates who could work on my problems.

I should mention that with all these duties my salary was $2,800 (the original $2,400 adjusted for the fact that I was now working a full twelve months), although I was directing others in the group whose salaries were two to three times greater. I doubt whether the University of Chicago ever realized a greater bargain!

As I have mentioned, academic positions involving tenure were very difficult to obtain in those years. At the University of Chicago, I had observed several young men who had remained in the Department as instructors for up to nine years and had then been released without achieving tenure. (Apparently, in those years tenure was achieved only after holding an appointment for ten years.) I gave myself five years including my postdoctoral year, and then in 1943 I asked Professor Schlesinger for the Department's decision. When the Department refused to commit itself, I resolved to leave.

At this point Professor M. S. Kharasch lent a helping hand. He contacted Professor Neil Gordon, who had just become Head of the Department at Wayne University. (Neil Gordon had been Chairman of the Department at the University of Maryland at the time Professor Kharasch had come there as a new Ph.D. from Chicago.) Professor Gordon invited me to come to Wayne.

Wayne University was operated by the Detroit Board of Education with city funds. It had a very small M.S. program in chemistry (there were two M.S. candidates at the time) and essentially no research facilities. The teaching load was eighteen contact hours per week. Professor Gordon wanted to initiate an active research program and promised me that my teaching load would be no greater than 12 hours if I came. My salary was to be $3,200. I accepted.

I was given a large corner room in the old central building. It had a small laboratory sink in one corner with hot and cold running water, two gas outlets, and a double electrical outlet. There were no other laboratory benches or utilities. I asked the maintenance people to place T's on the water and gas outlets with approximately six valves on each. I assembled a high-vacuum apparatus and mounted it on shock absorbers, permitting precise pressure measurements despite the vibrations of heavy trucks outside my windows. When I required liquid nitrogen, I had to drive to the Air Reduction plant and pick up my supply. In spite of these difficulties, my laboratory work got under way.

At this point a major difficulty appeared. Professor Gordon had failed to clear his promise of twelve contact hours with the other members of the staff. They asked him how he could give a new man a smaller teaching load than those with more seniority; they too wished to do research. He asked me how he could resolve this difficulty.

I told him that I could carry on research with twelve contact hours, but that eighteen hours plus preparation would make a viable research program almost impossible. It was equally impossible for him to cut everyone's teaching load suddenly by one-third. I suggested that he offer each person successively an opportunity to have a reduced load of twelve hours for one year. If that person worked effectively and published his work in a recognized journal, his load would be reduced from eighteen hours by two hours per publication. If he did not achieve any suitable publication, his load would return to eighteen hours, and another individual would be given the same opportunity. This plan was presented to the Department and accepted.

I worked very hard that year. During the summer I was offered an opportunity to teach, but elected instead to devote myself to research—without pay. I know that my decision caused my wife considerable anguish, but she never complained. As a result of my efforts, I was able to get out a total of seven publications that critical year. These included a review on sulfuryl chloride[6] (the library work had been largely completed during my postdoctoral year with Professor Kharasch) and an article on utilizing chlorination in undergraduate teaching.[7] Then there appeared five papers dealing with steric strains.[8] My teaching load for the following year was reduced to four contact hours, without complaint by the other members of the staff. Indeed, they were proud of my accomplishments, and it encouraged many of them to pursue their own research programs with new vigor.

I was promoted to Associate Professor in 1946 and was then invited to Purdue University as Professor in 1947. In 1959 I became the

[6] H. C. Brown, *Ind. Eng. Chem.*, **36,** 785 (1944).

[7] H. B. Cutter and H. C. Brown, *J. Chem. Educ.*, **21,** 443 (1944).

[8] H. C. Brown, *J. Amer. Chem. Soc.*, **67,** 374, 378, 503, 1452 (1945); H. C. Brown and H. Pearsall, *J. Amer. Chem. Soc.*, **67,** 1765 (1945).

R. B. Wetherill Professor and was appointed R. B. Wetherill Research Professor in 1960.

4. Academic Organic Chemistry in the U.S. in the 1930's

Today the organic chemist uses a wide variety of complex and expensive instruments in his research. He is deluged by journals and books. It is difficult to think back to a time when things were very different.

In the 1930's organic chemistry in the United States did not command much respect abroad. Very few postdoctoral graduates came from abroad for study here. Instead, it was more common for American postdoctoral students to go abroad.

Perhaps the status of American organic chemistry can best be described in terms of the journal and books used to report it. Professor M. S. Kharasch founded the *Journal of Organic Chemistry* in 1935. Prior to its appearance, organic papers were published only in the *Journal of the American Chemical Society*, a monthly; the 1934 volume totaled only 2,842 pages (much smaller pages than at present).

The first advanced organic chemistry textbook published by an American organic chemist appeared in 1937—Frank C. Whitmore's *Organic Chemistry* (D. Van Nostrand and Co., New York). The second advanced work was a two-volume treatise, that appeared in 1938—Henry Gilman was the editor: *Advanced Organic Chemistry* (John Wiley and Sons, New York). The third advanced work in organic chemistry appeared in 1940—Louis P. Hammett's *Physical Organic Chemistry* (McGraw-Hill and Co., New York). Thus over an entire decade American organic chemists produced a total of three advanced textbooks.

Compare that with the huge number of textbooks, monographs, and reviews published in the past ten years. In this connection the following quotation is of interest: "Scientific and technical literature has increased at such a rate as to be terrifying, and it is difficult to keep in touch with all that has been written even within a limited field. It is almost impossible for scientific journals and works of reference to do justice to their subjects. We all look forward to the future with anxiety. Things cannot be allowed to go on at this pace. Greater restraint is needed, both in the scope of the work and in publishing. This applies also to the excessive number of conventions and congresses, where often the same questions are discussed over and over again."

This is not a recent quotation. It comes from the 1932 Baker Lectures by Professor Alfred Stock![9] So a period that I look back upon as a relatively quiet and relaxed one, in which it was possible to keep up with the literature over a wide area, he found to be excessively active. Perhaps the younger generation does not find the present situation with regard to the literature so devastating.

Let us consider the situation as it existed for apparatus and equipment. Cork and rubber stoppers were still standard in the laboratories. In 1938 I required trichloroacetyl chloride in my research. I got the idea that I might be able to prepare such volatile acid chlorides conveniently by distilling them out from a mixture of the carboxylic acid and benzoyl chloride.[10] I asked Professor Schlesinger for permission to divert two or three weeks from my doctoral research to devote to this study, and he graciously gave it. I soon discovered that the hot acid chlorides were deadly to corks; I needed ground joints. These were not available from the storeroom. I had to go to Dr. Kenneth Adams, then the curator of the Department, and impress him with my need. Convinced, he went to the departmental safe, a huge affair, opened it, and gave me a pair of ground joints on loan for thirty days. At the end of that time I had to cut them off my apparatus and return them to him.

Later, after I had joined the staff in 1939 as Professor Schlesinger's assistant, with the rank of instructor and the right to take M.S. candidates, I had the idea of preparing an optically active deuterium compound by introducing deuterium without operating on the asymmetric center[11] (1.1).

$$CH_3\!-\!\overset{\displaystyle C_2H_5}{\underset{\displaystyle H}{\overset{|}{\underset{|}{C}}}}{}^{*}\!-\!CH_2OH \longrightarrow CH_3\!-\!\overset{\displaystyle C_2H_5}{\underset{\displaystyle H}{\overset{|}{\underset{|}{C}}}}{}^{*}\!-\!CH_2D \qquad 1.1$$

[9] *Hydrides of Boron and Silicon*, p. 3.

[10] H. C. Brown, *J. Amer. Chem. Soc.*, **60,** 1325 (1938).

[11] H. C. Brown and C. Groot, *J. Amer. Chem. Soc.*, **64,** 2563 (1942).

This experiment required construction of a fractioning column of approximately 100-theoretical plates to separate D-2-methyl-1-butanol from isoamyl alcohol in fusel oil. The column was constructed. It then required the purchase of a quantity of metal helices costing about $50. I requested the purchase. This purchase was considered of sufficient importance to be placed on the agenda of a departmental meeting. There I was asked whether I could not go down into the shop and turn the helices out by hand. Fortunately, I was able to persuade my senior colleagues that this would be a poor investment even of a lowly instructor's time, and the purchase was approved.

At Wayne University there were no major pieces of equipment. I found one mechanical oil pump in the Department. When I tested it, it pulled 100 mm, no more. Yet even here it proved possible to adapt my research program to the equipment I could either assemble or construct.

With this background, the reader may understand why I sometimes find myself disturbed by young men who tell me they cannot get their research under way without huge injections of costly equipment. It is nice to have all the equipment that one wants. It is still true, however, that many interesting and important research problems can be done with simple equipment. All that is necessary is the will to adapt one's research to the resources available and the imagination to select desirable research projects that can be handled with those resources.

PART ONE

EARLY STUDIES—
FROM HYDRIDES
TO FREE RADICALS

II. First Observations on Diborane as a Hydrogenating Agent

1. Origins of the Discovery that Diborane Is a Facile Hydrogenating Agent

In September 1936, just before I began my doctoral research on diborane, Schlesinger and Burg had observed that diborane reacts with carbon monoxide in a sealed tube at 100° under 20 atm pressure to give a product, now known to be the simple addition compound, borane carbonyl.[1] The product is a gas, bp $-64°$, and is largely dissociated into its components at atmospheric pressure (2.1).

$$\tfrac{1}{2}(BH_3)_2 + CO \rightleftharpoons H_3BCO \qquad 2.1$$

Initially there was considerable discussion as to whether the product was a simple addition compound (**2.1**), or whether the reaction had involved a migration of a hydride unit from boron to carbon (**2.2**).

$$
\begin{array}{cc}
\begin{array}{l}
H\; \_ \quad + \\[2pt]
\;\;\ddot{}\;\; \\[-6pt]
H\!:\!\ddot{B}\!:\!\ddot{C}\!:\!:\!:\!\ddot{O}\!: \\[-2pt]
\;\;\ddot{}\;\; \\[-6pt]
\ddot{H}
\end{array}
&
\begin{array}{l}
H\;\;H \\[-2pt]
\;\ddot{}\;\;\ddot{}\;\;\ddot{} \\[-6pt]
H\!:\!\ddot{B}\!:\!\ddot{C}\!:\!:\!\ddot{O}\!:
\end{array}
\\[20pt]
\mathbf{2.1} & \mathbf{2.2}
\end{array}
$$

It was noted that the product formed in the reaction was stable to the presence of excess diborane.[2] The question arose whether this observation might help decide between the two alternative structures, **2.1** and **2.2**. It was considered that **2.2** might exhibit some of the characteristics of an aldehyde and it was decided to examine the

---

[1] A. B. Burg and H. I. Schlesinger, *J. Amer. Chem. Soc.*, **59**, 780 (1937).

[2] This is not true in the presence of small, catalytic quantities of sodium borohydride. Thus, in the presence of such small quantities of the borohydride the reaction of carbon monoxide with diborane (in tetrahydrofuran as solvent) proceeds very rapidly at room temperature to yield trimethylboroxime (2.2).

$$\tfrac{1}{2}(BH_3)_2 + CO \xrightarrow{NaBH_4} \tfrac{1}{3}(H_3CBO)_3 \qquad 2.2$$

M. W. Rathke and H. C. Brown, *J. Amer. Chem. Soc.*, **88**, 2606 (1966).

behavior of diborane with typical organic compounds containing a carbonyl group. Accordingly, I undertook a study of the reaction of diborane with simple aldehydes, ketones, esters, and acid chlorides.

2. Reaction of Diborane with Organic Compounds Containing a Carbonyl Group

The first experiments revealed that diborane reacts very rapidly at room temperature, and even at $-80°$, with simple aldehydes and ketones, such as acetaldehyde and acetone, to yield the corresponding dialkoxy derivatives of borane (2.3, 2.4).[3]

$$2CH_3CHO + \tfrac{1}{2}(BH_3)_2 \longrightarrow (CH_3CH_2O)_2BH \qquad 2.3$$
$$2(CH_3)_2CO + \tfrac{1}{2}(BH_3)_2 \longrightarrow [(CH_3)_2CHO]_2BH \qquad 2.4$$

Treatment with water readily hydrolyzed the intermediate and provided the corresponding alcohol (2.5, 2.6).

$$(CH_3CH_2O)_2BH + 3H_2O \longrightarrow 2C_2H_5OH + H_2 + B(OH)_3 \quad 2.5$$
$$[(CH_3)_2CHO]_2BH + 3H_2O \longrightarrow$$
$$2(CH_3)_2CHOH + H_2 + B(OH)_3 \quad 2.6$$

In extending the reaction to other compounds containing a carbonyl group we concluded that we could classify these substances into three groups, depending on their rate of reaction with diborane, as follows:

I. *Rapid*—reaction complete within fifteen minutes, either at room temperature or at $-80°$. Examples: acetaldehyde, trimethylacetaldehyde, acetone.

II. *Slow*—several hours required to obtain measurable changes at room temperature; no reaction observed at low temperatures. Examples: methyl formate, ethyl acetate.

III. *No reaction*—no reaction observed over several hours at room temperature. Consequently, these compounds are less reactive than those of the preceding class. Examples: chloral, acetyl chloride, carbonyl chloride.

The fact that chloral (**2.3**) failed to react was initially very puzzling in view of the high reactivity of acetaldehyde. We con-

[3] H. C. Brown, H. I. Schlesinger, and A. B. Burg, *J. Amer. Chem. Soc.*, **61**, 673 (1939).

sidered the possibility that this might be a steric phenomenon. However, the steric requirements of methyl and chlorine are very similar. Consequently we tested trimethylacetaldehyde (**2.4**).

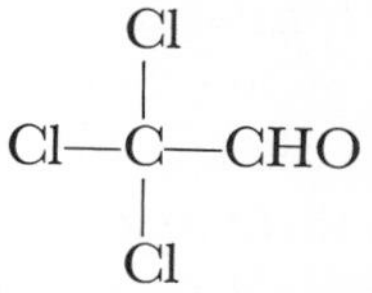

$$
\begin{array}{cc}
\text{Cl} & \text{CH}_3 \\
| & | \\
\text{Cl}-\text{C}-\text{CHO} \qquad & \text{H}_3\text{C}-\text{C}-\text{CHO} \\
| & | \\
\text{Cl} & \text{CH}_3 \\
\mathbf{2.3} & \mathbf{2.4}
\end{array}
$$

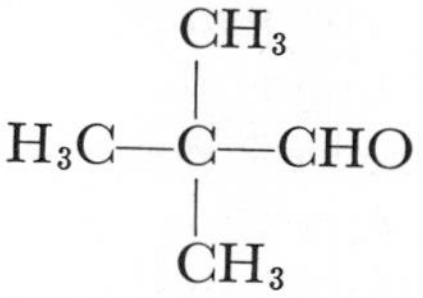

Its high reactivity ruled out the steric hypothesis. We concluded that the controlling factor must be the donor properties of the carbonyl oxygen, as indicated in the following proposed mechanism.[3]

1. Either spontaneously, or under the influence of the reagent, the diborane molecule is dissociated into borane moieties (2.7).

$$B_2H_6 \rightleftharpoons 2BH_3 \qquad\qquad 2.7$$

2. The resulting borane moieties coordinate with the oxygen of the carbonyl groups (2.8).

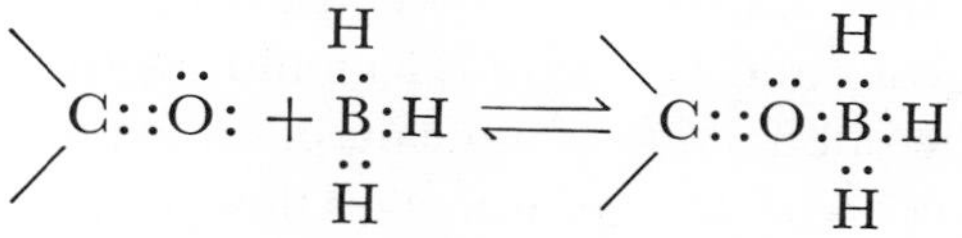

$$
\ce{>C::\overset{..}{\underset{..}{O}}: + \overset{H}{\underset{H}{\overset{..}{\underset{..}{B}}}}:H \rightleftharpoons >C::\overset{..}{\underset{..}{O}}:\overset{H}{\underset{H}{B}}:H} \qquad 2.8
$$

3. This intermediate undergoes a rapid hydride shift (2.9).

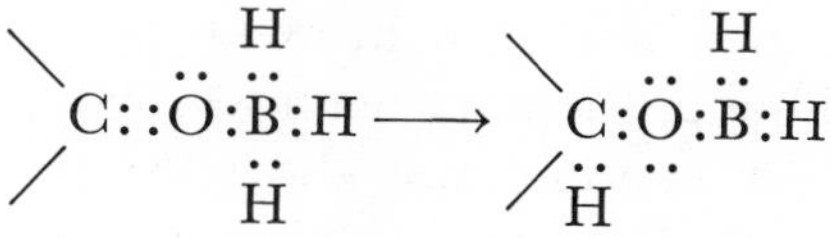

$$
\ce{>C::\overset{..}{\underset{..}{O}}:\overset{H}{\underset{H}{B}}:H \longrightarrow >C:\overset{..}{\underset{..}{O}}:\overset{H}{\underset{H}{B}}:H} \qquad 2.9
$$

4. The labile monoalkoxyborane reacts further to produce the dialkoxyborane, which is the product usually isolated.

In terms of this mechanism the failure of chloral to react was due to the greatly diminished donor properties of the carbonyl oxygen brought about by the electron-withdrawing ability of the three chlorine substituents.

This interpretation was tested by examining the addition compounds produced by the carbonyl group derivatives and boron trifluoride. The compounds that reacted rapidly with diborane (Group I) formed relatively stable addition compounds. The compounds that did not react with diborane (Group III) either did not form addition compounds with boron trifluoride or formed highly unstable compounds that existed only at low temperatures. These results confirmed the proposed interpretation.

This simple correlation between the rate of reaction with diborane and the stability of the addition compound with boron trifluoride broke down in the case of the esters (Group II). It was concluded that for these compounds the reaction mechanism must be different. It was proposed that both borane and boron trifluoride must coordinate preferentially with the alkoxy oxygen, and reduction may then involve a rupture of the carbon-oxygen bond rather than a simple addition to the carbonyl group.

This experience with molecular addition compounds later led me to explore their utility in studying steric strains (Chapter V).

3. Consequences of the Discovery

In 1939, when this study was published,[3] the organic chemist had available no really satisfactory method for reducing the carbonyl group of aldehydes and ketones under mild conditions. It might be thought that the publication of a means of reducing such compounds easily and rapidly at temperatures as low as $-80°$ would have aroused considerable interest and activity—but it did not. I am unable to find a single report by any organic chemist of a reduction utilizing diborane that was brought about by this discovery. Why?

At this time diborane was a chemical rarity. The best method available for its preparation[4] was the reaction of boron chloride with hydrogen in the silent electric discharge, producing hydrogen chloride and monochlorodiborane (2.10), followed by disproportionation of monochlorodiborane (2.11).

$$2BCl_3 + 5H_2 \longrightarrow B_2H_5Cl + 5HCl \qquad 2.10$$
$$6B_2H_5Cl \longrightarrow 5B_2H_6 + 2BCl_3 \qquad 2.11$$

To my knowledge, the only apparatus for the preparation of diborane in the United States was constructed by Anton B. Burg and

[4] H. I. Schlesinger and A. B. Burg, *J. Amer. Chem. Soc.*, **53**, 4321 (1931).

was located in Professor Schlesinger's personal laboratory. At first, Burg supplied investigators in other institutions with diborane for their studies. When this became burdensome, we began having visitors who would come to our laboratory to operate the unit to obtain sufficient diborane to take back with them for their measurements.

Each student in the group who required diborane would operate the unit. It required a 24-hour stint to produce approximately 500 cc of diborane gas, about 0.5 g. This may appear to be a tedious preparation of an essential reagent, but our experiments used Stock's high-vacuum-line techniques[5] and each experiment required only a few milligrams of the gas. Indeed, research involved in my entire Ph.D. thesis[3] required less than 500 ml of diborane gas. Many a Ph.D. candidate would be delighted if he could prepare an essential intermediate required for his doctoral program with a total expenditure of only 24 hours.

In the universities at this time there was no interest in synthetic inorganic chemistry—possibly because of the pervading influence of the Berkeley school with its emphasis on thermodynamics and kinetics. Students were given the impression that it was important to bring order to the facts and theories already present, rather than to add to the problem by discovering new phenomena.

At this time Professor Warren C. Johnson of the University of Chicago was engaged in a study of the phase relationships in the dissociation of lithium hydride[6] and calcium hydride.[7] I suggested that we explore the reaction of boron halides with these metal hydrides as a possible simple route to diborane, but could arouse no interest.

This situation was drastically altered by the requirements of our war research, as described in Chapter IV.

[5] R. T. Sanderson, *Vacuum Manipulation of Volatile Compounds* (John Wiley, New York, 1948).

[6] Unpublished data.

[7] W. C. Johnson, M. F. Stubbs, A. E. Sidwell, and A. Pechukas, *J. Amer. Chem. Soc.*, **61,** 318 (1939).

III. Free-Radical Substitution Reactions[1,2]

1. Entry into Free-Radical Chemistry

Having failed to obtain an industrial research position in 1938 after completing my requirements for the Ph.D. degree, I was more than happy to accept an appointment as a postdoctoral fellow with Professor M. S. Kharasch with funds provided by the Eli Lilly Co. I had by now married Sarah Baylen, so I asked for a stipend of $2,000 (considered in those days to be high). Professor Kharasch offered me $1,200. We compromised at $1,600. I shall relate later how he voluntarily made up the difference, a memorable experience in my life.

My research project was supposed to be the attempted isolation of an active principle from the pituitary gland. The work required a large stock of such glands from cattle, with a value of approximately $2,000, and Professor Kharasch was engaged in negotiations with one of the meat-packing concerns to donate the glands. The research was far removed from my previous experience with high-vacuum techniques, but it would broaden my experience and I was anxious to get started. Every morning and afternoon I dropped into Professor Kharasch's office to learn whether the problem of getting the glands had been resolved. Finally, in desperation, he suggested that until the glands could be obtained I should explore the possibility of achieving the chlorosulfonation of paraffin hydrocarbons with sulfuryl chloride.

2. Peroxide-Catalyzed Chlorinations with Sulfuryl Chloride

Shortly before this time C. F. Reed had discovered that a mixture of sulfur dioxide and chlorine react under photochemical activation

[1] H. C. Brown, *Record Chem. Progress*, **6,** 15 (1945).

[2] H. C. Brown in W. A. Waters, ed., *Vistas in Free Radical Chemistry* (Pergamon Press, New York, 1959), p. 196.

with paraffin hydrocarbons to introduce the sulfonyl chloride group[3] (3.1).

$$RH + SO_2 + Cl_2 \xrightarrow{h\nu} RSO_2Cl + HCl \qquad 3.1$$

The reaction was under active development at the DuPont organization as a possible route to detergents; Professor Kharasch, as a consultant, was familiar with these studies. The realization of optimum yields required careful control of the rate at which the gases, sulfur dioxide and chlorine, were introduced into the substrate undergoing substitution. This was not easy to achieve in the laboratory for isolated preparations, and he hoped that the use of sulfuryl chloride might overcome this difficulty.

Although I did not know it, several of Professor Kharasch's students had already tackled the problem without success.

I approached the problem eagerly. It did not take long to discover that a mixture of cyclohexane and sulfuryl chloride could be refluxed in the dark for long periods without significant reaction. The addition of small quantities of benzoyl peroxide resulted in a rapid reaction to produce chlorocyclohexane with the evolution of hydrogen chloride and sulfur dioxide (3.2).

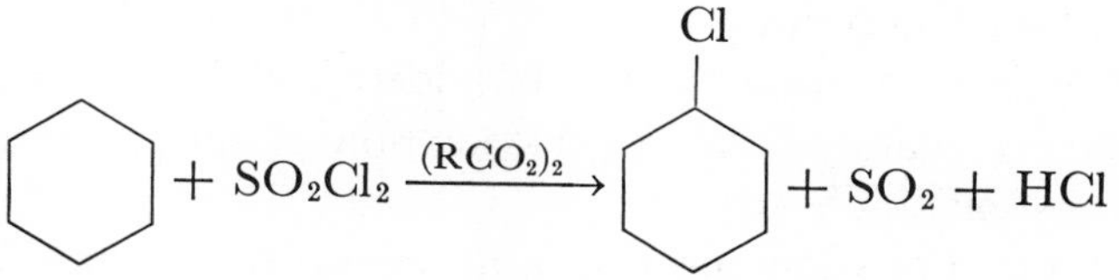

This was promising, and I rapidly applied this chlorination technique to representative hydrocarbons and chloro derivatives.[4] We then applied the reaction to ethylenic derivatives[5] and to aliphatic acids and acid chlorides.[6]

We proposed the following mechanism (3.3).

$$(R'CO_2)_2 \xrightarrow{\Delta} R'\cdot$$
$$R'\cdot + SO_2Cl_2 \longrightarrow R'Cl + \cdot SO_2Cl$$

[3] U.S. Pat. 2,174,494 (September 26, 1939).

[4] M. S. Kharasch and H. C. Brown, *J. Amer. Chem. Soc.*, **61**, 2142 (1939).

[5] *Ibid.*, 3432.   [6] *Ibid.*, **62**, 925 (1940).

$$\cdot SO_2Cl \longrightarrow SO_2 + Cl \cdot \qquad\qquad 3.3$$
$$RH + Cl \cdot \longrightarrow R \cdot + HCl$$
$$R \cdot + SO_2Cl_2 \longrightarrow RCl + \cdot SO_2Cl$$

Why had those who had already explored sulfuryl chloride at Professor Kharasch's request failed to achieve this development? The answer is that they had not. However, they were looking for chlorosulfonation and failed to recognize the possibilities of the peroxide-catalyzed chlorination reaction!

Interestingly, in exploring the reaction of carboxylic acids with the reagent we discovered that by omitting the peroxide catalyst while illuminating the reaction mixture we could achieve the desired chlorosulfonation[6,7] (3.4).

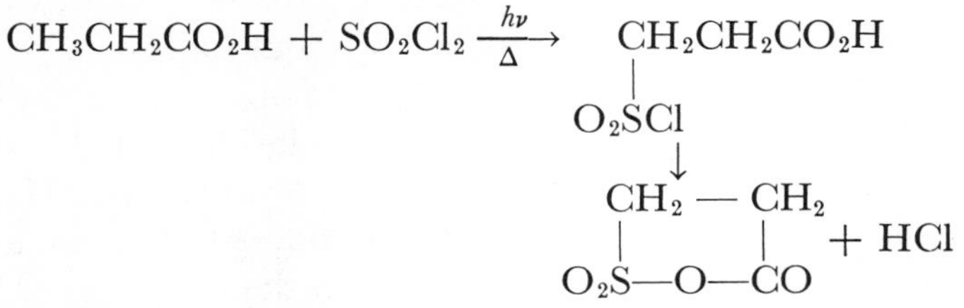

$$CH_3CH_2CO_2H + SO_2Cl_2 \xrightarrow[\Delta]{h\nu} \quad CH_2CH_2CO_2H \qquad 3.4$$
$$\underset{O_2SCl}{\big|}$$

$$\begin{array}{c} CH_2 - CH_2 \\ \big| \qquad \big| \\ O_2S{-}O{-}CO \end{array} + HCl$$

Apparently the carboxylic acid catalyzes the dissociation of the sulfuryl chloride into sulfur dioxide and chlorine and light catalyzes the Reed action.[3] A small amount of added pyridine proved to be a highly effective catalyst for the dissociation of sulfuryl chloride to achieve this reaction.[8]

The chlorosulfonation reaction appears to involve the following mechanism (3.5).

$$SO_2Cl_2 \xrightarrow[\text{cat.}]{\Delta} SO_2 + Cl_2 \qquad\qquad 3.5$$
$$Cl_2 \xrightarrow{h\nu} 2Cl \cdot$$
$$RH + Cl \cdot \longrightarrow R \cdot + HCl$$
$$R \cdot + SO_2 \rightleftharpoons RSO_2 \cdot$$
$$R \cdot + SO_2Cl_2 \longrightarrow RCl + SO_2Cl \cdot$$
$$R \cdot + Cl_2 \longrightarrow RCl + Cl \cdot$$
$$RSO_2 \cdot + Cl_2 \longrightarrow RSO_2Cl + Cl \cdot$$

[7] M. S. Kharasch, T. H. Chao, and H. C. Brown, *J. Amer. Chem. Soc.*, **62**, 2393 (1940).

[8] M. S. Kharasch and A. T. Read, *J. Amer. Chem. Soc.*, **61**, 3089 (1939).

The very high yields realized in the chlorination experiments evidently result from very low concentrations of sulfur dioxide in the reaction, achieved by vigorous refluxing of the reaction mixture, and the absence of free molecular chlorine. To achieve high yields in the chlorosulfonation reactions, the temperature is reduced to retain sulfur dioxide in solution.[8]

3. Chloroformylations with Oxalyl Chloride

In the Introduction (I–2) I mentioned that the Baker Lectures had twice played an important role in my life. First, the presentation of the volume containing the Baker Lectures by Alfred Stock by my girl friend, Sarah Baylen, had interested me in the hydrides of boron as a field for my doctorate research. Second, the lectures by Farrington Daniels[9] suggested trying the chloroformylation of a paraffin by a photochemical reaction with oxalyl chloride. This resulted in a financial bonanza for me that I still recall as one of the more pleasant experiences of my life.

While reading Daniels' book, I encountered the reference to the photochemical decomposition of oxalyl chloride (p. 180). At this time we had succeeded in achieving the photochemical chlorosulfonation of paraffins. Could we not similarly achieve the chloroformylation of paraffins with oxalyl chloride, I wondered (3.6)?

$$RH + (COCl)_2 \xrightarrow{h\nu} RCOCl + CO + HCl \qquad 3.6$$

I prepared some oxalyl chloride, dissolved it in cyclohexane, and illuminated it with a 300-watt electric lamp. Gases were evolved, which proved to consist of an equimolar mixture of carbon monoxide and hydrogen chloride. Distillation yielded cyclohexanecarboxylic acid chloride (3.7) characterized as the amide.[10]

$$\text{hexane} + (COCl)_2 \xrightarrow{h\nu} \text{hexane–}COCl + CO + HCl \qquad 3.7$$

[9] *Chemical Kinetics* (Cornell University Press, Ithaca, N.Y., 1938).
[10] M. S. Kharasch and H. C. Brown, *J. Amer. Chem. Soc.*, **62,** 454 (1940).

This was promising. I went to Professor Kharasch and described my experiment. He was deeply interested. He returned with me to the laboratory, traced each of my operations, smelled the product, and then sat at my desk. He said, "At the time you accepted the appointment you requested $2,000. I told you I could only pay $1,600. I believe you deserve the additional $400." With that he drew out his checkbook and wrote me a check for $400!

I do not know how Professor Kharasch's research funds were managed to allow him to write a check in this manner. Perhaps the payment came from personal funds. In any event, it was a memorable experience, even though I did not retain the money long—it went directly to the University to repay the balance of my wife's and my loans for tuition costs.

Phosgene also underwent the reaction[10] (3.8).

$$
\bigcirc + COCl_2 \xrightarrow{h\nu} \overset{\displaystyle COCl}{\underset{|}{\bigcirc}} + HCl \qquad 3.8
$$

The following mechanism was suggested[11] (3.9).

$$
\begin{aligned}
(COCl)_2 &\xrightarrow{h\nu} \cdot COCOCl + Cl\cdot \\
\cdot COCOCl &\longrightarrow 2CO + Cl\cdot \\
RH + Cl\cdot &\longrightarrow R\cdot + HCl \\
R\cdot + (COCl)_2 &\longrightarrow RCOCl + \cdot COCl \\
\cdot COCl &\longrightarrow CO + Cl\cdot
\end{aligned} \qquad 3.9
$$

This mechanism predicted that the addition of benzoyl chloride or other free-radical initiator to the reaction mixture should induce the reaction. The prediction was tested and confirmed.[11]

It appeared that it should also be possible to achieve the chloroformylation of paraffins by using a mixture of carbon monoxide and chlorine activated by light (3.10).

$$
\begin{aligned}
Cl_2 &\xrightarrow{h\nu} 2Cl\cdot \\
RH + Cl\cdot &\longrightarrow R\cdot + HCl \\
R\cdot + CO &\rightleftharpoons RCO\cdot \\
RCO\cdot + Cl_2 &\longrightarrow RCOCl + Cl\cdot
\end{aligned} \qquad 3.10
$$

[11] *Ibid.*, **64**, 329 (1942).

The analogy with the Reed reaction[3] is clear. Unfortunately, we could not achieve this reaction. The free radicals reacted with the chlorine in solution in preference to the small amount of carbon monoxide dissolved. Our analysis indicated that we needed to dissolve the carbon monoxide in the paraffin under high pressure to realize a satisfactory concentration. Chlorine should then be introduced at a very slow controlled rate to maintain its concentration low. Finally, light had to be introduced in this system at high pressure to achieve the desired reaction. I tried several experiments, but our equipment was simply too crude to offer any hope.

I mention this because the problem was solved in an elegant manner and some twenty years later by the use of carbon tetrachloride as the chain-transfer agent[12] (3.11).

$$RH + CCl_3 \cdot \longrightarrow R \cdot + HCCl_3 \qquad 3.11$$
$$R \cdot + CO \rightleftharpoons RCO \cdot$$
$$RCO \cdot + CCl_4 \longrightarrow RCOCl + CCl_3 \cdot$$

I have always taken a special pleasure in such simple solutions. Finally, I might point out that this reaction has lately been applied to norbornane, a system in which I have taken a close interest[13] (3.12).

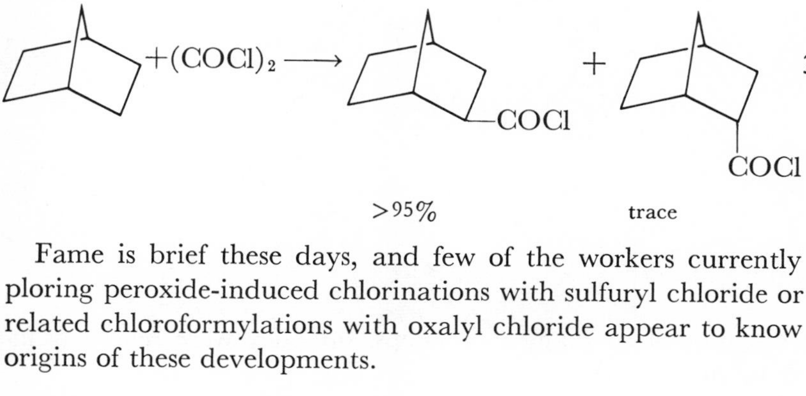

$$>95\% \qquad\qquad\qquad \text{trace}$$

Fame is brief these days, and few of the workers currently exploring peroxide-induced chlorinations with sulfuryl chloride or the related chloroformylations with oxalyl chloride appear to know the origins of these developments.

4. Properties of Free Radicals

The rapid proliferation of new free-radical substitution reactions of alkanes and their derivatives led to a consideration of the precise

[12] W. A. Thaler, *J. Amer. Chem. Soc.*, **88**, 4278 (1966).

[13] I. Tabushi, T. Okada, and R. Oda, *Tetrahedron Letters*, 1605 (1969).

properties of free radicals. Were they planar or pyramidal? Did they isomerize during reactions, or did the substituent appear at the point where the hydrogen atom had been removed? We undertook to answer some of these questions.

This was shortly after the elegant study by Bartlett and Knox on the remarkable inertness of apobornyl chloride, presumably attributable to the difficulty of forming a highly strained planar carbonium ion at the bridgehead.[14] It appeared of interest to undertake a study of the photochemical chlorination of bicyclo[2.2.2]octane in order to ascertain whether chlorine could be introduced at the bridgehead. Accordingly, I undertook the synthesis of the hydrocarbon.

Two procedures had been described in the literature. One involved only a few steps but depended on the relatively new Diels-Alder reaction, with which I was unfamiliar[15] (3.13).

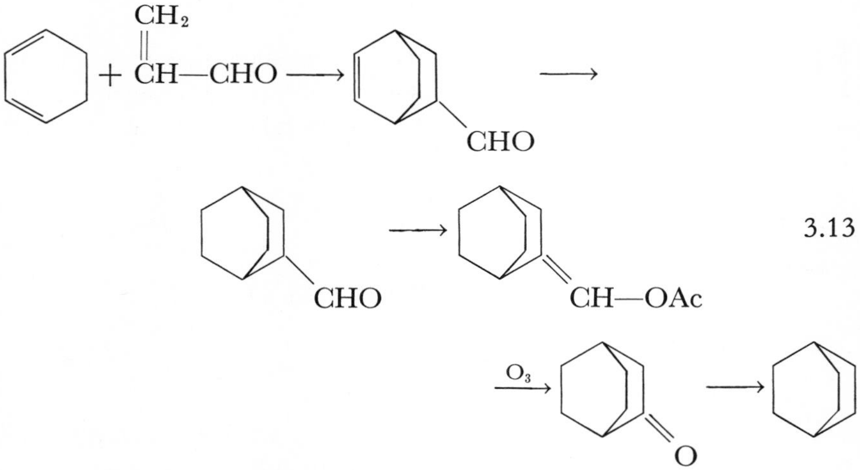

3.13

The other involved many more steps, but they were relatively familiar to me[16] (3.14).

[14] P. D. Bartlett and L. H. Knox, *J. Amer. Chem. Soc.*, **61**, 3184 (1939).

[15] O. Diels and K. Alder, *Ann.*, **478,** 137 (1930); K. Alder and G. Stein, *Ann.*, **514,** 1 (1934).

[16] G. Komppa, *Ber.*, **68,** 1267 (1935).

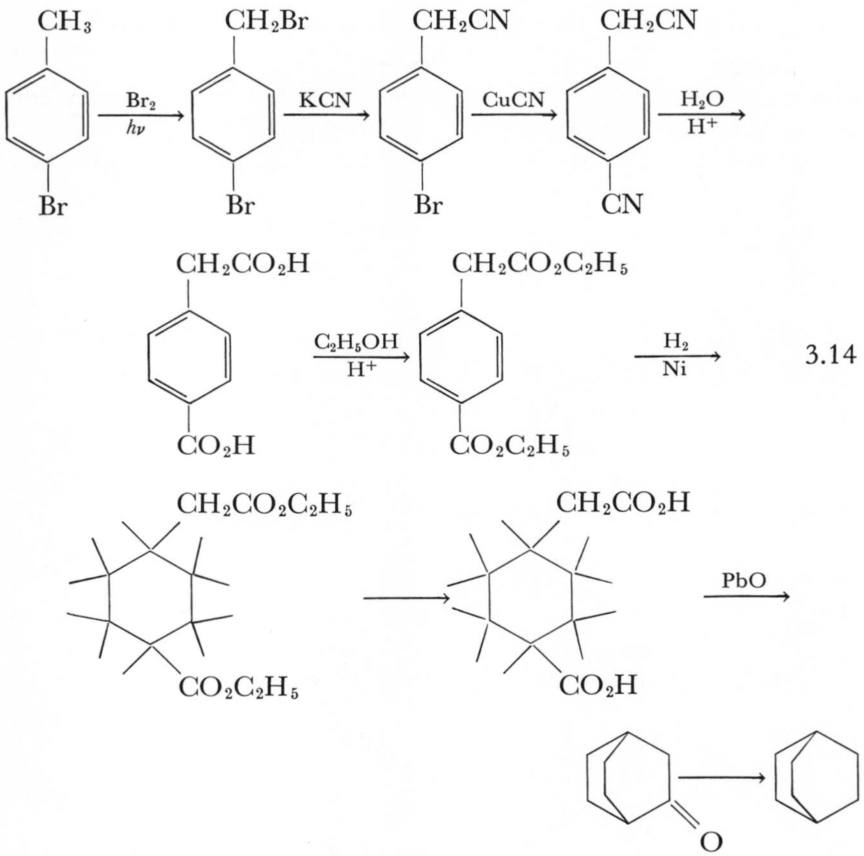

3.14

I undertook the synthesis and carried it all the way down to the hydrogenated dicarboxylic acid. The closure failed. By this time I had so little material that even success in the synthesis would have been of little avail. In those days, without gas chromatography, we required considerable quantities of the products to achieve a satisfactory separation and identification. Consequently, the project was abandoned.

This was a most instructive experience for me. I had never previously calculated what my overall yield would be in a 10-step synthesis with yields as high as 80 percent in each step. Ever since, I have sought for simple, few-step syntheses.

We had more success with our next project, the chlorination of optically active 1-chloro-2-methylbutane.[17] One liter of active amyl alcohol was obtained by the fractionation of 20 l of fusel oil. It was transformed into active 1-chloro-2-methylbutane and the latter chlorinated both photochemically and by the peroxide-catalyzed reaction with sulfuryl chloride to produce the dichloride. This was fractionated to separate the five dichlorides (3.15).

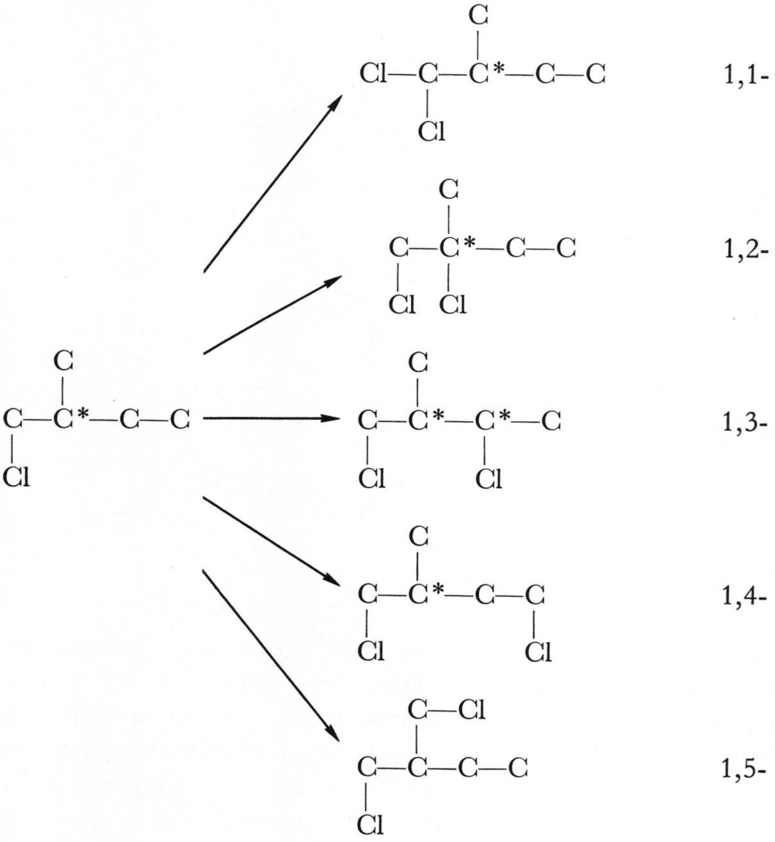

$$
\begin{array}{ccc}
 & \text{C} & \\
 & | & \\
\text{Cl—C—C*—C—C} & & \quad 1,1\text{-} \\
 & | & \\
 & \text{Cl} & \\
\end{array}
$$

$$
\begin{array}{ccc}
 & \text{C} & \\
 & | & \\
\text{C—C*—C—C} & & \quad 1,2\text{-} \\
 | & | & \\
\text{Cl} & \text{Cl} & \\
\end{array}
$$

$$
\begin{array}{ccc}
 & \text{C} & \\
 & | & \\
\text{C—C*—C—C} \longrightarrow \quad \text{C—C*—C*—C} & & \quad 1,3\text{-} \quad 3.15 \\
| \qquad\qquad\qquad | & | & \\
\text{Cl} \qquad\qquad\; \text{Cl} & \text{Cl} & \\
\end{array}
$$

$$
\begin{array}{ccc}
 & \text{C} & \\
 & | & \\
\text{C—C*—C—C} & & \quad 1,4\text{-} \\
 | & | & \\
\text{Cl} & \text{Cl} & \\
\end{array}
$$

$$
\begin{array}{ccc}
 & \text{C—Cl} & \\
 & | & \\
\text{C—C—C—C} & & \quad 1,5\text{-} \\
 | & & \\
\text{Cl} & & \\
\end{array}
$$

To our amazement we observed not five, but six distinct plateaus. Today one flings this sort of question at a Ph.D. candidate and

[17] H. C. Brown, M. S. Kharasch, and T. H. Chao, *J. Amer. Chem. Soc.*, **62,** 3435 (1940).

expects an instant reply. It is amusing to recall that it took Professor Kharasch, Dr. Chao, and myself most of one afternoon to deduce that we had successfully separated the 1,3-dichloro-2-methylbutane into its diasteriomers.

The important point was whether the 1,2-dichloro-2-methylbutane was optically active or inactive. The results revealed that it was inactive. Clearly the free radical must either be planar or else involve pyramidal free radicals that equilibrate rapidly compared to the rate of attachment of chlorine.

More recently Skell has reported that the bromination of either optically active 1-chloro- or 1-bromo-2-methylbutane gives some retention of optical activity in the products.[18] It has been argued that the results support the incursion of a bridged-radical intermediate. Alternatively, the much faster capture by the free radical of a bromine atom from the molecule may be responsible for the retention of activity in the bromination reaction in contrast to the chlorination results.

We next explored the question whether isomerization of free radicals occurred during these substitution reactions.[19] We carried out the decomposition of di-*n*-butyryl peroxide in carbon tetrachloride and identified *n*-propyl chloride and hexachloroethane as the products. Similarly, diisobutyryl peroxide yielded only isopropyl chloride and hexachloroethane. Consequently, there did not appear to be any isomerization of *n*-propyl free radicals produced in the decomposition of *n*-butyryl peroxide to the isomeric isopropyl free radicals, nor any isomerization of the isopropyl free radical to the *n*-propyl intermediate.

In later studies of directive effects in chlorination it became important to examine the question of nonisomerization of free radicals under chlorination conditions. It was observed that the photochlorination of 2-methylpropane-2-*d* at −15° produces equimolar amounts of *t*-butyl chloride and deuterium chloride[20] (3.16).

[18] P. S. Skell, D. L. Tuleen, and P. D. Readio, *J. Amer. Chem. Soc.*, **85,** 2849 (1963).

[19] M. S. Kharasch, S. S. Kane, and H. C. Brown, *J. Amer. Chem. Soc.*, **63,** 526 (1941).

[20] H. C. Brown and G. A. Russell, *J. Amer. Chem. Soc.*, **74,** 3995 (1952).

$$
\begin{array}{ccccc}
& \overset{\displaystyle CH_3}{\underset{\displaystyle CH_3}{|}} & & & \overset{\displaystyle CH_2\cdot}{\underset{\displaystyle CH_3}{|}} \\
H_3C-\overset{|}{\underset{|}{C}}-D + Cl\cdot & \longrightarrow & H_3C-\overset{|}{\underset{|}{C}}-D + HCl & & 3.16
\end{array}
$$

$$
\begin{array}{ccc}
\overset{\displaystyle CH_3}{\underset{\displaystyle CH_3}{|}} & & \overset{\displaystyle CH_3}{\underset{\displaystyle CH_3}{|}} \\
H_3C-\overset{|}{\underset{|}{C}}-D + Cl\cdot & \longrightarrow & H_3C-\overset{|}{\underset{|}{C}}\cdot + DCl
\end{array}
$$

The correspondence between the amount of deuterium chloride and the amount of t-butyl chloride in the product argues for the position that no significant rearrangement of free radicals or hydrogen exchange between radicals and hydrocarbon occurs during photochemical chlorination. The results realized in the competitive photochemical chlorination of a mixture of cyclohexane and α-d_1-toluene support the same conclusion. Consequently, for such chlorinations it appears safe to relate the isomer distribution observed in the products to the free radicals formed in the attack of the chlorine atoms on the hydrocarbons.

5. Directive Effects in Aliphatic Substitution [2, 21]

At Wayne University I had my hands full pursuing my program on steric effects, while initiating my program on aromatic complexes with Friedel-Crafts catalysts. I had no plans to continue any work in the homolytic substitution area.

In 1945, just after we had initiated a Ph.D. program at Wayne, an evening school student, Arthur B. Ash, approached me about beginning work toward a Ph.D. degree. He was only the second such candidate in the Department, so I was very interested. He was employed at the Wyandotte Chemical Corporation, where much of his work involved chlorination, and could not leave his industrial job for full-time academic work for at least one year. However, if I could suggest a suitable problem involving chlorination, he might be able to persuade his superiors to permit him to combine his research with his normal activities.

[21] A. B. Ash and H. C. Brown, *Record Chem. Progress*, **9,** 81 (1948).

In the course of my research on peroxide-catalyzed chlorinations I had been impressed by the strong directive influence exhibited by the chlorine substituent in *n*-butyl chloride[4] and by the carboxyl group in *n*-butyric acid.[6] I suggested a systematic exploration of such directive effects. Ash was delighted with the suggestion and he experienced no difficulty in initiating the program at Wyandotte, later coming to Wayne full time to complete his thesis.

Typical results realized with the 1-chlorobutanes are indicated below[22] (3.17).

$$
\begin{array}{ccccl}
\text{C} & \text{C} & \text{C} & \text{C} & X_n \qquad\qquad 3.17 \\
\uparrow & \uparrow & \uparrow & \uparrow & \\
24 & 47 & 22 & 7 & Cl \\
\\
37 & 49 & 12 & 2 & Cl_2 \\
\\
51 & 49 & 0 & & Cl_3
\end{array}
$$

Similarly, *n*-butyryl chloride exhibits an exceedingly powerful directive effect of the —COCl group[22] (3.18).

$$
\begin{array}{cccl}
\text{C} & \text{C} & \text{C} & \text{COCl} \qquad\qquad 3.18 \\
\uparrow & \uparrow & \uparrow & \\
48 & 49 & 3 &
\end{array}
$$

These studies required the preparation of large quantities of starting material, followed by chlorination at high dilution to minimize polychlorination. Then the products had to be subjected to careful fractionation to establish the isomeric distribution. It is heart-wrenching to realize that with modern gas-chromatographic methods the research could have been accomplished with about one-tenth the effort.

6. Selectivity of the Attacking Species [2]

Unlike chlorination, photochemical bromination is highly selective and constitutes an excellent method for the preparation of tertiary bromides from hydrocarbons containing only primary hydrogen atoms adjacent to the tertiary hydrogen.[23] The presence of

[22] H. C. Brown and A. B. Ash, *J. Amer. Chem. Soc.*, **77**, 4019 (1955).

[23] M. S. Kharasch, W. Hered, and F. R. Mayo, *J. Org. Chem.*, **6**, 818 (1941).

secondary or tertiary hydrogen atoms in the adjacent position leads to the formation of considerable quantities of dibromides.[24] These apparently arise from a dark reaction between elementary bromine and the tertiary bromide initially formed.

The competitive photohalogenation of cyclohexane and toluene at 80 ° revealed that toluene is sixty times as reactive as cyclohexane toward bromine atoms, whereas toward chlorine atoms cyclohexane is eleven times as reactive as toluene.[25] This indicates that in photobromination a cyclohexane hydrogen atom is only 0.004 times as reactive as a toluene hydrogen, whereas in photochlorination the cyclohexane hydrogen atom is 2.8 times as reactive as a toluene side-chain hydrogen atom.

The marked difference in the relative reactivities of toluene and cyclohexane in these two reactions was explicable in terms of a marked difference in the degree of bond breaking in the respective transition states. It was proposed that in bromination the carbon-hydrogen bond is largely broken in the transition state, so that the activated complex is strongly stabilized by resonance contributions from the partially formed benzyl free radical. On the other hand, the attack of the chlorine atom involves an exothermic process. Here the transition state would be expected to resemble reactants. Consequently, the carbon-hydrogen bond will be but slightly broken in the activated complex, and resonance stabilization from the incipient benzyl free radical will contribute very little to the stability of the transition state.[25]

The photochlorination of isobutane, 2,3-dimethylbutane, 2,2,3-trimethylbutane, and 2,3,4-trimethylbutane revealed an essentially constant ratio of attack of the primary and tertiary hydrogen atoms.[26] Assuming that the reactivities of the primary hydrogen atoms in these compounds do not vary significantly, it appears that the reactivities of the tertiary hydrogen atoms in these aliphatic hydrocarbons are remarkably constant. The observation that the reactivity of the tertiary hydrogen toward attack by the chlorine atom is not altered appreciably by the accumulation of bulky groups is con-

[24] G. A. Russell and H. C. Brown, *J. Amer. Chem. Soc.*, **77,** 4025 (1955).
[25] *Ibid.*, 4578.   [26] *Ibid.*, 4031.

sistent with the conclusion that the transition state in the attack of the chlorine atom on hydrogen involves very little bond breaking, so that the geometrical arrangement is still essentially tetrahedral.[26]

In the course of his doctoral thesis Glen Russell observed indications that the selectivity of the chlorine in attacking hydrogen could be modified by the presence of aromatic rings. He later explored this phenomenon in a brilliant series of investigations. He demonstrated that photochemical chlorination of 2,3-dimethylbutane at 25° results in the formation of the two isomers, 2-chloro- and 1-chloro-2,3-dimethylbutane in a ratio of 40:60. If the photochemical chlorination is carried out in the presence of benzene, the ratio changes to 90:10.[27] Indeed, comparison of the relative effects of a number of aromatic derivatives resulted in a good correlation with the ability of these aromatics to form π-complexes. He therefore concluded that the chlorine atom forms such π-complexes with the aromatic ring, and that the selectivity of the chlorine atom in attacking carbon-hydrogen bonds is thereby modified.

This concept of selectivity of the attacking species, originally developed to account for these interesting phenomena in free-radical reactions, later helped support a new understanding of the factors influencing electrophilic aromatic substitution. It led to the development of the selectivity relationship, recognition of the possibility of treating aromatic substitution data quantitatively through the development of a new set of substituent constants, and the development of such a set of σ^+ constants. Unfortunately, these topics could not be included in the Baker Lectures, and they are not discussed in the present book.

7. The Reactive Paraffins

In 1941 Professor Kharasch and I were stimulated to publish a brief discussion of the implications of the uncovering of these new substitution reactions of the paraffins. We pointed out that the term paraffin (from "parum affinis," little affinity) was a misnomer. "The hallowed belief of the organic chemist in the inertness of the paraffin hydrocarbons is probably to be ascribed to the fact that these sub-

[27] G. A. Russell, *J. Amer. Chem. Soc.*, **79**, 2977 (1957).

stances undergo reaction most readily by means of mechanisms involving atomic and free-radical intermediates, and the study of reactions of this type is a recent development."[28]

At about this time there was a conference at the University of Chicago, at which one of the speakers was Professor Louis P. Hammett of Columbia University. He made the point that the chemistry of organic chemistry is predominantly the chemistry of ionic intermediates. As a brash young man, I stood up to disagree. Although this was true of the reactions currently known, I said, one should not extrapolate present knowledge into the future. I believed there were probably as many homolytic processes as heterolytic processes and it merely required imaginative exploration to uncover them.

My response was the same twenty years later when many chemists asked me why I was spending so much effort exploring the hydroboration reaction. These reactions yielded organoboranes. Such organoboranes had been known for a century, yet very little interesting had been described. I shall leave it to the reader of Chapters XVI–XIX to decide whether my optimism has been justified.

[28] H. C. Brown and M. S. Kharasch, *J. Chem. Ed.*, **18,** 589 (1941).

IV. The Alkali Metal Hydride Route to Diborane and the Borohydrides[1]

1. Background in 1940

In 1940 there were available two synthetic routes to diborane. The first, developed by Stock in 1912, involved the preparation and hydrolysis of magnesium boride followed by the thermal decomposition of the higher boron hydrides thus obtained.[2] The second, developed by Schlesinger and Burg in 1931, utilized the interaction of hydrogen and boron chloride in the silent electric discharge.[3] A brief review of these methods will illustrate some of the difficulties facing the research worker who wished to utilize diborane in his investigations.

The first step in the Stock synthesis involved the preparation of magnesium boride by the reaction of carefully selected magnesium and boric oxide.[4] Since the reaction is violent when the stoichiometric quantities of the reactants are used, it was recommended that excess magnesium be utilized to moderate the vigor of the reaction. The reaction was carried out in thin-walled iron crucibles under a hydrogen atmosphere. Best results were realized by carrying out the preparation in individual lots of 10 g. Since the procedure utilized as much as 4000 g of the magnesium boride, some 400 individual batches of the boride were required for a single preparation of the boron hydride.

The magnesium boride (2000 to 4000 g) was then added slowly at a rate of not more than 100 g per hour to a large volume of 10 percent hydrochloric acid or phosphoric acid. Mechanical devices were used to introduce the boride at a uniform rate to the acid over the several days required for the process. The main product of the reaction was hydrogen, presumably from hydrolysis of borane as the chief product

---

[1] H. C. Brown, *Congress Lectures, XVII International Congress of Pure and Applied Chemistry* (Butterworth & Co. Publishers Ltd., London, 1960), p. 167.

[2] A. Stock and C. Massenez, *Ber.*, **45**, 3539 (1912).

[3] *J. Amer. Chem. Soc.*, **53**, 4321 (1931).

[4] A. Stock, *Hydrides of Boron and Silicon.*

of the reaction. The hydrogen was passed through traps cooled in liquid air and a condensate obtained that contained SiH_4, H_2S, CO_2, PH_3, Si_2H_6, B_4H_{10}, Si_3H_8, B_5H_9, B_6H_{10}, Si_4H_{10}, $B_{10}H_{14}$, and higher materials. This mixture was then subjected to laborious fractionation in the high-vacuum apparatus.

Tetraborane, B_4H_{10}, was the chief product. From 4200 g of magnesium boride there was realized a yield of 6 g of "sufficiently pure" B_4H_{10},[4] a yield of 0.5 percent based on boron. By warming from 90° to 95° for five hours B_4H_{10} was converted into diborane along with some higher boron hydrides.

Stock and his students had available only soft glass for their traps and Dewar vessels and only liquid air (not nitrogen) for their refrigerant. On occasion a trap broke, the boron hydride product encountered the liquid air, and a violent explosion ensued.[4] One can only admire the courage and perseverance of these pioneers.

The Schlesinger-Burg process represented a major improvement. Boron chloride was prepared by passing chlorine over commercial calcium boride, CaB_6, maintained at approximately 500° in an electric tube furnace. A mixture of hydrogen and boron chloride at a pressure of about 10 mm was passed through a silent electric discharge of 15,000 volts maintained between water-cooled cooper electrodes. In the discharge partial hydrogenation of the boron chloride occurred to a partially hydrogenated boron chloride, possibly $BHCl_2$, and hydrogen chloride.

The unconverted boron chloride, the partially hydrogenated intermediate, and the hydrogen chloride were condensed with liquid nitrogen and thereby separated from the hydrogen. Hydrogen chloride could be pumped off from the product at −80°. Distillation then yielded boron chloride and chlorodiborane, B_2H_5Cl. The latter material was fractionated at a pressure of approximately two atmospheres to yield diborane and boron chloride.

This preparation required a relatively complex apparatus. However, once the apparatus had been constructed and was available, it could be used to produce some 500 ml of diborane in a 24-hour operation. This was ample for our requirements in those days, when experiments were performed on the Stock high-vacuum apparatus on a micro scale. Indeed, I used less than 500 ml of diborane for my entire Ph.D. research. Moreover, synthetic inorganic chemistry was in relatively low esteem in the universities in those days, so there was no incentive to undertake the development of more convenient synthetic methods (Section II-3). We were jarred out of this rut by the requirements of the research that we undertook as part of the defense effort, beginning late in 1940.

2. The Uranium Borohydride Project

Late in 1940 Professor Schlesinger received a request from the National Defense Agency to investigate the synthesis of new volatile compounds of uranium. He was not told the precise application to be made of these compounds, but was merely informed that a volatility as low as 0.1 mm at a temperature at which the material would be stable for a considerable period of time would be satisfactory. (Uranium hexafluoride was available, but there was some doubt whether technology could overcome the problems of handling this highly reactive material.)

I had recently completed the laboratory work involved in our synthesis of lithium borohydride and our study of its properties[5] and was in the midst of a study of gallium borohydride.[6] Professor Schlesinger asked me to drop my normal academic research and devote myself to this Defense Project.

We assembled a small group, including some remarkable individuals, and began a study of various possible compounds of uranium with emphasis on uranium (IV) acetonylacetonate and related derivatives.[7] The volatility of the acetonylacetonate was too low, so we explored structural changes that might improve it. We observed a major improvement with the fluorine derivative **4.1** and were well our way to synthesizing **4.2** when the blow fell.

[5] H. I. Schlesinger and H. C. Brown, *J. Amer. Chem. Soc.*, **62,** 3429 (1940).

[6] H. I. Schlesinger, H. C. Brown, and G. W. Schaeffer, *J. Amer. Chem. Soc.*, **65,** 1786 (1943).

[7] H. I. Schlesinger, H. C. Brown, J. J. Katz, S. Archer, and R. A. Lad, *J. Amer. Chem. Soc.*, **75,** 2446 (1953).

$$\begin{array}{cc} \text{F}_3\text{C} & \text{F}_3\text{C} \\ \qquad\qquad\text{C—O} & \qquad\qquad\text{C—O} \\ \text{HC} \qquad\qquad)_4\text{U} & \text{HC} \qquad\qquad)_4\text{U} \\ \qquad\quad\text{C=O} & \qquad\quad\text{C=O} \\ \text{H}_3\text{C} & \text{F}_3\text{C} \\ \quad\textbf{4.1} & \quad\textbf{4.2} \end{array}$$

We were advised that there was an important property about which we had not been informed. It was important that the molecule have a low molecular weight, preferably not greater than 238! Since **4.2** would have a molecular weight of 1066, it was obviously unsuitable.

In desperation, I undertook to prepare uranium (IV) borohydride, U(BH$_4$)$_4$, by treatment of uranium (IV) fluoride with aluminum borohydride. The first experiment was successful[8] (Section I-3). We were immediately requested to assemble a large group (1) to prepare uranium borohydride in quantity for testing, (2) to devise simple methods for its synthesis adaptable to large-scale production, and (3) to establish the volatility, stability, and characteristics of uranium borohydride and other promising derivatives.

Professor Schlesinger assumed charge of this last phase of the project. I was placed in charge of the other two. This was a fortunate division as far as I was concerned, since it gave me an opportunity to explore new synthetic methods, a type of research in which I have always taken especial pleasure.

We first had to get a group going to prepare uranium borohydride utilizing procedures already known to us. It is difficult for a chemist familiar with the present ready availability of many intermediates to realize how far back we had to go in our synthesis. We decided to utilize boron tribromide[9] instead of the trichloride for the preparation of diborane because some of the intermediates would be easier to handle. This required the synthesis of boron tribromide from calcium

[8] H. I. Schlesinger and H. C. Brown, *J. Amer. Chem. Soc.*, **75,** 219 (1953).
[9] A. Stock and W. Sutterlin, *Ber.*, **67,** 407 (1934).

42

boride (4.1) and reduction of boron tribromide by the silent electric discharge method[10] (4.2). (A battery of six individual units was operated to produce the diborane required.)

$$CaB_6 + 10Br_2 \xrightarrow{\ 900°\ } CaBr_2 + 6BBr_3 \qquad 4.1$$

$$BBr_3 + H_2 \xrightarrow[\text{12,000 volts}]{\text{15 mm}} [BHBr_2] + HBr \qquad 4.2$$

$$\downarrow \Delta$$

$$B_2H_6 + BBr_3 \longleftarrow B_2H_5Br$$

Aluminum borohydride was produced by treatment of trimethylaluminum with excess diborane[11] (4.5). The trimethylaluminum was prepared by treating dimethylmercury with aluminum (4.4). In turn, the dimethylmercury was synthesized via mercuric chloride and methylmagnesium halide (4.3).

$$2CH_3MgX + HgX_2 \longrightarrow (CH_3)_2Hg + 2MgX_2 \qquad 4.3$$

$$3(CH_3)_2Hg + 2Al \longrightarrow 2(CH_3)_3Al + 3Hg \qquad 4.4$$

$$(CH_3)_3Al + 2B_2H_6 \longrightarrow Al(BH_4)_3 + B(CH_3)_3 \qquad 4.5$$

Uranium (IV) fluoride was available from government sources. It was converted to uranium (IV) borohydride by treatment with aluminum borohydride[8] (4.6).

$$UF_4 + 2Al(BH_4)_3 \longrightarrow U(BH_4)_4 + 2AlF_2(BH_4) \qquad 4.6$$

Finally, treatment of the uranium borohydride with trimethyl-boron yielded a monomethyl derivative of uranium (IV) borohydride that was even more volatile than the parent compound.[12]

Most of our effort had to be devoted to the synthesis of sufficient uranium borohydride and the methyl derivative for detailed study of their volatilities, stabilities, and properties. However, it was evident that if these materials proved useful and were required in large

---

[10] H. I. Schlesinger, H. C. Brown, B. Abraham, N. Davidson, A. E. Finholt, R. A. Lad, J. Knight, and A. M. Schwartz, *J. Amer. Chem. Soc.*, **75**, 191 (1953).

[11] H. I. Schlesinger, R. T. Sanderson, and A. B. Burg, *J. Amer. Chem. Soc.*, **62**, 3421 (1940).

[12] H. I. Schlesinger, H. C. Brown, L. Horwitz, A. C. Bond, L. D. Tuck, and A. O. Walker, *J. Amer. Chem. Soc.*, **75**, 222 (1953).

quantities, the known synthetic route would not be satisfactory for large-scale industrial application. Accordingly, a relatively small group began exploring new synthetic routes. We had almost immediate success.

We discovered that boron trifluoride etherate reacted readily with a suspension of finely divided lithium hydride suspended in ether to give diborane[13] (4.7). Moreover, in the same solvent diborane and lithium hydride readily reacted to produce lithium borohydride[14] (4.8).

$$6\text{LiH} + 8\text{BF}_3 \xrightarrow{\text{Et}_2\text{O}} \text{B}_2\text{H}_6 + 6\text{LiBF}_4 \qquad\qquad 4.7$$

$$2\text{LiH} + \text{B}_2\text{H}_6 \xrightarrow{\text{Et}_2\text{O}} 2\text{LiBH}_4 \qquad\qquad 4.8$$

In turn lithium borohydride reacted readily with boron trifluoride (4.9) or boron trichloride to produce diborane (4.10) and with aluminum trichloride (in the absence of solvent) to produce aluminum borohydride[15] (4.11).

$$3\text{LiBH}_4 + 4\text{BF}_3 \xrightarrow{\text{Et}_2\text{O}} 2\text{B}_2\text{H}_6 + 3\text{LiBF}_4 \qquad\qquad 4.9$$

$$3\text{LiBH}_4 + \text{BCl}_3 \xrightarrow{\text{Et}_2\text{O}} 2\text{B}_2\text{H}_6 + 3\text{LiCl} \qquad\qquad 4.10$$

$$3\text{LiBH}_4 + \text{AlCl}_3 \longrightarrow \text{Al(BH}_4)_3 + 3\text{LiCl} \qquad\qquad 4.11$$

Consequently, for the synthesis of aluminum borohydride we could now replace the five stages, 4.1 through 4.5, with three much simpler stages, 4.7, 4.8, and 4.11.

When we reported this to headquarters our delight was severely dampened. There was a serious shortage of lithium hydride and none could be spared for this application. (Every plane going over water carried two 1-pound charges of lithium hydride, which could be used to inflate a balloon with hydrogen to carry aloft an antenna for distress signals.) On the other hand, restricted production of

---

[13] H. I. Schlesinger, H. C. Brown, J. R. Gilbreath, and J. J. Katz, *J. Amer. Chem. Soc.*, **75,** 195 (1953).

[14] H. I. Schlesinger, H. C. Brown, H. R. Hoekstra, and L. R. Rapp, *J. Amer. Chem. Soc.*, **75,** 199 (1953).

[15] H. I. Schlesinger, H. C. Brown, and E. K. Hyde, *J. Amer. Chem. Soc.*, **75,** 209 (1953).

tetraethyllead meant that excess sodium capacity was available. Could we adapt these methods to sodium hydride?

At this time the highly useful solvents, tetrahydrofuran and diglyme, were not available. Some time later, using these solvents, we were able to duplicate the reactions 4.7, 4.8, 4.9, and 4.10 without difficulty.[16] However, in ethyl ether, or the other solvents we tried, the reactions with sodium hydride were not satisfactory.

We solved this problem by using an addition compound of sodium hydride with methyl borate, sodium trimethoxyborohydride.[17] This new addition compound was readily prepared by heating sodium hydride with excess methyl borate under reflux (4.12). The reaction is an interesting one. The reaction proceeds with a remarkable expansion in the volume of the solid in the reaction vessel and is readily followed by observing the increase in volume.

$$NaH + B(OCH_3)_3 \longrightarrow NaBH(OCH_3)_3 \qquad 4.12$$

The product behaves as a form of active or soluble sodium hydride. Reaction readily occurs with diborane to form sodium borohydride[14] (4.13) and with boron halides to form diborane[13] (4.14, 4.15).

$$2NaBH(OCH_3)_3 + B_2H_6 \longrightarrow 2NaBH_4 + 2B(OCH_3)_3 \qquad 4.13$$
$$6NaBH(OCH_3)_3 + 8BF_3 \longrightarrow B_2H_6 + 6NaBF_4 + 6B(OCH_3)_3 \qquad 4.14$$
$$6NaBH(OCH_3)_3 + 2BCl_3 \longrightarrow B_2H_6 + 6NaCl + 6B(OCH_3)_3 \qquad 4.15$$

Sodium borohydride could be used in the reaction with aluminum chloride to prepare aluminum borohydride.[15]

Our results indicated that uranium (IV) borohydride possessed sufficient volatility and stability to meet the specifications. A supply of uranium borohydride was prepared, and I received priority to fly to New York for testing at Columbia University (a four-stop flight aboard a DC3). At Columbia I worked with Dr. Willard E. Libby, subjecting the material to the metal barriers that would be used in the diffusion plants. Alas! Uranium borohydride proved unstable to these metal barriers.

[16] H. C. Brown and P. A. Tierney, *J. Amer. Chem. Soc.*, **80**, 1552 (1958).

[17] H. C. Brown, H. I. Schlesinger, I. Sheft, and D. M. Ritter, *J. Amer. Chem. Soc.*, **75**, 192 (1953).

3. Project for the Field Generation of Hydrogen

At this point we received a visit from representatives of the Signal Corps. They had heard we had prepared a material having promising properties for the field generation of hydrogen, and they wanted to review the question with us.

They told us of their difficulties. Their standard method of generating hydrogen for field use involved the action of sodium hydroxide on ferrosilicon. Into a large steel cylinder, about twice the size of the usual hydrogen cylinder in the laboratory, was placed the charge of ferrosilicon. Sodium hydroxide pellets were added, then water, and the top was quickly screwed on. The sodium hydroxide dissolved in the water and raised the temperature to above 100°. At that temperature the strong caustic began to act on the ferrosilicon, liberating hydrogen. The temperature rose to perhaps 200°. The cylinder was cooled and hydrogen drawn from it until it was exhausted.

This procedure had four difficulties. First, the large and bulky cylinder was difficult to transport to battle areas around the world. Second, the soldiers had to handle sodium hydroxide pellets. This was hazardous; a number of men had been hospitalized with caustic burns. Third, the residue, a mixture of sodium silicate and ferrous hydroxide, set to a glassy material in the cylinder. To chip out this deposit, so that the cylinder could be used again, took the average private a full day. Fourth, this residue presented a disposal problem. The soldiers had orders to dig a deep hole and bury it, but often they simply dropped it behind a bush. Here it was found and eaten by the cattle, and as a consequence the Army was facing numerous damage suits.

We told them about sodium borohydride. It should hydrolyze in water to produce a mole of hydrogen per 9.5 g of borohydride, almost as good as lithium hydride and far better than calcium hydride or ferrosilicon. The product of the reaction would be sodium borate, a water-soluble, easily disposable material.

Interested, they asked for a demonstration. This led to one of the greatest shocks of my life.

In research involving the hydrogen compounds of carbon one

customarily oxidizes them to carbon dioxide and water and collects and weighs these to get an analysis. One does not normally anticipate that a given hydrocarbon will fail to burn. Similarly, in research involving the hydrides of boron it is customary to treat the compound with water to form boric acid and hydrogen. It is ingrained in one always to protect such compounds from air and water. We had always so protected sodium borohydride.

Since our visitors wanted a demonstration, I weighed out a sample of sodium borohydride in a dry box and placed the sample in a flask fitted with a gas outlet tube connected to a gas meter. A dropping funnel containing water was attached to the flask. The entire assembly was mounted behind a safety screen. (I expected a violent reaction, similar to that which occurs with lithium aluminum hydride.)

With the several colonels and civilians of the visiting party surrounding me, I cautiously allowed the water to flow from the dropping funnel into the flask. To my amazement, the sodium borohydride simply dissolved and no significant gas evolved.

This was embarrassing indeed!

I quickly added acid and achieved the rapid evolution of hydrogen.

In spite of this *faux pas* the visitors were interested, and a project was initiated to find a simpler, more economical route to sodium borohydride and a means of facilitating its hydrolysis. We soon discovered that we could readily prepare sodium borohydride by adding the calculated quantity of methyl borate to well-stirred sodium hydride maintained at 250° (4.16).[18]

$$4NaH + B(OCH_3)_3 \xrightarrow{250°} NaBH_4 + 3NaOCH_3 \qquad 4.16$$

This is the basis of the present industrial method for the production of sodium borohydride.

Since we required a solvent to separate the sodium borohydride from the sodium methoxide, we began to determine the solubility of sodium borohydride in various readily available solvents. One of the solvents we tested was acetone. In this material the borohy-

[18] H. I. Schlesinger, H. C. Brown, and A. E. Finholt, *J. Amer. Chem. Soc.*, **75**, 205 (1953).

dride dissolved with evolution of heat. Hydrolysis with dilute acid produced no hydrogen. Analysis showed the presence of four moles of isopropyl alcohol per mole of borohydride introduced. Obviously, sodium borohydride was a facile hydrogenating agent capable of reducing the carbonyl groups of aldehydes and ketones in the manner previously demonstrated for diborane (Chapter II).

The incorporation of an equal weight of boric oxide into the sodium borohydride process gave satisfactory rates of hydrolysis. However, it was discovered that the introduction of a few percent of cobalt chloride into the sodium borohydride pellet offered an even better solution. When the pellet contacted water, the cobalt salt was reduced by the borohydride to a black, finely divided cobalt boride. This cobalt boride was an excellent catalyst for the hydrolysis of sodium borohydride.[19,20]

The Signal Corps was interested in this product, and the Ethyl Corporation was commissioned to do the pilot plant work. Late in 1944 a contract was finally given for the construction of a plant; however, instructions followed shortly from Washington that no new war plants were to be built—so this development, like that of uranium borohydride before it, failed to be utilized. Nevertheless, the research opened up major new areas. It completely revolutionized the methods used by the organic chemists for the reduction of functional groups. And sodium borohydride, a product developed under the exigencies of war research, later found its main application in the pharmaceutical industry.

4. Publication

Before leaving this topic, it might be of interest to recount why publication of these papers did not take place until 1953. The re-

[19] H. I. Schlesinger, H. C. Brown, A. E. Finholt, J. R. Gilbreath, H. R. Hoekstra, and E. K. Hyde, *J. Amer. Chem. Soc.*, **75,** 215 (1953).

[20] Much later, my son, as an undergraduate at Purdue University, reopened this study and discovered that platinum chloride is reduced to give an even more active catalyst for the hydrolysis of borohydride. Even more important, he discovered that the black powder produced was a highly active hydrogenation catalyst that could be utilized *in situ* for such hydrogenations. H. C. Brown and C. A. Brown, *J. Amer. Chem. Soc.*, **84,** 1493, 1494, 2827 (1962).

search described above was largely completed by the time I left the University of Chicago for Wayne University in September 1943, and Professor Schlesinger insisted that I write up the papers for publication. I have explained the difficulties under which I was working at Wayne (I-3). I felt it was unfair to take time at Wayne to write up a set of eleven papers when it was essential for me to initiate a research program there. I urged Professor Schlesinger, as the senior man, to do the writing, but he declined. This impasse continued even after I had moved to Purdue.

It finally became evident that if I did not yield, the papers would not be published. Indeed, many papers were already appearing using results from our unpublished studies. In 1951, during a summer as Visiting Professor at U.C.L.A., I devoted my free time to the preparation of the eleven manuscripts that—after so long a delay— ultimately appeared in the January 5, 1953, issue of the *Journal of the American Chemical Society*.

PART TWO

STERIC STRAINS

V. Steric Effects — Important or Not

1. Origins of the Program on Steric Effects

In September 1939, I became assistant to Professor H. I. Schlesinger at the University of Chicago with the rank of Instructor. My obligations in these positions would occupy most of my time; however, I would be permitted to direct the research of M.S. candidates, provided I could attract them. Consequently I considered striking out in a new direction in my own research program.

I had observed that certain compounds containing carbonyl groups, such as acetaldehyde, acetone, and trimethylacetaldehyde, reacted rapidly with diborane, whereas other derivatives, such as chloral, acetyl chloride, and phosgene, failed to react under the same mild conditions (Section II-2). I had accounted for the difference in terms of a major difference in the donor properties of the carbonyl oxygen atom in the two groups of compounds. Those that were capable of coordinating with BH_3 were postulated to undergo rapid reduction; those that failed to coordinate in the initial stage failed to react. I had tested this hypothesis by comparing the relative stabilities of the molecular addition compounds formed by these substances with boron trifluoride. Indeed, acetaldehyde, acetone, and trimethylacetaldehyde formed relatively stable addition compounds, whereas the compounds from chloral, acetyl chloride, and phosgene were either highly unstable or showed no evidence of formation. This experience convinced me of the value of utilizing such molecular addition compounds to explore chemical behavior, and I began to consider specific problems where the tool might be applied. One fertile area might be the testing of the possible importance of steric effects in chemical behavior.

2. The Place of Steric Effects in Organic Theory in 1940

Steric effects as a factor in chemical behavior appear to have originated with Kehrmann's observations on the chemical inertness of quinones containing *ortho* substituents.[1] Investigation of similar phenomena in the esterification of hindered aromatic acids was later pursued with considerable success by Victor Meyer[2] and numerous contemporary workers.[3] Indeed, the topic proved to be exceptionally popular, as indicated by the nearly 200 pages that were used to review it in 1928.[3]

Evidence continued to accumulate that steric effects were of widespread importance in chemical behavior. Conant and his coworkers noted that the tendency of the dialkyltetraphenylethanes (5.1) and the dialkylxanthates (5.2) to dissociate into free radicals increased with the increasing bulk of the alkyl groups (Me < Et < *i*-Pr < *t*-Bu).[4]

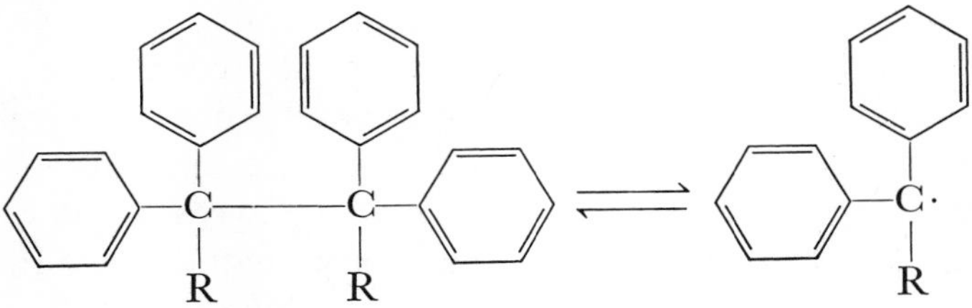

5.1

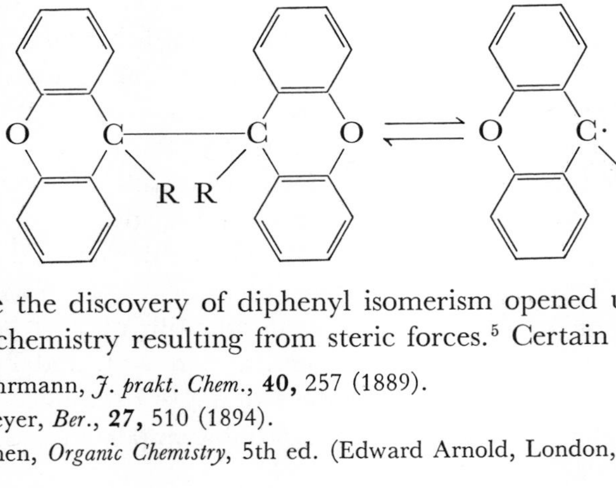

5.2

Likewise the discovery of diphenyl isomerism opened up an entire area of chemistry resulting from steric forces.[5] Certain peculiarities

[1] F. Kehrmann, *J. prakt. Chem.*, **40**, 257 (1889).

[2] V. Meyer, *Ber.*, **27**, 510 (1894).

[3] J. Cohen, *Organic Chemistry*, 5th ed. (Edward Arnold, London, 1928), Vol. I, Chap. 5.

[4] J. B. Conant and N. M. Bigelow, *J. Amer. Chem. Soc.*, **50**, 2041 (1928); J. B. Conant, L. F. Small, and A. W. Sloan, *J. Amer. Chem. Soc.*, **48**, 1743 (1926).

[5] E. E. Turner and R. J. W. LeFèvre, *Chem. and Ind.*, **45**, 831 (1926); R. Adams and H. C. Yuan, *Chem. Rev.*, **12**, 261 (1933).

in *ortho* substitution led Holleman to suggest that steric effects were an important factor in aromatic substitution,[6] and LeFèvre later utilized this explanation to account for the observed directive effects in the nitration and sulfonation of *p*-cymene[7] (5.3).

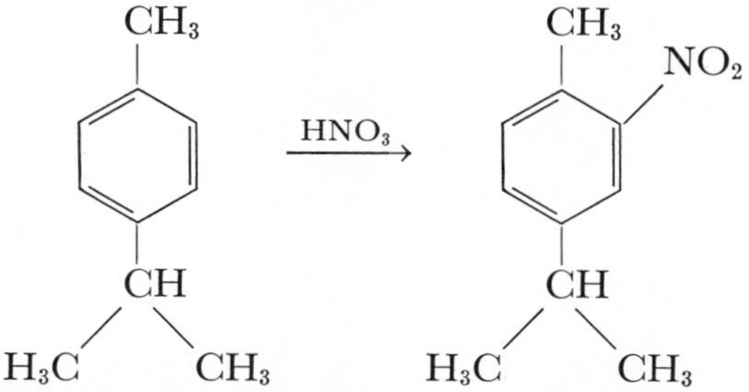

$$5.3$$

Moreover, Polanyi utilized the concept to account for the decreasing rates of the reactions of iodide ion with methyl, ethyl, and isopropyl chloride[8] (5.4).

$$I^- + CH_3Cl \gg I^- + \underset{H_3C}{\overset{H_3C}{\diagdown}} CHCl \qquad 5.4$$

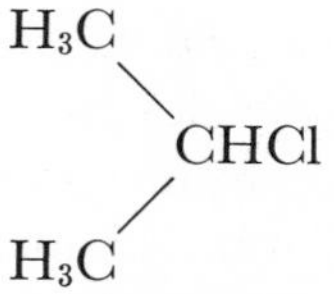

However, the immense success achieved by the electronic theory in the 1930's led to attempts to account for essentially all chemical behavior in terms of electronic effects. For example, Wheland suggested that the increased dissociation of the dialkyltetraphenylethanes (5.1) and the dialkyldixanthyls (5.2) with increasing bulk of the alkyl groups might be due to the increased possibility for resonance in which the alkyl groups participate.[9] Likewise, it was suggested that the directive effects in *p*-cymene were due, not to the relative steric requirements of the methyl and isopropyl groups, but instead to their relative ability to participate in hyperconjugation[10] (5.5).

[6] A. F. Holleman, *Chem. Rev.*, **1**, 187 (1924).

[7] R. J. W. LeFèvre, *J. Chem. Soc.*, 977, 980 (1933); *ibid.*, 1501 (1934).

[8] N. Meer and M. Polanyi, *Z. phys. Chem.*, **19B**, 164 (1932); A. G. Evans, *Trans. Faraday Soc.*, **42**, 719 (1946).

[9] G. W. Wheland, *J. Chem. Phys.*, **2**, 474 (1934).

[10] J. W. Baker and W. S. Nathan, *J. Chem. Soc.*, 1844 (1935).

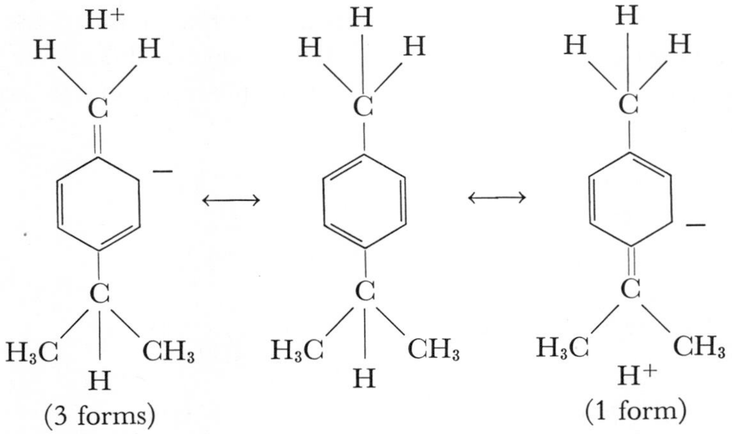

(3 forms)   (1 form)

Similarly, the decreased reactivity toward bimolecular displacement reactions in the series MeX, EtX, i-PrX, and t-BuX was attributed to the polar effects of the increasing number of the methyl groups attached to the central carbon.[11]

By 1940 these developments had engendered a widespread scepticism as to the importance of steric effects in chemical behavior. The excellent book published in that year by Professor Louis P. Hammett played a major role in the development of physical organic chemistry in the United States; steric effects are conspicuously absent from its discussions.[12] Indeed, he coined a new term, proximity effects, to discuss the influence of *ortho* substituents without committing himself to an interpretation of the origin of that influence. Similarly, the book published in 1944 by my colleague at Wayne University, Professor Remick, offers no discussion of the role of steric effects in organic theory.[13] Perhaps the prevalent view is best expressed in the words of a textbook of the day: "Steric hindrance . . . has become the last refuge of the puzzled organic chemist."[14]

If steric effects were again to receive serious consideration as a major factor in chemical behavior, a technique of study was re-

[11] C. K. Ingold, *Chem. Rev.*, **15**, 225 (1934); C. N. Hinshelwood, K. J. Laidler, and E. W. Timm, *J. Chem. Soc.*, 848 (1938).

[12] L. P. Hammett, *Physical Organic Chemistry* (McGraw-Hill, New York, 1940).

[13] A. E. Remick, *Electronic Interpretation of Organic Chemistry* (John Wiley, New York, 1944).

[14] F. E. Ray, *Organic Chemistry* (J. B. Lippincott, Philadelphia, 1941), p. 522.

quired that would provide quantitative, unambiguous data on the magnitudes of the steric effects in the systems under investigation. The dissociation of molecular addition compounds offered promise of providing such a tool, and I therefore undertook their study with my associates.

3. Qualitative Studies of Steric Strains

The initial investigations involved essentially qualitative studies of the relative stabilities of pairs of addition compounds.[15] Since the later quantitative studies proved much more satisfying, these initial studies will be described only briefly to give a taste of the approach.

Triethylamine ($pK_a = 10.65$) is a considerably stronger base than trimethylamine ($pK_a = 9.76$). This means that in aqueous solution triethylamine would compete effectively for an equivalent of protons (5.6).

$$Me_3N:H^+ + Et_3N \rightleftharpoons Me_3N + Et_3N:H^+ \qquad 5.6$$

Yet when these two bases are compared against trimethylborane as the reference acid, the equilibrium lies strongly to the left (5.7).

$$Me_3N:BMe_3 + Et_3N \rightleftharpoons Me_3N + Et_3N:BMe_3 \qquad 5.7$$

Similarly, the introduction of methyl groups into the pyridine molecule increases the strength of the base. Thus pyridine exhibits a pK_a of 5.17 and 2,6-lutidine a pK_a of 6.75. In the absence of steric effects one would expect 2,6-lutidine to form a more stable addition compound with trimethylborane than does pyridine. Yet the results reveal that pyridine reacts readily to form a stable addition compound at room temperature (5.8), whereas 2,6-lutidine fails to react even at $-80°$ (5.9).

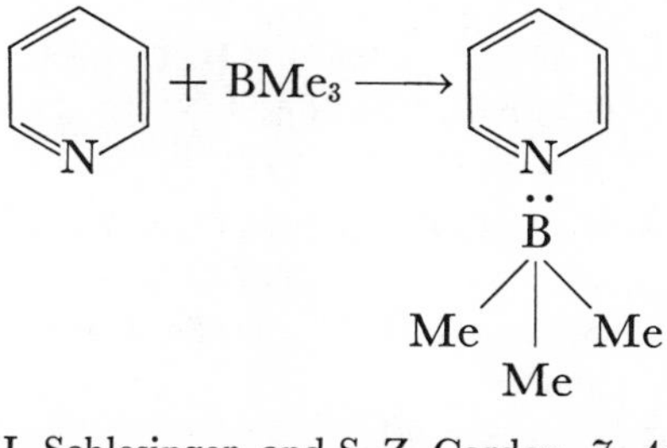

$$5.8$$

[15] H. C. Brown, H. I. Schlesinger, and S. Z. Cardon, *J. Amer. Chem. Soc.*, **64**, 325 (1942).

$$\text{(2,6-lutidine structure)} + BMe_3 \longrightarrow \text{No reaction} \qquad 5.9$$

4. Quantitative Studies of Steric Strains

The success achieved in these early qualitative experiments encouraged the development of quantitative methods of determining thermodynamic data for the dissociation of addition compounds. One approach has been the measurement of the dissociation constants for the gaseous systems over a range of temperatures.[16] The experimental results[17] are summarized in Table V-1.

This dissociation technique is applicable only to molecular addition compounds of a relatively restricted range of stability. In order to study additional compounds that lie outside this range, we had to go to calorimetric methods.[18] The experimental results[19] are summarized in Table V-2.

Though we cannot consider here all the problems that were explored, a discussion of four representative studies will indicate the general approach.

The introduction of a methyl group into the ammonia molecule increases the heat of dissociation for the trimethylborane addition

[16] H. C. Brown, M. D. Taylor, and M. Gerstein, *J. Amer. Chem. Soc.*, **66,** 431 (1944); H. C. Brown and M. Gerstein, *J. Amer. Chem. Soc.*, **72,** 2923 (1950).

[17a] H. C. Brown, H. Bartholomay, Jr., and M. D. Taylor, *J. Amer. Chem. Soc.*, **66,** 435 (1944). [b] H. C. Brown and M. D. Taylor, *J. Amer. Chem. Soc.*, **69,** 1332 (1947). [c] H. C. Brown, M. D. Taylor, and S. Sujishi, *J. Amer. Chem. Soc.*, **73,** 2464 (1951). [d] H. C. Brown and G. K. Barbaras, *J. Amer. Chem. Soc.*, **75,** 6 (1953). [e] H. C. Brown and S. Sujishi, *J. Amer. Chem. Soc.*, **70,** 2878 (1948). [f] H. C. Brown and G. K. Barbaras, *J. Amer. Chem. Soc.*, **69,** 1137 (1947). [g] H. C. Brown and M. Gerstein, *J. Amer. Chem. Soc.*, **72,** 2926 (1950). [h] S. Sujishi, Ph.D. Thesis, Purdue U., 1949. [i] E. A. Fletcher, Ph.D. Thesis, Purdue U., 1952.

[18a] H. C. Brown and R. H. Horowitz, *J. Amer. Chem. Soc.*, **77,** 1730 (1955). [b] H. C. Brown and D. Gintis, *J. Amer. Chem. Soc.*, **78,** 5378 (1956). [c] J. S. Olcott, Ph.D. Thesis, Purdue U., 1957. [d] S. Bank, Ph.D. Thesis, Purdue U., 1960. [e] J. G. Koelling, Ph.D. Thesis, Purdue U., 1962.

[19] See footnotes 20–32 in Chapter VII.

Table V–1. Thermodynamic Data for the Dissociation of Molecular Addition Compounds in the Gaseous Phase

Compound	K_{100}	ΔF°_{100}	ΔH	ΔS
Ammonia-trimethylboron[17a]	4.6	−1134	13.75	39.9
Methylamine-trimethylboron[17a]	0.0350	2472	17.64	40.6
Dimethylamine-trimethylboron[17a]	0.0214	2885	19.26	43.6
Trimethylamine-trimethylboron[17a]	0.472	557	17.62	45.7
Ethylamine-trimethylboron[17b]	0.0705	1965	18.00	43.0
Diethylamine-trimethylboron[17b]	1.22	−147	16.31	44.1
Triethylamine-trimethylboron[17b]	Unstable	—	~10	—
Methylamine-trimethylboron[17a]	0.0350	2472	17.64	40.6
Ethylamine-trimethylboron[17b]	0.0705	1965	18.00	43.0
n-Propylamine-trimethylboron[17c]	0.0598	2088	18.14	43.0
n-Butylamine-trimethylboron[17c]	0.0470	2266	18.41	43.2
n-Pentylamine-trimethylboron[17c]	0.0415	2359	18.71	43.9
n-Hexylamine-trimethylboron[17c]	0.0390	2404	18.53	43.2
Methylamine-trimethylboron[17a]	0.0350	2472	17.64	40.6
Ethylamine-trimethylboron[17b]	0.0705	1965	18.00	43.0
isoPropylamine-trimethylboron[17d]	0.368	740	17.42	44.7
sec-Butylamine-trimethylboron[17d]	0.373	732	17.26	44.3
tert-Butylamine-trimethylboron[17d]	9.46	−1665	12.99	39.3
Triethylamine-trimethylboron[17b]	Unstable	—	~10	—
Quinuclidine-trimethylboron[17e]	0.0196	2916	19.94	45.6
Pyridine-trimethylboron[17f]	0.301	890	17.00	43.2
2-Picoline-trimethylboron[17f]	Unstable	—	~10	—
3-Picoline-trimethylboron[17f]	0.138	1468	17.81	43.9
4-Picoline-trimethylboron[17f]	0.105	1671	(19.40)	(47.5)
Dimethylamine-trimethylboron[17a]	0.0214	2885	19.26	43.6
Ethyleneimine-trimethylboron[17g]	0.0284	2640	17.59	40.1
Trimethyleneimine-trimethylboron[17g]	0.000332	5960	22.48	44.3
Pyrrolidine-trimethylboron[17g]	0.00350	4190	20.43	43.5
Piperidine-trimethylboron[17g]	0.0210	2864	19.65	45.0
Methylphosphine-trimethylboron[17h]	Too highly dissociated to be measured			
Dimethylphosphine-trimethylboron[17h]	9.8	−1690	11.41	35.1
Trimethylphosphine-trimethylboron[17h]	0.128	1525	16.47	40.0
Methylphosphine-boron trifluoride[17i]	Too highly dissociated to be measured			
Dimethylphosphine-boron trifluoride[17i]	10.5	−1740	14.7	44.1
Trimethylphosphine-boron trifluoride[17i]	0.0669	1986	18.9	45.3

Table V–2.   Calorimetric Heats of Reaction of Pyridine Bases with Reference Acids

Substituent in Pyridine Base	pK_a, 25°	Heat of Reaction, $-\Delta H$, kcal/mole				$10^5 k_2$[a,f]	
		CH_3SO_3H[a,b]	$\frac{1}{2}(BH_3)_2$[a,c]	BF_3[a,d]	BMe_3[a,e]	CH_3I	E_{act}
Hydrogen	5.17	17.1	20.2	33.3	21.4	34.3	13.9
4-Me	6.02	18.4	20.8	33.8	22.0	76.0	13.6
4-Et	6.02	18.3		33.6		77.7	
4-*i*-Pr	6.02	18.4		33.7		76.7	
4-*t*-Bu	5.99	18.3		33.6		75.7	13.7
3-Me	5.68	17.8	20.7	33.6	21.7	71.2	13.6
3-Et	5.70	18.1		33.5		76.1	
3-*i*-Pr	5.72	18.0		33.6		81.0	
3-*t*-Bu	5.82	18.2		33.9		95.0	
2-Me	5.97	18.3	19.7	31.6	16.1	16.2	14.0
2-Et	5.92	18.2	19.2	31.0	15.1	7.64	14.2
2-*i*-Pr	5.83	18.1	19.0	30.2	13.7	2.45	14.8
2-*t*-Bu	5.76	18.0	10.3	23.1	0	0.008	17.5
2,6-Me$_2$	6.75	19.5	16.4	25.8		1.45	15.1
2,4,6-Me$_3$	7.59	20.7	17.2	26.8		3.75	14.8

[a] Nitrobenzene solution.   [b] Reaction: $CH_3SO_3H(soln) + Py(soln) \rightarrow PyH^{+-}O_3SCH_3$.
[c] Reaction: $\frac{1}{2}B_2H_6(g) + Py(soln) \rightarrow Py:BH_3(soln)$.   [d] Reaction: $BF_3(g) + Py(soln) \rightarrow Py:BF_3(soln)$.   [e] Reaction: $BMe_3(g) + Py(soln) \rightarrow Py:BMe_3(soln)$.   [f] Reaction: $CH_3I + Py \rightarrow Py:CH_3^{+}I^{-}$. 25°, l mole^{-1} sec^{-1}.

compound from 13.75 kcal mole^{-1} to 17.6. This increase of 3.9 kcal mole^{-1} is presumably the result of the inductive effect of the methyl group. The effect of a second methyl group is less, resulting in an increment of 1.6 kcal mole^{-1}. The third methyl group in trimethylamine results not in a further increase, but in a decrease to 17.6 kcal mole^{-1}.[17a] How can we account for this reversal?

If we make the reasonable assumption that the inductive effect of the methyl group should be additive, the predicted value of ΔH for trimethylamine is 25.3 kcal mole^{-1}. Thus there is a discrepancy of 7.7 kcal mole^{-1}—no small factor—to be accounted for. This has been attributed to steric interactions between the six methyl groups of the addition compound (**5.1**), resulting in strains that are relieved in the dissociation.

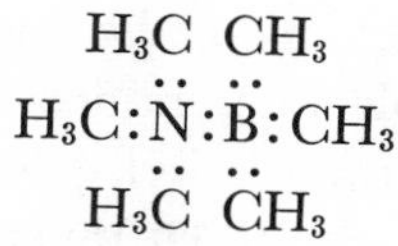

$$H_3C \quad CH_3$$
$$H_3C : \overset{..}{\underset{..}{N}} : \overset{..}{\underset{..}{B}} : CH_3$$
$$H_3C \quad CH_3$$

5.1

There is no great difference in the base strengths of methylamine, ethylamine, isopropylamine, and *t*-butylamine. Consequently, if steric effects were not a factor, we should expect to find similar heats of dissociation for the corresponding trimethylborane addition compounds. The value increases slightly from 17.6 for methylamine to 18.0 for ethylamine. Then there is a slight drop to 17.42 for isopropylamine. However, a major decrease is observed for *t*-butylamine, 13.0 kcal mole^{-1}.[17d] An examination of molecular models reveals major steric interactions if the molecule is to retain reasonable bond angles (**5.2**).

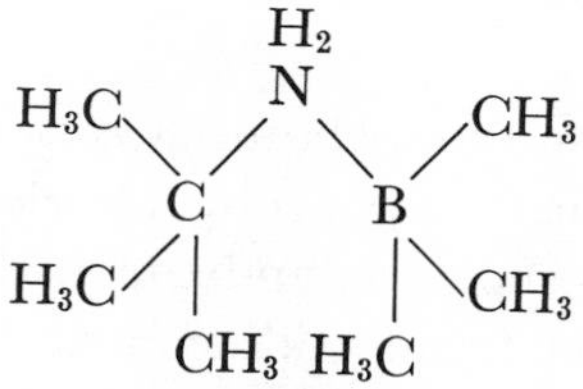

5.2

The heat of dissociation of ammonia-trimethylborane is 13.75 kcal mole^{-1}. This rises to 17.6 for methylamine- and to 18.0 for ethylamine-trimethylborane. These increases are readily attributable to the inductive effects of the methyl and the ethyl groups. The heat of dissociation of trimethylamine-trimethylborane is 17.62, not significantly different from that of the methylamine derivative. However, triethylamine-trimethylborane is a highly unstable substance with a heat of dissociation in the neighborhood of 10 kcal mole^{-1}. How can we account for the observation that one ethyl group has about the same effect as one methyl group, and that three methyl groups exert an effect not significantly different from the effect of one methyl group, but that three ethyl groups bring about an enormous decrease in the stability of the addition compound?

A possible explanation is that the steric requirements of three ethyl groups are far larger than those of three methyl groups, resulting in much larger steric strains in the addition compound. Indeed, an examination of the molecular models suggests that it is possible to rotate only two of the three ethyl groups out of the path of the trimethylboron molecule—the third must project in such a way as to interfere with the adding molecule (**5.3**).

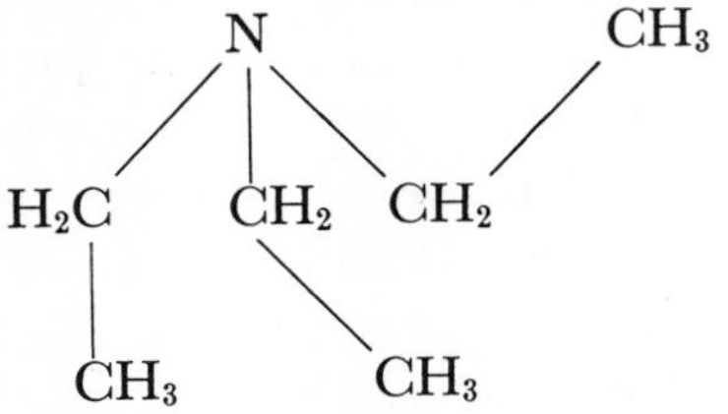

5.3 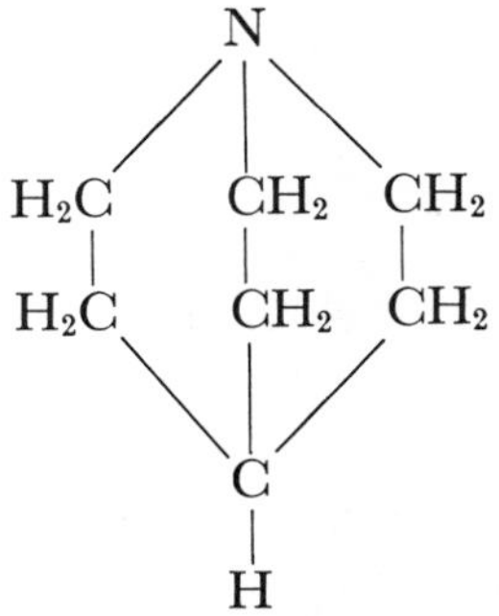   **5.4**

   To test this explanation quinuclidine was synthesized and tested. In this molecule the third ethyl group of triethylamine has been rotated to the rear of the molecule and effectively held there as part of the cage structure (**5.4**). Quinuclidine forms an exceedingly stable addition compound with trimethylborane.[17e] The heat of dissociation, 20.0 kcal mole^{-1}, is the highest observed for any tertiary amine (Table V-1). It does not appear possible to account for the enormous difference between quinuclidine and triethylamine in terms of polar effects, whereas the steric interpretation provides a simple, reasonable explanation for the observed phenomena.

   Finally, the behavior of the pyridine bases toward a family of reference acids of increasing steric requirements provides convincing evidence for the need to consider steric effects. The introduction of a methyl group in the 3-position of pyridine results in an increase of the pK_a value from 5.17 for pyridine to 5.68 for 3-picoline. This increase is attributed to the inductive effect of the methyl group. 3-Picoline likewise exhibits a consistent increase over pyridine in its heat of reaction with methanesulfonic acid, diborane, boron trifluoride, and trimethylborane, all in nitrobenzene solution (Table V-2).

The effects of the methyl group in the 2-position do not exhibit the same consistency. The pK_a value and the heats of reaction with methanesulfonic acid show a regular increment with one and two methyl groups in the series, pyridine, 2-pyridine, 2,6-lutidine, suggesting that we are observing the polar effects of the methyl groups uncomplicated by their steric requirements. However, with borane, boron trifluoride, and trimethylborane the heats of reaction exhibit a decrease rather than the expected increase. Moreover, the discrepancy increases quite sharply with the increasing steric requirements of the reference acid (Table V-3).

Table V-3. Comparison of Pyridine, 2-Picoline, and 2,6-Lutidine with Various Reference Acids

Reaction			
pK_a	5.17	5.97	6.75
$-\Delta H, CH_3SO_3H$	17.1	18.3	19.5
$-\Delta H, \frac{1}{2}(BH_3)_2$	20.2	19.7	16.4
$-\Delta H, BF_3$	33.3	31.6	25.8
$-\Delta H, BMe_3$	21.4	16.1	No reaction

The results are represented graphically in Fig. V-1. It does not appear possible to account for these results in terms of any known electronic effects of the different reference acids. However, a simple explanation is afforded in terms of the acids' steric requirements. In the case of the proton in water (pK_a) and methanesulfonic acid, the steric requirements are so small that a regular increase in base strength results from the presence of one and two methyl groups in the pyridine base. Borane is large enough to cause steric interaction with the methyl groups in the 2-position. The resulting strains cause a small decrease in the heat of reaction of diborane with 2-picoline and a larger decrease in the case of 2,6-lutidine. With the increasing steric requirements of boron trifluoride and trimethylborane, the strains become much larger, and the observed heats of reaction drop sharply below those to be expected. Finally, trimethylborane fails

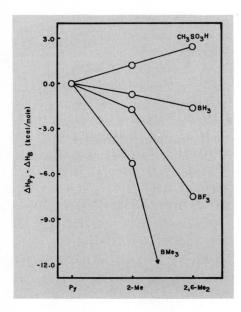

Figure V-1. Relative strengths of pyridine, 2-picoline, and 2,6-lutidine with reference acids of increasing steric requirements.

to add to 2,6-lutidine, even though it is a stronger base than pyridine itself.

5. Conclusion

These results appeared to provide convincing evidence for the importance of steric effects in organic chemistry.[20] We proceeded to apply this technique to a number of problems, such as exploring buttressing and rotational effects (Chapter VI), secondary isotope effects as a steric phenomenon (Chapter VI), the effect of molecular shape on chemical behavior (Chapter VII), steric strains in the transition state of S_N2 displacement reactions (Chapter VIII), and the chemical effects of steric strains in highly branched carbon compound and ring systems (Chapter VIII).

[20] H. C. Brown, *J. Chem. Soc.*, 1248 (1956).

VI. Exploration of Steric Phenomena with Molecular Addition Compounds

1. Molecular Addition Compounds—A Powerful Tool [1]

The ionization of simple acids and bases has played a major role in the development of modern theories of organic chemistry. Indeed, many of the currently accepted theoretical concepts receive their most direct and best confirmation from acid-base phenomena.[2-4] It is probable that the major role this reaction has played in the development of organic theory is due not to any inherent importance of the ionization reaction, but rather to the lack of other simple reversible reactions of wide applicability.

G. N. Lewis pointed out that the essential process in the ionization of classical acids and bases is the coordination of the proton with the free electron pair of the base.[5] He suggested, therefore, that molecules such as boron trichloride and stannic chloride, which also possess the ability to accept the electron pair of a base, should also be considered to be acids. Whether or not this definition is accepted, it is apparent that the coordination of such Lewis acids with bases represents a simple reversible process of far greater versatility than the original reaction involving the transfer of a proton (6.1).

$$H^+ + :Base \rightleftharpoons {}^+H:Base$$
$$\tfrac{1}{2}(BH_3)_2 + :Base \rightleftharpoons H_3B:Base$$
$$BF_3 + :Base \rightleftharpoons F_3B:Base \qquad 6.1$$
$$BMe_3 + :Base \rightleftharpoons Me_3B:Base$$
$$AlMe_3 + :Base \rightleftharpoons Me_3Al:Base$$
$$GaMe_3 + :Base \rightleftharpoons Me_3Ga:Base$$

[1] H. C. Brown, *J. Chem. Educ.*, **36,** 424 (1959).

[2] H. B. Watson, *Modern Theories of Organic Chemistry*, 2d ed. (Oxford University Press, New York, 1941).

[3] G. W. Wheland, *The Theory of Resonance*, 2d ed. (John Wiley, New York, 1944).

[4] H. C. Brown, D. H. McDaniel, and O. Hafliger, "Dissociation Constants," Chap. 14 in E. A. Braude and F. C. Nachod, eds., *Determination of Organic Structures by Physical Methods* (Academic Press, New York, 1955).

[5] *J. Franklin Inst.*, **226,** 293 (1938).

In studying molecular addition compounds, we are able to introduce wide variations in the structures of both components, so we can study in detail the effect of structure and of substituents upon the stability of the molecular addition compounds. In this way it is possible to evaluate the importance of the classical electronic effects, induction and resonance, as well as steric effects arising from the conflicting steric requirements of the two components.

There is an important additional advantage in the study of these systems. Classical acid-base reactions must be studied in water or related solvents. Interpretation of the results is frequently rendered difficult by the complexities of separating solvation effects from the structural interactions under examination. On the other hand, molecular addition compounds can be examined in the vapor phase or in nonhydroxylic solvents, greatly simplifying the analysis of the data.

The proton is a charged particle of negligible steric requirements. As long as organic theory concerned itself primarily with the factors controlling the addition or removal of a proton from an organic moiety, it naturally emphasized the role of the electrical factor and neglected the steric factor. Extension of the study of acid-base phenomena to Lewis acids with varying steric requirements led to a clearer appreciation of the role of the steric factor in organic theory.

2. Steric Requirements of Alkyl Substituents

Earlier (Section IV-4) we discussed the phenomena observed in the interaction of a series of reference acids of increasing steric requirements, $CH_3SO_3H < BH_3 < BF_3 < BMe_3$, with the series of bases of increasing steric requirements, pyridine $<$ 2-picoline $<$ 2,6-lutidine. This series is a particularly simple one, since rotation of the methyl substituents in the pyridine bases does not significantly alter the steric requirements of the individual bases. It was of interest to explore the behavior of the 2-alkylpyridines (R = Me, Et, i-Pr, t-Bu) in order to establish how much effect rotation of the alkyl substituent will have on the behavior of the base.[6] The experimental data are

[6] H. C. Brown, D. Gintis, and L. Domash, *J. Amer. Chem. Soc.*, **78**, 5387 (1956).

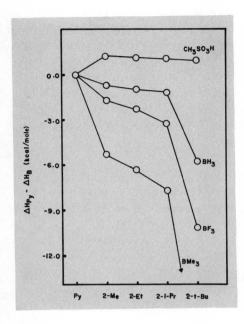

Figure VI-1. Relative strengths of pyridine and the 2-alkylpyridines with reference acids of increasing steric requirements.

reported in Table V-2 and reported graphically in Figure VI-1.

A simplistic approach might have led to the prediction that the steric requirements of the bases in question, pyridine, 2-methyl-, 2-ethyl-, 2-isopropyl-, and 2-*t*-butylpyridine, should increase monotonically. However, with reference acids of significant steric requirements there is observed a marked increase in steric effect with the introduction of the methyl group in the 2-position, but little further effect as the alkyl group is varied to ethyl and isopropyl. Only with *t*-butyl is a large additional effect observed (Figure VI-1).

This apparent large difference in the steric requirements of the four alkyl substituents is attributed to the ability of the ethyl and isopropyl groups to rotate in such a way as to minimize the steric interactions with the reference acid. Consequently, the steric effects of these two groups in these systems are but little larger than that of the methyl substituent. On the other hand, the *t*-butyl group possesses essentially a spherical symmetry and cannot minimize its steric requirements significantly through rotation.

3. Effect of Hindered Rotation on the Steric Requirements of Alkyl Substituents

The small decrease in the heat of reaction of a typical reference acid, such as boron trifluoride, with 2-ethylpyridine and 2-isopropylpyridine over that exhibited by 2-methylpyridine is reasonably accounted for in terms of a facile rotation of the ethyl and isopropyl groups in the addition compounds away from the adding acid (6.2–6.4).

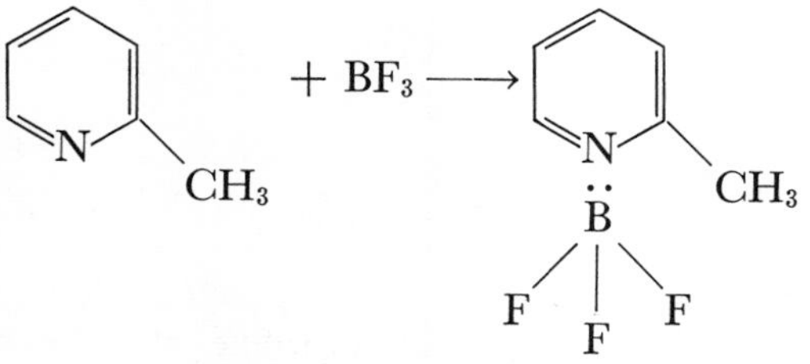

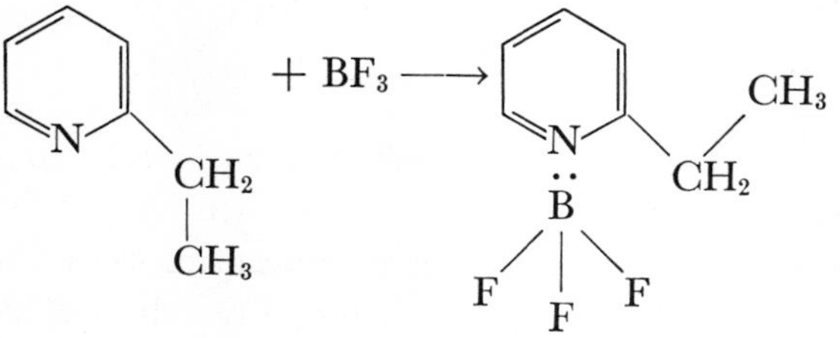

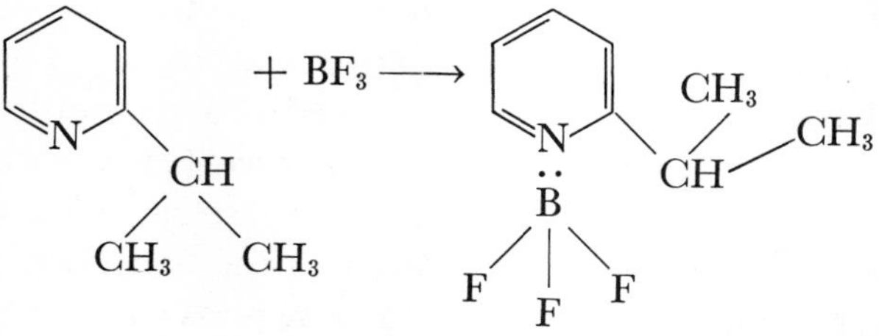

Consequently, in the addition compound the steric effects of the 2-ethyl and 2-isopropyl groups can be only slightly larger than that of the 2-methyl group. In the case of the 2-t-butyl group, rotation will have relatively little influence on the steric requirements (6.5).

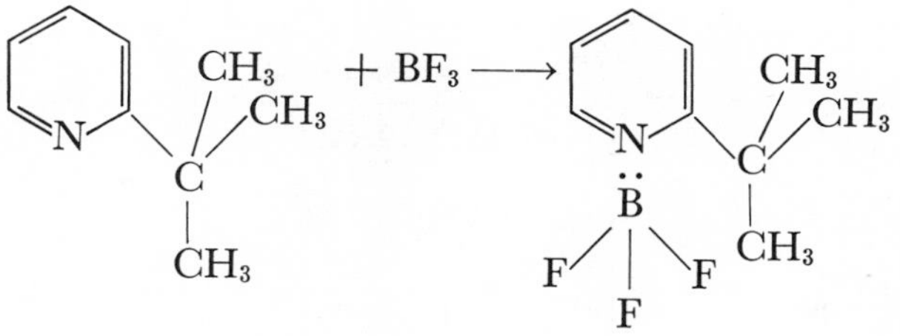

6.5

Indeed, the heat of reaction of boron trifluoride with 2-*t*-butylpyri-dine is approximately 8 kcal mole^{-1} lower than the values exhibited by the previous 2-alkylpyridines.

If this explanation is valid, a methyl group in the 3-position should hinder rotation of the 2-ethyl and 2-isopropyl group, greatly increasing their apparent steric requirements (6.7). On the other hand, a methyl group in the 5-position should have relatively little effect (6.6). We undertook to test this prediction.[7]

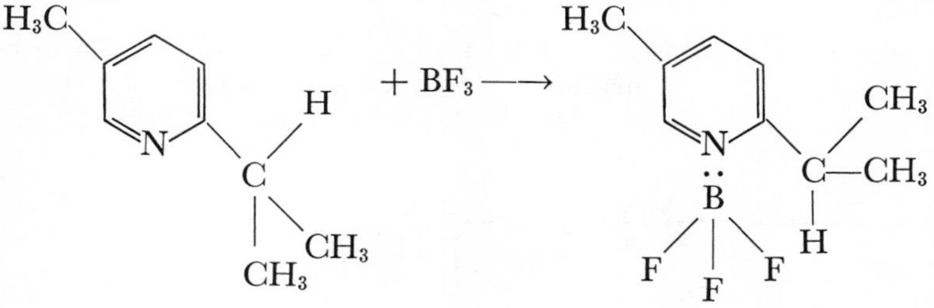

6.6

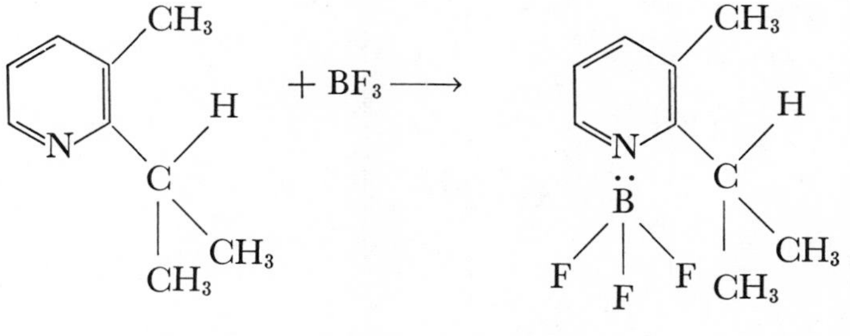

6.7

The results are given in Table VI-1 and Figure VI-2. It is quite clear that the methyl group in the 3-position greatly enhances the steric requirements of the isopropyl group in the 2-position.

[7] J. A. Donahue, Ph.D. Thesis, Purdue U., 1957.

Table VI-1. Calorimetric Heats of Reaction of Dialkylpyridine Bases with Reference Acids

Substituent in Pyridine Base	pK_a, 25°	Heat of Reaction, $-\Delta H$, kcal mole^{-1}			$10^5 k_2^{a,e}$
		$CH_3SO_3H^{a,b}$	$\frac{1}{2}(BH_3)_2^{a,c}$	$BF_3^{a,d}$	CH_3I
2,3-Me$_2$	6.56	19.6	19.6	32.0	19.4
2,4-Me$_2$	6.72	19.7	20.6	32.5	38.7
2,5-Me$_2$	6.42	19.4	20.3	32.5	37.6
2,6-Me$_2$	6.75	19.5	16.4	25.8	1.45
2-Me-3-Et	6.59			31.7	16.4
2-Me-3-i-Pr	6.63			31.3	17.4
2-Me-3-t-Bu	6.81		18.9	30.2	11.2
2-Me-5-Et	6.45			32.6	36.2
2-Me-5-i-Pr	6.50			32.4	40.3
2-Me-5-t-Bu	6.60		20.2	32.5	43.7
2-Et-3-Me	6.39			30.9	8.13
2-Et-5-Me	6.39			31.8	18.4
2-i-Pr-3-Me	6.36		14.6	26.8	0.105
2-i-Pr-5-Me	6.32		18.9	30.8	5.89

[a] Nitrobenzene solution. [b] Reaction: $CH_3SO_3H(soln) + Py(soln) \rightarrow PyH^{+-}O_3SCH$. [c] Reaction: $\frac{1}{2}B_2H_6(g) + Py(soln) \rightarrow Py{:}BH_3(soln)$. [d] Reaction: $BF_3(g) + Py(soln) \rightarrow Py{:}BF_3(soln)$. [e] Reaction: $CH_3I + Py \rightarrow Py{:}CH_3^{+}I^{-}$. 25°, l mole^{-1} sec^{-1}.

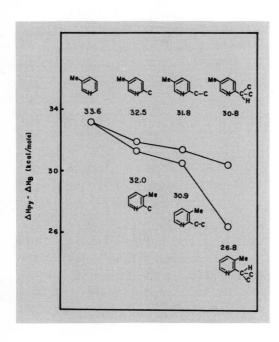

Figure VI-2. Effect of rotation and hindered rotation on the steric requirements of alkyl groups.

4. Effect of Buttressing on the Steric Requirements of Alkyl Substituents

This experimental approach offered an opportunity to explore the importance of buttressing effects.[8] Accordingly, we synthesized the entire series of 2-methyl-3-alkylpyridines (6.8) and 2-methyl-5-alkyl-pyridines (alkyl = Me, Et, i-Pr, and t-Bu) (6.9) and determined their heats of reaction with boron trifluoride[9,10] (Table VI-1).

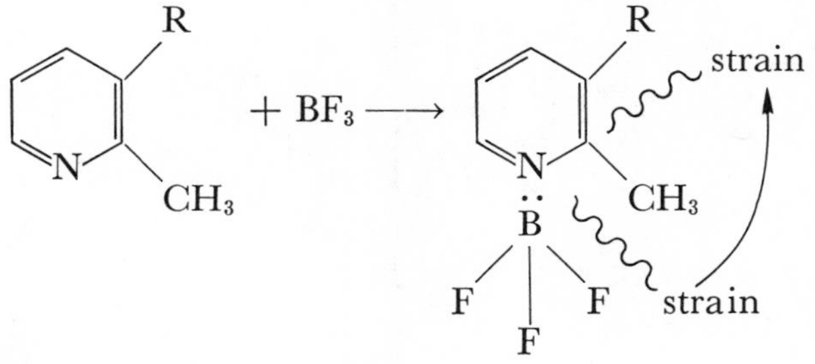

6.8

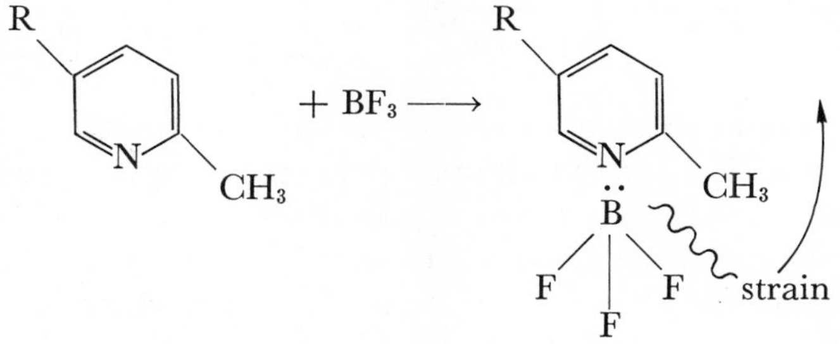

6.9

As anticipated, the data reveal only a small increase in the effect of the 3-alkyl group as one proceeds from 3-Me to 3-Et to 3-i-Pr. However, a 3-t-Bu group exhibits a significantly larger effect (Figure VI-3).

5. Secondary Isotope Effects as a Steric Phenomenon

Numerous secondary isotope effects observed in organic reactions have been interpreted in terms of classical effects that have long been

[8] M. Rieger and F. H. Westheimer, *J. Amer. Chem. Soc.*, **72,** 19 (1950).
[9] H. Podall, Ph.D. Thesis, Purdue U., 1955.
[10] M. S. Howie, Ph.D. Thesis, Purdue U., 1960.

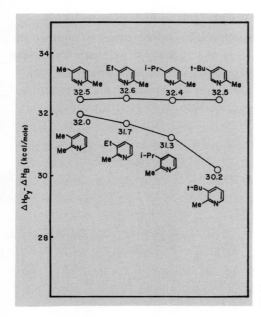

Figure VI-3. Effect of buttressing in the 2-methyl-3-alkyl- and 2-methyl-5-alkylpyridines.

attributed to the more usual substituents,[11] hyperconjugative, inductive, and steric.[12] The steric interpretation of β-deuterium isotope effects has been severely criticized,[13] principally on the basis of a sensitivity to conformational effects. However, the concept has been applied to account for the enhancement in the rates of racemization of 9,10-dihydro-4,5-dimethylphenanthrene that accompanies the introduction of deuterium into the methyl groups and the bridge positions,[14] and for the partial asymmetric alcoholysis of α-phenylbutyric anhydride in pyridine with (+)-S-2-propanol-1-d_3.[15]

It appeared desirable to apply the present technique to test the

[11] For an excellent review and discussion of secondary isotope effects, see E. A. Halevi, *Progr. Phys. Org. Chem.*, **1,** 109 (1963).

[12] L. S. Bartell, *J. Amer. Chem. Soc.*, **83,** 3567 (1961).

[13] V. J. Shiner, Jr. and J. S. Humphrey, Jr., *J. Amer. Chem. Soc.*, **85,** 2416 (1963).

[14] K. Mislow, R. Graeve, A. J. Gordon, and G. H. Wahl, Jr., *J. Amer. Chem. Soc.*, **86,** 1733 (1964).

[15] A. Horeau, A. Nouaille, and K. Mislow, *J. Amer. Chem. Soc.*, **87,** 4957 (1965).

utility of the steric model to account for the results.[16,17] Accordingly, 2-, 3-, 4-picoline, and 2,6-lutidine were synthesized with deuterium in the methyl groups. The k_D/k_H rate ratios for the reaction with methyl iodide (and other derivatives) at 25.0° were determined[16] (**6.1, 6.2, 6.3,** and **6.4**).

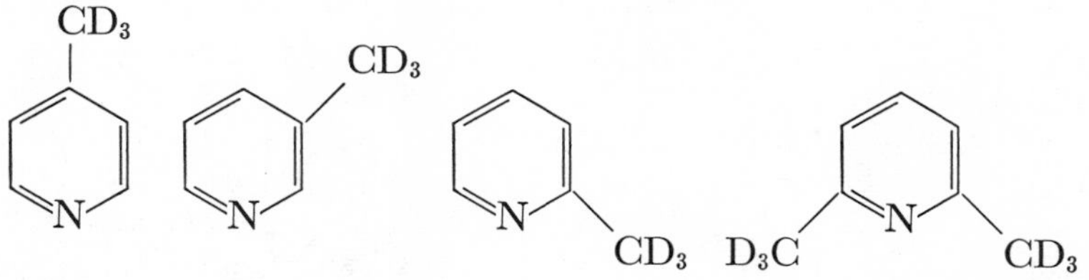

k_D/k_H	1.001	1.009	1.030	1.095
	± 0.003	± 0.002	± 0.003	± 0.003
	6.1	**6.2**	**6.3**	**6.4**

The results reveal that the —CD_3 groups in the 2-position cause a small increase in the rate with methyl iodide, in accord with the postulated smaller steric requirements of the —CD_3 group as compared to the —CH_3 group.

The heats of reaction of these bases with boron trifluoride and diborane (for the 2,6-derivatives) were also determined;[17] the results are summarized in Table VI-2. The data will be discussed in considerable detail to clarify the application of the present method.

It is apparent from the data that a 4-CD_3 substituent exerts no significant effect ($\Delta H_H - \Delta H_D = -0.04$ kcal mole^{-1}) upon the heat of reaction of 4-picoline with boron trifluoride. In the case of the 3-CD_3 substituent the observed difference is larger ($\Delta H_H - \Delta H_D = -0.10$ kcal mole^{-1}), but still within the experimental uncertainty. Supporting the conclusion that this difference is not significant, only the inductive effect of the substituent should be operative from the 3-position, and the observed change is in the wrong direction for such an effect.

In the 2-position the CD_3 substituent results in an enhanced heat

[16] H. C. Brown and G. J. McDonald, *J. Amer. Chem. Soc.*, **88**, 2514 (1966).

[17] H. C. Brown, M. E. Azzaro, J. G. Koelling, and G. J. McDonald, *J. Amer. Chem. Soc.*, **88**, 2520 (1966).

Table VI–2. Isotope Effects in the Heats of Reaction of the Methyl-d_3-pyridines with Boron Trifluoride and Diborane

		Heat of Reaction, $-\Delta H^a$ (kcal/mole)			
Pyridine	Ref. Acid	Previous value	Present Value	Average deviation[c]	$\Delta H_H - \Delta H_D$
4-Methyl	BF$_3$	33.8[b]	33.61	± 0.26(11)	
4-Methyl-d_3	BF$_3$		33.57	± 0.24(12)	−0.04
3-Methyl	BF$_3$	33.6[b]	33.52	± 0.11(9)	
3-Methyl-d_3	BF$_3$		33.42	± 0.23(11)	−0.10
2-Methyl	BF$_3$	31.6[b]	31.12	± 0.10(15)	
2-Methyl-d_3	BF$_3$		31.28	± 0.17(14)	0.16
2,6-Dimethyl	BF$_3$	25.8[b]	25.61	± 0.06(7)	
2,6-Dimethyl-d_6	BF$_3$		25.84	± 0.09(7)	0.23
2,6-Dimethyl	$\frac{1}{2}$(BH$_3$)$_2$	16.3[b]	16.42	± 0.08(9)	
2,6-Dimethyl-d_6	$\frac{1}{2}$(BH$_3$)$_2$		16.40	± 0.10(6)	−0.02

[a] Nitrobenzene solution at 25° for the reactions: Py(soln) + BF$_3$(g) → Py:BF$_3$(soln) and Py(soln) + $\frac{1}{2}$B$_2$H$_6$(g) → Py:BH$_3$(soln). [b] See Table V–2. The values are all corrected by 0.4 kcal/mole on the basis of more recent determinations of the heat of reaction of gasious boron trifluoride with pyridine. [c] The figures in the parentheses give the number of individual measurements that were averaged.

of reaction of 0.16 kcal mole^{-1}. Again this is near the limit of the experimental uncertainty, and no conclusion can safely be drawn. (However, the statistical analysis[17] indicates a 96 percent confidence level that the observed difference is significant.)

On the other hand, the reaction of 2,6-dimethyl-d_6-pyridine is more exothermic than the protium base by 0.23 kcal mole^{-1}. This appears to be significantly larger than the combined average deviation in the two measurements. (The statistical analysis[17] indicates a 99 percent confidence level that the observed difference is significant.)

No isotope effect was observed in the heat of reaction of diborane with 2,6-dimethyl-d_6-pyridine. It may appear surprising that the secondary isotope effect should not merely decrease, but apparently vanish with the smaller reference acid borane. However, the estimated steric strain for the reaction of diborane with 2,6-lutidine is approximately one-fourth of the corresponding strain for boron trifluoride.[6] A decrease of this magnitude would reduce the secondary isotope effect from 0.23 kcal mole^{-1} observed with boron trifluoride, down to 0.06 kcal mole^{-1}, within the range of the experimental variation.

The liberation of more energy in the formation of boron trifluoride addition compounds of methylpyridines in which the 2-methyl groups are labeled with deuterium indicates that such deuteration must stabilize the products. On the other hand, deuteration of the methyl groups more remote from the reaction center has no effect.

It has been pointed out in this discussion that the stabilities of the addition compounds of alkylpyridines with Lewis acids vary markedly and in a predictable manner with the steric requirements of alkyl groups in the 2-position of the pyridine ring. Consequently, it is reasonable to examine whether the results are likewise consistent with the operation of a steric effect. Indeed, in terms of this interpretation the smaller steric requirements of a —CD$_3$ group as compared to a normal methyl group[18] would result in diminished steric interactions with the boron trifluoride group, giving rise to greater stability of the adduct. The absence of any observable effect from the presence of —CD$_3$ groups in the 3- and 4-positions is likewise consistent with the steric interpretation.

The results rule out any significant hyperconjugative contributions in the present case. In the addition compounds there must be considerable positive charge on the nitrogen atom. Hyperconjugation with the methyl substituents should provide electron density to stabilize the charge. Thus, in terms of this explanation, a methyl group in the 4-position should be just as effective as a methyl group in the 2-position, and both should stabilize the normal addition over the deuterium analogs. Clearly this is contrary to the experimental results.

The proposal of a differential inductive effect of hydrogen and deuterium[11] also predicts an enhanced stability of the boron trifluoride adducts of the methyl-d_3-pyridines as a consequence of the proposed greater electron-donating capability of deuterium relative

[18] At the triple point deuterium and methane-d_4 have smaller molar volumes than hydrogen and normal methane: K. Clusius and K. Weigand, *Z. physik. Chem.*, **B46,** 1 (1940).

75

to hydrogen. However, an inductive effect would be expected to operate from the 3- and 4-positions nearly as effectively as from the 2-position. For example, a methyl substituent in the 2-, 3-, or 4-position of the pyridine ring results in a large, quite similar enhancement of the strength of the base (Table V-2). Consequently, the failure to observe any effect of the —CD_3 group in the 3- or 4-position, either in the rate of reaction with methyl iodide[16] or in the heat of reaction with boron trifluoride, argues against a significant inductive contribution in the present system.

In conclusion, the observation that significant isotope effects occur only with those labeled bases having —CD_3 groups located in a position where a steric effect can operate, together with the sensitivity of the isotope effect to the steric requirements of the reagent, strongly supports the conclusion that the fundamental origin of the secondary isotope effect in these reactions must reside in the decreased steric requirements of the —CD_3 substituents.

6. Application of a Test for Steric vs. Polar Basis

An important test of steric vs. polar contributions is provided by the relative effect of a substituent on the acidic and basic portion of a molecular addition compound. For example, the introduction of an electronegative substituent, such as halogen ($-I$), into the boron component (**6.5**) results in a stronger acid and an enhanced stability (in the absence of large steric effects). In the amine component (**6.6**) such electronegative substituents reduce the base strength and decrease the stability of the addition compound.

$$
\begin{array}{cc}
\begin{array}{c}
\text{H}_3\text{C} \quad\quad \text{CH}_3 \\
| \quad\quad\quad | \\
\text{H}_3\text{C}-\text{N} \;:\; \text{B}-\text{X} \\
| \quad\quad\quad | \\
\text{H}_3\text{C} \quad\quad \text{CH}_3 \\
\textbf{6.5}
\end{array}
&
\begin{array}{c}
\text{H}_3\text{C} \quad\quad \text{CH}_3 \\
| \quad\quad\quad | \\
\text{X}-\text{N} \;:\; \text{B}-\text{CH}_3 \\
| \quad\quad\quad | \\
\text{H}_3\text{C} \quad\quad \text{CH}_3 \\
\textbf{6.6}
\end{array}
\end{array}
$$

Thus, a polar effect manifests itself through opposing influences that accompany its presence in the acidic or basic component. On the other hand, if X is a group with a low polar effect but large steric requirements (such as t-butyl), a decrease in stability would be ob-

Table VI-3. Predicted and Observed Isotope Effects in Molecular Addition Compounds

Compound	Predicted isotope effects[a]			Observed effect
	Inductive theory	Hyper-conjugative theory	Steric theory	
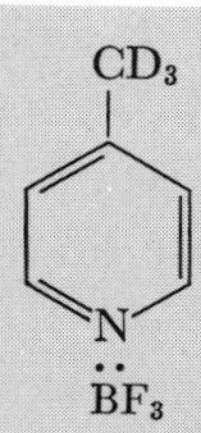 CD₃—pyridine·BF₃	Inverse	Normal	None	None
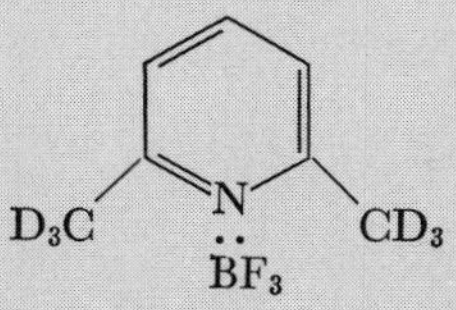 D₃C, CD₃ pyridine·BF₃	Inverse	Normal	Inverse	Inverse
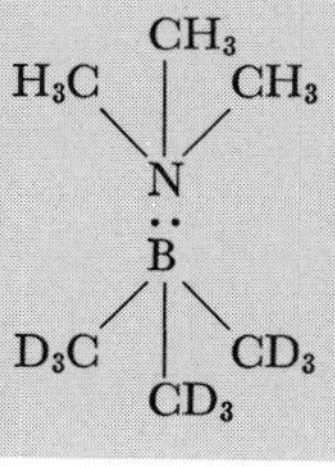 (H₃C)(CH₃)N·B(CD₃)(D₃C)(CD₃)	Normal	Inverse	Inverse	Inverse

[a] Predicted for formation of addition compound from its components.

served to accompany its introduction in either the amine or the boron component.

Fortunately, we can now apply this test to the present problem of secondary isotope effects. In the study described[17] the effect of deuterium substitution in the amine component of the molecular addition compound was established. The effect of deuterium substitution in the boron component has been examined earlier.[19] It was observed that trimethylborane-*d₉* forms a more stable addition com-

[19] P. Love, R. W. Taft, Jr., and T. Wartik, *Tetrahedron,* **5,** 116 (1959).

pound with trimethylborane. In other words, the following equilibrium (6.10) tends to the right with an equilibrium constant of 1.25.

$$(CD_3)_3B + (CH_3)_3B:N(CH_3)_3 \rightleftharpoons$$
$$(CH_3)_3B + (CD_3)_3B:N(CH_3)_3 \quad 6.10$$

Thus, deuterium substitution in the basic component brings about an increase in the stability of the addition compound. Likewise, substitution of deuterium in the acidic component results in an increase in the stability of the molecular addition compound. Clearly this is contrary to the predicted operation of a polar influence and in accordance with the predicted effect of a steric influence.

The summary in Table VI-3 supports the steric interpretation.

VII. Strained Homomorphs

1. Strains in Carbon Compounds from Molecular Addition Compounds

In the original study using this approach to steric effects[1] it was suggested that the study of molecular addition compounds might provide a convenient tool for the estimation of steric strains in related carbon compounds. The dimensions and geometry of the boron-nitrogen bond are almost identical with those of the carbon-carbon bond. It appeared reasonable to expect comparable strains in the addition compounds, trimethylamine-trimethylborane (**7.1**), *t*-butylamine-trimethylborane (**7.2**), and 2-picoline-trimethylborane (**7.3**), and in the related carbon compounds, hexamethylethane (**7.4**), di-*t*-butylmethane (**7.5**), and *o-t*-butyltoluene (**7.6**).

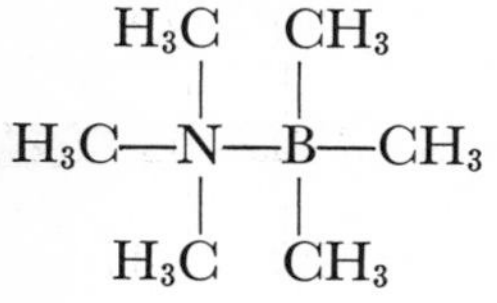

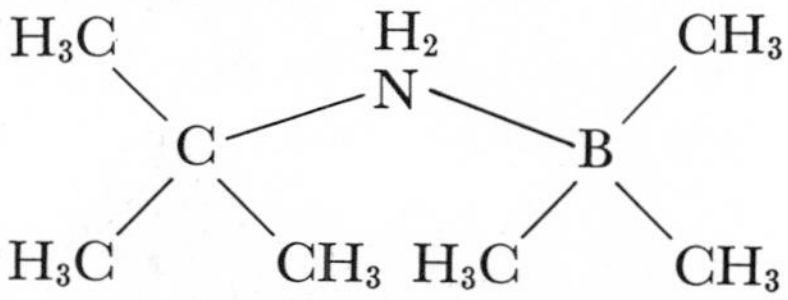

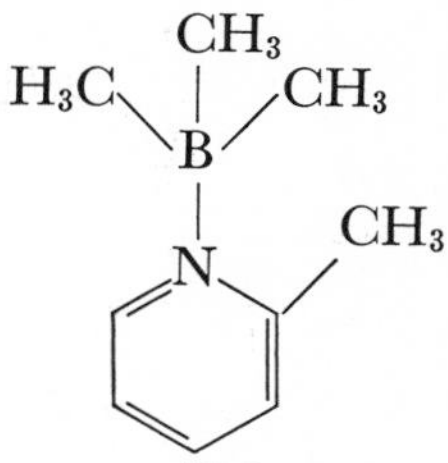

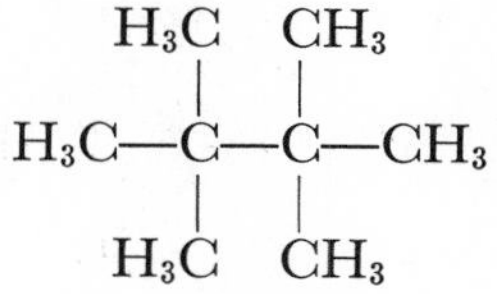

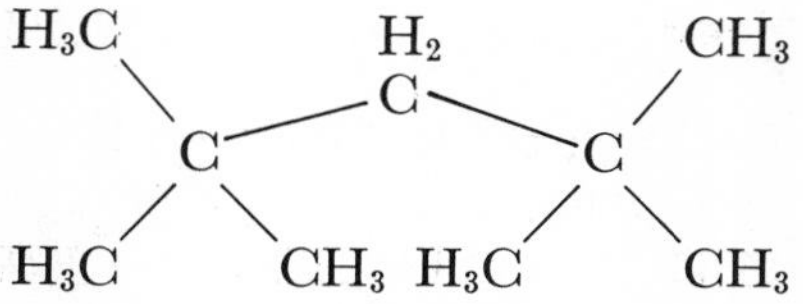

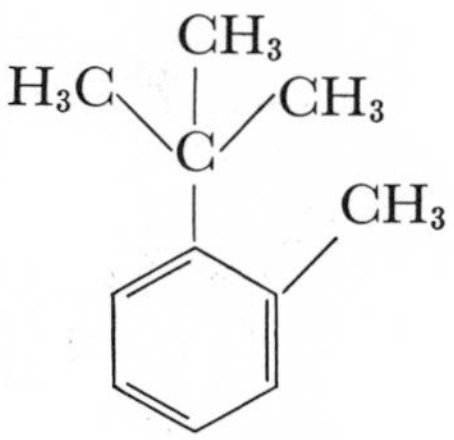

[1] H. C. Brown, H. I. Schlesinger, and S. Z. Cardon, *J. Amer. Chem. Soc.*, **64**, 325 (1942).

This possibility was explored and evidence obtained to support the thesis that strains, estimated from the stability of molecular addition compounds, persist in related organic molecules of similar sizes and shapes and exert a profound effect upon the chemical behavior of these substances.[2] It is convenient to have a term to designate molecules that have similar molecular dimensions; the term "homomorph" was proposed.[2]

2. Homomorphs of Di-t-Butylmethane

Typical homomorphs of di-t-butylmethane are indicated in Figure VII-1. If the values of the van der Waals radii for methyl, chlorine, borane, and the $-^{+}NH_3$ group are considered, it is evident that there must be considerable strain in these homomorphic molecules.

An estimate of this strain is provided by the dissociation data for t-butylamine-trimethylborane (Figure VII-1, A). The heat of dissociation of this compound is 13.0 kcal mole^{-1}, as compared to the value of 18.4 for the corresponding n-butylamine derivative (Table V-1). Since the amines do not differ significantly in base strength (n-butylamine, pK_a 10.61; t-butylamine, pK_a 10.82), the difference

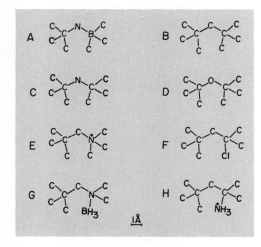

Figure VII-1. Homomorphs of di-t-butylmethane.

[2] H. C. Brown, G. K. Barbaras, H. L. Berneis, W. H. Bonner, R. B. Johannesen, M. Grayson, and K. L. Nelson, *J. Amer. Chem. Soc.*, **75**, 1 (1953).

in the heats of dissociation may be attributed to steric strain in the
t-butylamine derivative.

Combustion data permit an estimate of the strain present in di-t-
butylmethane (Figure VII-1, B). This approach leads to an estimated
value of 5.2 kcal mole^{-1} for the strain in the parent hydrocarbon.[2]

Similarly, heat-of-combustion data have been used to estimate the
strain energy present in di-t-butyl ether[3] (Figure VII-1, D). The
value realized, 7.6 kcal mole^{-1}, is moderately higher. However, all
three systems indicate that strains of appreciable magnitude should
be present in the homomorphs of di-t-butylmethane and should exert
a strong influence on their chemical behavior, especially in reactions
involving the rupture or making of bonds that would alter the strain.

From the similar dimensions of the chlorine atom and the methyl
group, it would be anticipated that strains of this magnitude should
also be present in neopentyldimethylcarbinyl chloride (Figure VII-1,
F). Such strain should be partially relieved in the transition state for
solvolysis (Chapter VIII). Consequently, the rate of solvolysis of
this chloride should be enhanced (7.1).

$$
\begin{array}{c}
\mathrm{H_3C}\diagdown\qquad\underset{\mathrm{C}}{\overset{\mathrm{H_2}}{\mathrm{C}}}\qquad\diagup\mathrm{Cl}\\
\mathrm{C}\qquad\qquad\mathrm{C}\\
\mathrm{H_3C}\diagup\quad\mathrm{CH_3}\ \mathrm{CH_3}\diagdown\mathrm{CH_3}
\end{array}
\ \longrightarrow\
\begin{array}{c}
\mathrm{H_3C}\diagdown\qquad\underset{\mathrm{C}}{\overset{\mathrm{H_2}}{\mathrm{C}}}\qquad\diagup\mathrm{CH_3}\\
\mathrm{C}\qquad\qquad\mathrm{C^+}\\
\mathrm{H_3C}\diagup\quad\mathrm{CH_3}\qquad\mathrm{CH_3}
\end{array}
\ +\ \mathrm{Cl^-}\qquad 7.1
$$

In agreement with this prediction, neopentyldimethylcarbinyl chloride undergoes solvolysis in 80 percent ethanol at a rate some sixteen
times that of n-butyldimethylcarbinyl chloride.[4]

This approach leads to the prediction that the reaction of neo-
pentyldimethylamine with methyl iodide, leading to the strained
homomorph neopentyltrimethylammonium ion (Figure VII-1, E)
should be strongly hindered. Indeed, the reaction (7.2) proceeds at
1 percent of the rate of the corresponding reaction of n-butyldi-
methylamine with methyl iodide.[5]

[3] E. J. Smutny and A. Bondi, *J. Phys. Chem.*, **65,** 546 (1961).
[4] H. C. Brown and H. L. Berneis, *J. Amer. Chem. Soc.*, **75,** 10 (1953).
[5] H. C. Brown and W. H. Bonner, *J. Amer. Chem. Soc.*, **75,** 14 (1953).

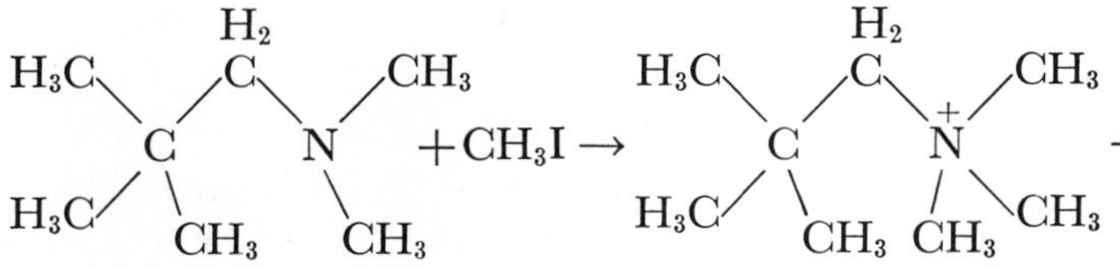

$$\underset{\substack{H_3C \\ H_3C}}{\overset{H_2}{C}} \underset{CH_3}{\overset{CH_3}{N}} + CH_3I \rightarrow \underset{\substack{H_3C \\ H_3C}}{\overset{H_2}{C}} \underset{CH_3\ CH_3}{\overset{+}{N}} \underset{CH_3}{\overset{CH_3}{}} + I^- \quad 7.2$$

3. Homomorphs of 2,6-Dimethyl-t-Butylbenzene

Trimethylborane forms a stable addition compound with pyridine with a heat of dissociation of 17.0 kcal mole^{-1} (Table V-1). No reaction occurs with 2,6-lutidine, a considerably stronger base in aqueous solution. Indeed, even at $-80°$ there is no evidence for combination.[6] The strain present in homomorphs of this structure (Figure VII-2, A) has been estimated to be in the neighborhood of 24 kcal mole^{-1}.[7]

The early attempts to synthesize homomorphs of 2,6-dimethyl-t-butylbenzene (Figure VII-2, C) failed. However, the chemical results are still of interest in revealing the influence of the large strains

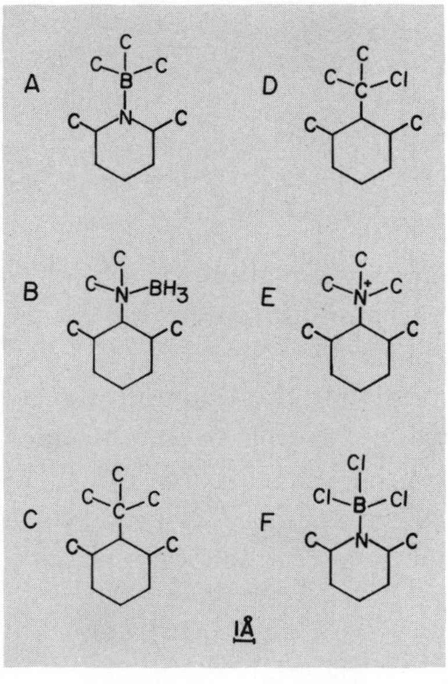

Figure VII-2. Homomorphs of 2,6-dimethyl-t-butylbenzene.

[6] H. C. Brown and R. B. Johannesen, *J. Amer. Chem. Soc.*, **75,** 16 (1953).
[7] H. C. Brown, D. Gintis, and L. Domash, *J. Amer. Chem. Soc.*, **78,** 5387 (1956).

present in the homomorphic structures. Thus, over a period of several months no significant reaction of 2,6-N,N-tetramethylaniline with methyl iodide (Figure VII-2, E) was observed.[8] Similarly, 2,6-N,N-tetramethylaniline adds boron trifluoride with difficulty. Moreover, the compound is unstable and readily loses boron trifluoride, suggesting a heat of reaction in the neighborhood of 10 kcal mole^{-1}.[6] In contrast, pyridine, a base of comparable strength, reacts readily with boron trifluoride, with $-\Delta H° = 33.3$ kcal mole^{-1} (Table V-2).

2,6-Dimethylphenyldimethylcarbinol and mesityldimethylcarbinol were synthesized. We attempted to convert these tertiary alcohols into the corresponding chlorides by treatment with hydrogen chloride. However, the alcohols were merely converted to the olefins, which neither add bromine nor decolorize cold permanganate solution. At temperatures of -30 to $-80°$ the olefins appeared to absorb considerable hydrogen chloride and were converted into red solids. However, on solution in alcohol at low temperatures the hydrogen chloride was instantly released, so that it was concluded that the tertiary chlorides (Figure VII-2, D) are not formed.[8]

Although the estimated strain is large, it is still only approximately one-third the energy of a carbon-carbon bond. Consequently, we thought it might be possible to synthesize the parent hydrocarbon by introducing a *t*-butyl group between two methyl groups through a Friedel-Crafts operation on mesitylene. However, under conditions that gave an excellent yield of 1,3-dimethyl-5-*t*-butylbenzene with *m*-xylene, no *t*-butylation of mesitylene occurred.[8] Relatively recently the successful preparation of 2,6-dimethyl-*t*-butylbenzene in a yield of 2.5 percent was achieved by the direct reaction of 2,6-dimethylphenylmagnesium bromide with *t*-butyl chloride in tetrahydrofuran.[9]

4. Homomorphs of o-Di-t-Butylbenzene

Trimethylborane also fails to add to 2-*t*-butylpyridine at low temperatures,[6] so the strain in this derivative (Figure VII-3, A) and in

[8] H. C. Brown and M. Grayson, *J. Amer. Chem. Soc.*, **75,** 20 (1953).

[9] A. W. Burgstahler, D. J. Malfer, and M. O. Abdel-Rahman, *Tetrahedron Letters*, 1625 (1965).

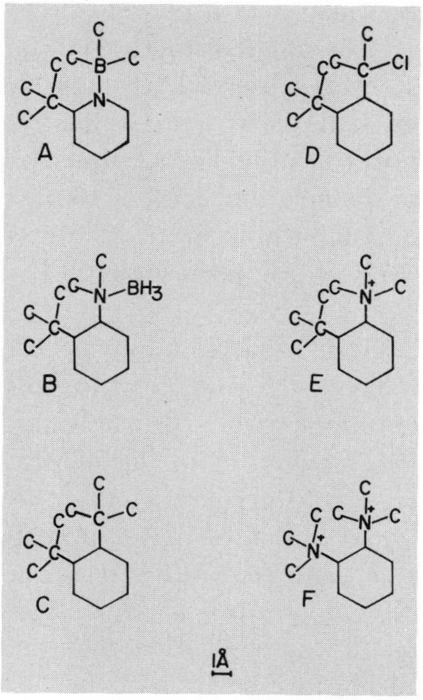

Figure VII-3. Homomorphs of *o*-di-*t*-butylbenzene.

the related homomorphs of *o*-di-*t*-butylbenzene (Figure VII-3) must be relatively high. Indeed, a value of 25.5 kcal mole^{-1} has been estimated for the strain present in these homomorphs.[7] Again we find an interesting chemistry of these derivatives, which reflects the operation of this large strain.

Thus treatment of *o*-*t*-butyl-N,N-dimethylaniline with methyl iodide for several months failed to yield even traces of the homomorphic *o*-*t*-butyltrimethylanilinium ion[10] (Figure VII-3, E). Similarly, *m*- and *p*-phenylenediamine readily react with six molar equivalents of methyl iodide to form the bis-quaternary ammonium salts, whereas *o*-phenylenediamine reacts with a maximum of five molar equivalents. The homomorphic structure (Figure VII-3, F) does not form.

In spite of the large strain estimated for the *o*-di-*t*-butylbenzene molecule, considerable success has been realized in recent years in

[10] H. C. Brown and K. L. Nelson, *J. Amer. Chem. Soc.*, **75**, 24 (1953).

developing syntheses for it,[11] and its chemistry has been explored.[12] Moreover, Arnett has recently measured the heat of isomerization of this hydrocarbon and has arrived at an estimate of 22 kcal mole^{-1} for the strain in this system,[13] a value in reasonable agreement with the earlier estimate.[7]

5. Homomorphs of Hemimellitene

Representative homomorphs of hemimellitene are shown in Figure VII-4. From the heat of reaction of diborane with 2,6-lutidine (Table V-2), a strain of approximately 5 kcal mole^{-1} is estimated for 2,6-lutidine-borane (Figure VII-4, A). On the other hand, the heats

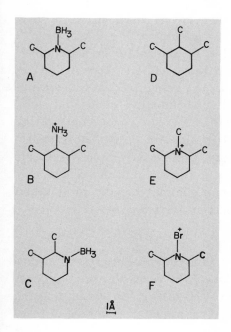

Figure VII-4. Homomorphs of hemimellitene.

[11] For leading references, see A. W. Burgstahler and M. O. Abdel-Rahman, *J. Amer. Chem. Soc.*, **85,** 173 (1963).

[12] A. W. Burgstahler, P.-L. Chien, and M. O. Abdel-Rahman, *J. Amer. Chem. Soc.*, **86,** 5281 (1964).

[13] E. M. Arnett, J. C. Sanda, J. M. Bollinger, and M. Barber, *J. Amer. Chem. Soc.*, **89,** 5389 (1967).

of combustion of the three trimethylbenzenes leads to an estimated strain of 1.2 kcal mole^{-1} for hemimellitene[14] (Figure VII-4, D). The steric requirements of the borane group may be significantly larger than those of a methyl group, and this may be responsible for the relatively large difference in the estimated strains.

The observed energy of activation for the reaction of methyl iodide with 2,6-lutidine (Figure VII-4, E) is 15.1 kcal mole^{-1} (Table V-2). This is 1.8 kcal mole^{-1} higher than the value estimated for an unhindered molecule (Section VII-7). Experience indicates that strains in the transition state are smaller than strains in the parent homomorph. Consequently, the strain estimated from the heat of combustion, involving as it does very small differences between large heats, may be low. In any event, the evidence is that strains in these homomorphs are relatively small, in the range of 3 ± 2 kcal mole^{-1}.

m-2-Xylidinium ion (Figure VII-4, B), the conjugate acid of m-2-xylidine, is also homomorphic with hemimellitene. It is of interest that this basis is unexpectedly weak, pK_a being 3.42 vs. 4.25 for aniline, whereas the operation of both the inductive effect and steric inhibition of resonance should tend to increase the strength of the first base. The postulate that such strains are present in other related *ortho* substituted aromatic bases provides a simple reasonable explanation of certain peculiarities in the strength of such bases.[15] This topic of possible steric effects accompanying the protonation of an amine will be discussed in Section VIII-5.

6. Homomorphs of o-t-Butyltoluene

Representative homomorphs of o-t-butyltoluene are shown in Figure VII-5. The heats of combustion of the isomeric t-butyltoluenes reveal that the *ortho* isomer (Figure VII-5, B) is less stable than the *meta* and *para* isomers by 5.6 ± 0.8 kcal mole^{-1}.[16] The heat of reaction of trimethylborane with 2-picoline (Figure VII-5, E) is 5.9 kcal

[14] W. H. Johnson, E. J. Prosen, and F. D. Rossini, *J. Research Natl. Bur. Standards*, **35**, 141 (1945).

[15] H. C. Brown and A. Cahn, *J. Amer. Chem. Soc.*, **72**, 2939 (1950).

[16] I. Jaffe and E. J. Prosen, *Abstracts of the American Chemical Society Meeting, Minneapolis, 1955*, p. 3R.

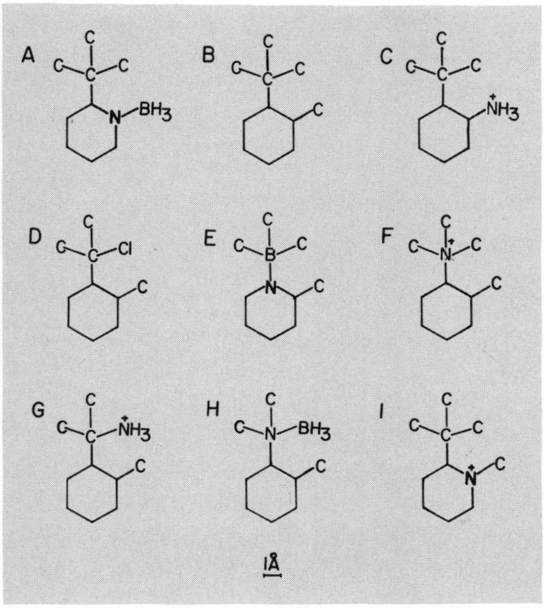

Figure VII-5. Homomorphs of *o-t*-butyltoluene.

mole^{-1} less than the value estimated in the absence of steric interactions.[7] Consequently, there is excellent agreement between these estimates. On the other hand, the heat of reaction of 2-*t*-butylpyridine with diborane (Table V-2) indicates the presence of strain amounting to 10.5 kcal mole^{-1}. This is considerably larger. As in the hemimellitene case previously mentioned, it may indicate that the steric requirements of a borane group are significantly greater than those of the methyl group.

The energies of activation for the reaction of methyl iodide with pyridine and 2-*t*-butylpyridine in nitrobenzene solution are 13.9 and 17.5 kcal mole^{-1}, respectively (Table V-2). This suggests a strain in the transition state leading to 2-*t*-butyl-N-methylpyridinium ion (Figure VII-5, I) of approximately 4 kcal mole^{-1}. Similarly, the energy of activation for the reaction of N,N-dimethyl-*o*-toluidine with methyl iodide to form the homomorph, *o*-tolyltrimethylammonium ion (Figure VII-5, F), is reported to be 21.1 kcal mole^{-1}, as compared to a value of 15.2 for dimethylaniline,[17] leading to an

[17] D. P. Evans, H. B. Watson, and R. Williams, *J. Chem. Soc.*, 1345, 1348 (1939).

Table VII–1. Steric Strains in Related Homomorphs

Parent Homomorph	Structures (strains in kcal/mole)		
Di-*t*-butylmethane	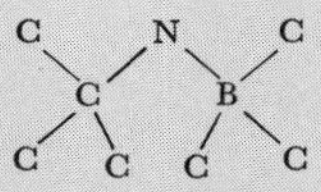 (5.4)	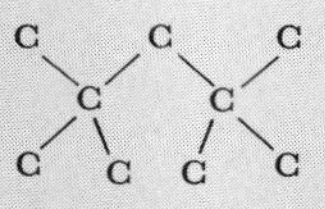 (5.2)	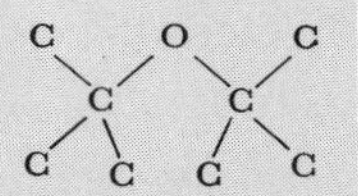 (7.6)
o-t-Butyltoluene	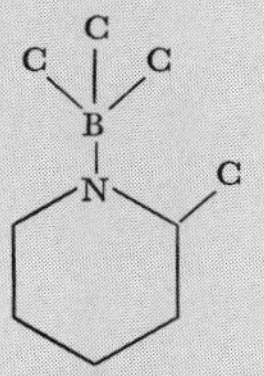 (5.9)	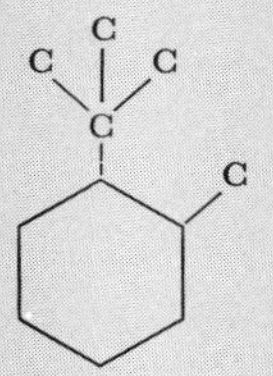 (5.6)	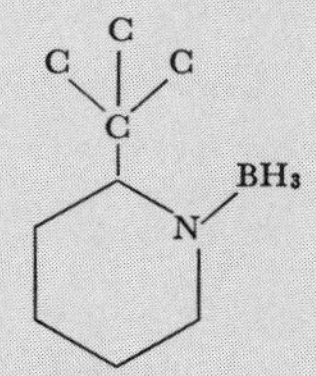 (10.5)
Hemimellitene	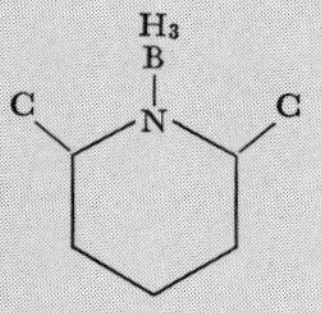 (2.7)	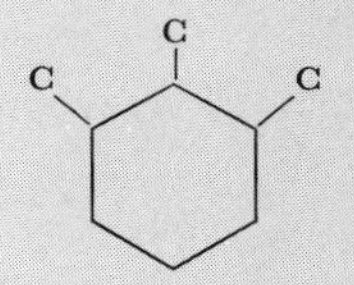 (1.6)	
o-Di-*t*-butylbenzene	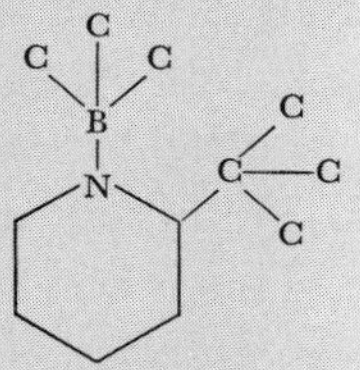 (25.5)	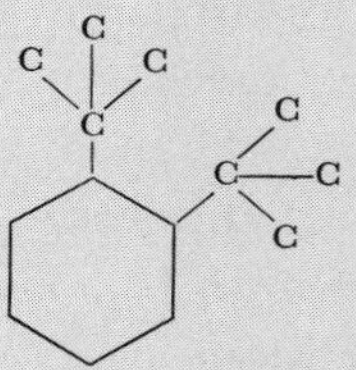 (22)	
2,6-Dimethyl-*t*-butylbenzene	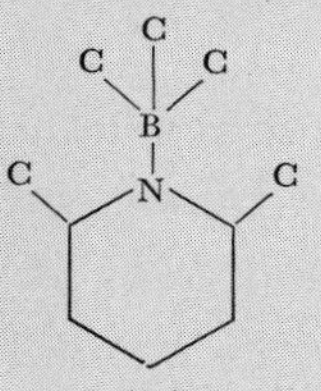 (24)		

estimate for the strain of 5.9 kcal mole^{-1}. Finally, it is significant that *o-t*-butylaniline is quite weak in water, pK_a being 3.78 vs. 4.58 for aniline itself.[18]

These results are all consistent with the influence of a moderate

[18] B. M. Wepster, *Rec. tran. Chim.*, **76,** 357 (1957).

amount of strain in these related homomorphs of *o-t*-butyltoluene.

Table VII-1 summarizes the data for steric strains in related homomorphs.

7. Steric Effects in the Transition State

The phenomena observed in typical displacement reactions resemble closely those noted in the behavior of related addition compounds. It was therefore suggested that the steric strains in the transition states of such displacement reactions (**7.7**) might be very similar to the strains in addition compounds of related structures (**7.8**).

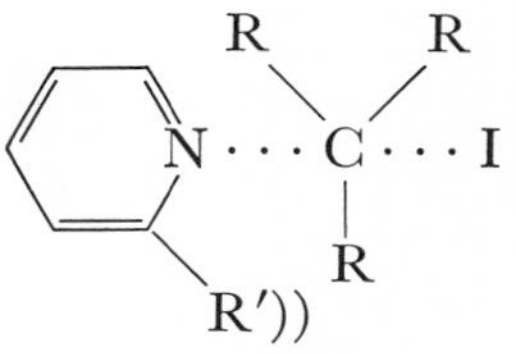

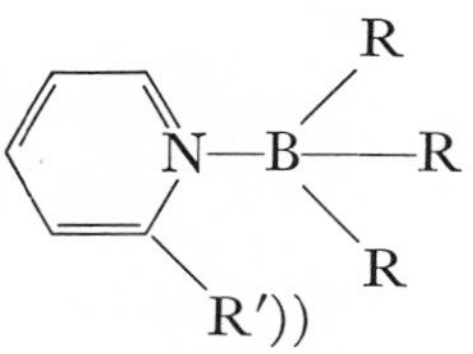

To test this proposal required considerable data. Over a period of some fifteen years the following investigations were carried out.

(1) Development of synthetic methods and the synthesis of the 2-, 3-, and 4-monoalkylpyridines[19] and selected dialkylpyridines.[20-22]

(2) Development of a spectroscopic method for the measurement of pK_a and its application to the monoalkylpyridines[23] and dialkylpyridines.[21]

(3) Measurement of the heats of reaction of methanesulfonic acid with the monoalkylpyridines.[24]

(4) Development of a calorimetric technique and its application to the determination of the heats of reaction of boron trifluoride-tetrahydropyran with the monoalkylpyridines.[25]

---

[19] H. C. Brown and W. A. Murphey, *J. Amer. Chem. Soc.*, **73**, 3308 (1951).

[20] H. Podall, Ph.D. Thesis, Purdue U., 1955.

[21] J. A. Donahue, Ph.D. Thesis, Purdue U., 1957.

[22] M. S. Howie, Ph.D. Thesis, Purdue U., 1960.

[23] H. C. Brown and X. R. Mihm, *J. Amer. Chem. Soc.*, **77**, 1723 (1955).

[24] H. C. Brown and R. R. Holmes, *J. Amer. Chem. Soc.*, **77**, 1727 (1955).

[25] H. C. Brown and R. H. Horowitz, *J. Amer. Chem. Soc.*, **77**, 1733 (1955).

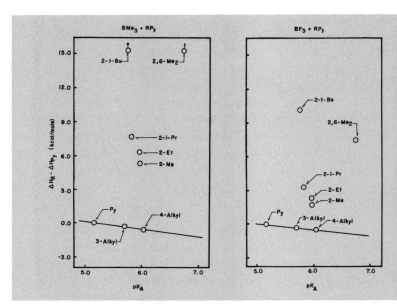

Figure VII-6. Relationship between the strengths of the pyridine bases and their heats of reaction with reference acids of varying steric requirements.

(5) Development of a vacuum-line calorimeter and its application to the determination of the heats of reaction of the monoalkylpyridines with trimethylborane[26] and with diborane.[27]

(6) Development of an improved vacuum-line calorimeter and its application of the heats of reaction of the mono- and selected dialkylpyridines with gaseous boron trifluoride[28,29] and with diborane.[30]

(7) Development of a precise kinetic technique and its application to the reaction of alkyl iodides with the monoalkylpyridines[31] and selected dialkylpyridines.[20-22]

The data, the results of thirteen Ph.D. theses, are summarized in Tables V-2 and VI-1. It is difficult to describe in mere words the effort required in carrying through a closely integrated program of this kind.

[26] H. C. Brown and D. Gintis, *J. Amer. Chem. Soc.*, **78,** 5378 (1956).
[27] H. C. Brown and L. Domash, *J. Amer. Chem. Soc.*, **78,** 5384 (1956).
[28] J. S. Olcott, Ph.D. Thesis, Purdue U., 1957.
[29] S. Bank, Ph.D. Thesis, Purdue U., 1960.
[30] J. G. Koelling, Ph.D. Thesis, Purdue U., 1962.
[31] H. C. Brown and A. Cahn, *J. Amer. Chem. Soc.*, **77,** 1715 (1955).

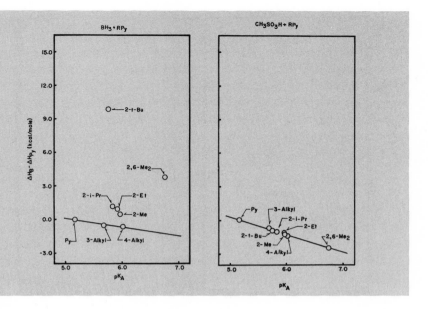

To observe the effect of decreasing steric requirements of the reference acid on the achievement of a linear energy relationship, the values for the heats of reaction of the various reference acids with the pyridine bases are plotted against their pK_a values (Figure VII-6). In the case of trimethylboron, pyridine and the 3- and 4-alkyl derivatives define a straight line, but the 2-alkyl derivatives deviate from the line. Moreover, the extent of the deviation varies with the steric requirements of the substituent.

Use of a reference acid of smaller steric requirements, boron trifluoride, moves the points toward the line, and the deviation decreases further with borane as the reference acid. Finally, an excellent linear relationship is achieved with methanesulfonic acid as the reference acid.

A similar plot of the log k_2 values for the reaction with methyl iodide against the pK_a values gives very similar results (Figure VII-7). This suggests that the steric effects in the transition state must be related to the steric effects in the related molecular addition compounds.

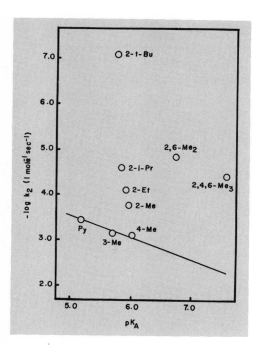

Figure VII-7. Relationship between the strengths of the alkylpyridine bases and their rates of reaction with methyl iodide.

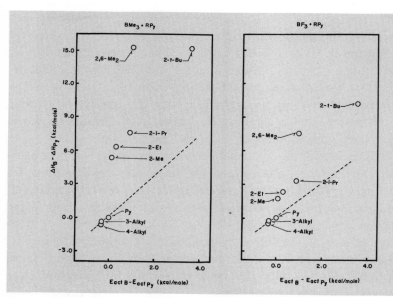

Figure VII-8. Relationship between the strengths of the pyridine bases and the energies of activation for their reaction with methyl iodide.

This conclusion is confirmed by Figure VII-8. Here also the 2-alkyl derivatives deviate from the line with trimethylborane. The deviations diminish with boron trifluoride as reference acid. A reasonably good linear relationship is achieved with borane as the reference acid. Finally, the use of methanesulfonic acid results in deviations in the opposite direction.

Consequently, there appears to be no significant difference between the rate and equilibrium data. When the steric requirements of the reaction on the vertical axis are greater than those on the horizontal axis, deviations from a linear relationship are observed, and the magnitude of the deviations increases with increasing steric requirements of the reference acid. Only when the steric requirements are similar (ΔH, CH_3SO_3H vs. pK_a, and ΔH, BH_3 vs. E_{act}, CH_3I) do the data define satisfactory linear relationships including the 2-alkyl substituent.

These results suggest that molecular addition compounds of the pyridine bases may serve as reasonably good models for transition states involving these bases and alkyl halides.

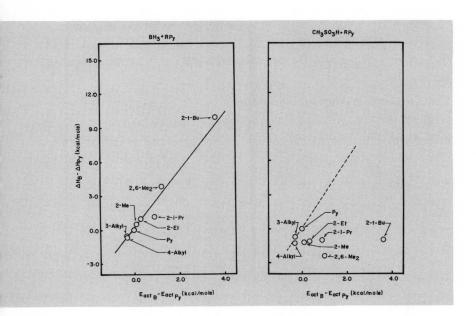

VIII. Chemical Effects
of Steric Strains[1]

1. Background

In Section V-2 I pointed out the unusual historical development of the concept of steric hindrance in organic chemistry. Originally developed by Kehrmann in 1889 and by Victor Meyer in 1894, the concept was actively explored and applied for many years, until in 1928 it constituted a major section of the advanced textbook by Cohen.[2] Yet by 1940 the concept had practically disappeared from discussions of the theory of organic chemistry.

Our original experiments with molecular addition compounds (Chapters V and VI) provided convincing evidence for the importance of steric effects in these derivatives. Our explorations of the chemistry of strained homomorphs (Chapter VII) persuaded us that such steric effects must also be important in the related carbon compounds. Accordingly, we undertook to explore suitable systems of carbon compounds for experimental evidence of the operation of steric effects.

2. Steric Effects in Aromatic Substitution

LeFèvre had pointed out that preferential substitution of 1-methyl-4-isopropylbenzene in the 2-position (8.1) could reasonably be ac-

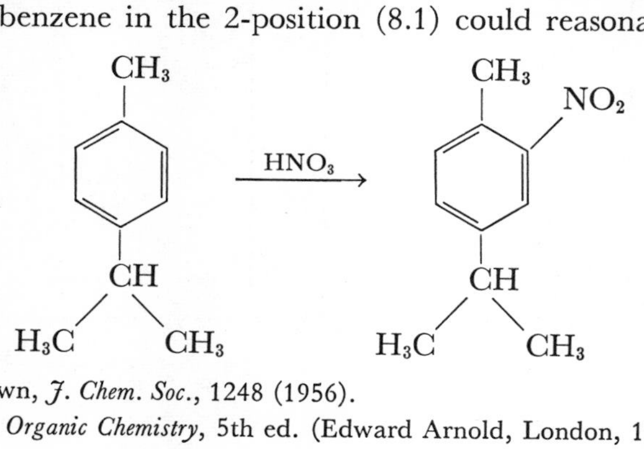

8.1

---

[1] H. C. Brown, *J. Chem. Soc.*, 1248 (1956).

[2] J. Cohen, *Organic Chemistry*, 5th ed. (Edward Arnold, London, 1928), Vol. I, Chap. 5.

counted for in terms of the difference in the steric requirements of the two alkyl groups.[3]

However, at the time that steric effects were being relegated to the background, it was suggested that this result could be accounted for in terms of the relative hyperconjugative contributions of the two groups.[4,5] It was considered that the greater possibilities for hyperconjugation involving the three α-hydrogen atoms of the methyl group as compared to the one α-hydrogen atom of the isopropyl group would strongly favor electrophilic substitution *ortho* to the methyl group (8.2).

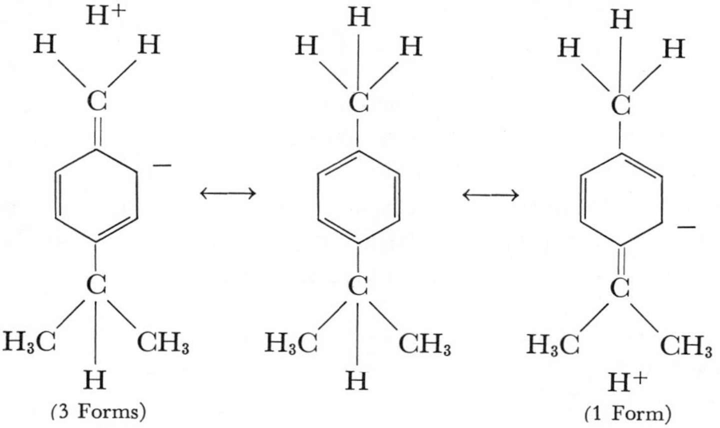

It appeared that the ambiguity might be resolved by a study of the mononitration of the monoalkylbenzenes, RC_6H_5 (R = Me, Et, *i*-Pr, *t*-Bu). Hyperconjugative effects should be transmitted to both the *ortho* and *para* positions. Consequently, if steric effects were unimportant, the *ortho/para* ratio should remain essentially constant. If steric effects were important, the ratio should decrease with increasing steric requirements of the substituent.[6]

[3] R. J. W. LeFèvre, *J. Chem. Soc.*, 977, 980 (1933); *ibid.*, 1501 (1934).

[4] J. W. Baker and W. S. Nathan, *J. Chem. Soc.*, 1844 (1935).

[5] A. E. Remick, *Electronic Interpretation of Organic Chemistry*, 2d ed. (John Wiley, New York, 1949), p. 367.

[6] K. L. Nelson and H. C. Brown, *J. Amer. Chem. Soc.*, **73,** 5605 (1951); H. C. Brown and W. H. Bonner, *J. Amer. Chem. Soc.*, **76,** 605 (1954).

The data are summarized in Table VIII-1. The results clearly support the steric interpretation.

Table VIII–1. Isomer Ratios for the Mononitration of the Monoalkylbenzenes

Alkylbenzene	Isomer distribution			Isomer ratios		
	ortho	*meta*	*para*	o/p	o/m	p/m
Toluene	58.5	4.4	37.1	1.57	13.3	8.5
Ethylbenzene	45.0	6.5	48.5	0.93	6.9	7.5
Isopropylbenzene	30.0	7.7	62.3	0.48	3.9	8.1
t-Butylbenzene	15.8	11.5	72.7	0.22	1.4	6.3

More recently I became interested in the general problem of electrophilic substitution and undertook a detailed study of the phenomena and the development of a quantitative theory. Unfortunately, it was not possible to include this topic in the Baker Lectures.[7] In this study we obtained considerable quantitative data on the partial rate factors for aromatic substitution. Typical data for bromination in acetic acid at 25° are summarized in Table VIII-2.

Table VIII–2. Isomer Distribution and Partial Rate Factors for the Bromination of the Monoalkylbenzenes

Alkylbenzene	Isomer distribution			Rate ratio, k_R/k_B	Partial rate factors[a]		
	ortho	*meta*	*para*		o_f	m_f	p_f
Toluene	32.9	0.3	66.8	605	600	5.5	2420
Ethylbenzene	33.8		66.2	460	465		1800
Isopropylbenzene	23		77	260	180		1200
t-Butylbenzene	1.20	1.47	97.3	138	5.0	6.1	806

[a] Rate of substitution of one position of the alkylbenzene relative to the rate of substitution of one position of benzene.

It should be noted that the value of p_f decreases by a factor of three from toluene to t-butylbenzene. This may reflect decreased

[7] For a detailed review, see L. M. Stock and H. C. Brown, *Adv. Phys. Org. Chem.*, **1,** 35 (1963).

hyperconjugative contributions from the alkyl substituent. However, for the *ortho* position the value of o_f decreases by a factor of 120. This much sharper decrease must include the influence of a large steric factor hindering substitution in the position *ortho* to the alkyl substituent.

3. Steric Effects in Aliphatic Substitution—The Methyl, Ethyl, Isopropyl Problem

Bimolecular displacement reactions normally show a regular decrease in rate in the order:

$$CH_3X > CH_3CH_2X > (CH_3)_2CHX > (CH_3)_3CX$$

It was originally suggested by Polanyi that this decrease was primarily the result of steric forces.[8] However, adherents of the electronic theory maintained that the decrease was primarily the result of the inductive effect of the methyl substituents. By increasing the electron density at the central carbon atom these substituents would increase the difficulty for the approach of an electronegative reagent.[9,10]

This viewpoint was used to account for the larger increase in activation energy observed in the reaction of triethylamine with methyl and isopropyl iodide, 6.3 kcal mole^{-1}, as compared with the smaller increment observed in the reaction of pyridine with these two halides, 3.8 kcal mole^{-1}. According to the proposed interpretation, the nitrogen atom of the stronger base, triethylamine, would be strongly electronegative and would be greatly affected by the increasing negative charge (or decreasing positive charge) on the central carbon atom of the isopropyl group. On the other hand, the nitrogen atom in the weaker base, pyridine, would be less electronegative and would be affected to a much lesser degree by changes in the electronegative character of the central atom of the alkyl group.[10]

In fact, the alternate steric interpretation accounts for the phe-

[8] N. Meer and M. Polanyi, *Z. phys. Chem.*, **19B,** 164 (1932); A. G. Evans, *Trans. Faraday Soc.*, **42,** 719 (1946).

[9] C. K. Ingold, *Chem. Rev.*, **15,** 225 (1934).

[10] C. N. Hinshelwood, K. J. Laidler, and E. W. Timm, *J. Chem. Soc.*, 848 (1938).

nomenon equally well. According to the steric interpretation, the reaction of a base with large steric requirements, such as triethylamine, with an alkyl halide with large steric requirements, such as isopropyl iodide, should result in large steric strains in the transition state. Consequently, such a reaction should exhibit a relatively large energy of activation. Pyridine is a base with relatively low steric requirements. It should be much less sensitive than triethylamine to the increase in steric requirements in proceeding from methyl iodide to isopropyl iodide.

To resolve this question, we obviously require a base with the high strength of triethylamine and the low steric requirements of pyridine. If the polar interpretation were correct, such a base would show a large increment in the energies of activation for the reactions with methyl and isopropyl iodides. On the other hand, the steric interpretation would require such a base to exhibit a smaller increment in the energies of activation.

We previously pointed out that quinuclidine (**8.1**), with $pK_a = 10.58$,[11] was such a base, comparable in strength to triethylamine, $pK_a = 10.76$, but with much smaller steric requirements (**8.2**).

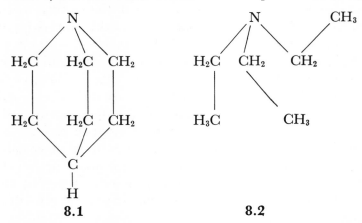

8.1   8.2

The experimental results[10,12] are summarized in Table VIII-3. The increment for quinuclidine ($E_{act}^{i\text{-}PrI} - E_{act}^{MeI}$), 4.1 kcal mole^{-1},

[11] B. M. Wepster, *Rec. trav. chim.*, **71**, 1159, 1171 (1952).
[12] H. C. Brown and N. R. Eldred, *J. Amer. Chem. Soc.*, **71**, 445 (1949).

Table VIII–3. Rates and Energies of Activation for the Reaction of Tertiary Amines with Alkyl Iodides in Nitrobenzene Solution

Base	Alkyl iodide	$k_{25°}$, l. mole^{-1}sec^{-1}	Energy of activation, kcal mole^{-1}
Triethylamine[10]	methyl	3.29×10^{-2}	9.7
	ethyl	1.92×10^{-4}	12.5
	isopropyl	1.13×10^{-6}	16.0
Pyridine[13]	methyl	3.43×10^{-4}	13.9
	ethyl	1.83×10^{-5}	16.0
	isopropyl	0.94×10^{-6}	17.7
Quinuclidine[12]	methyl	1.88	9.5
	ethyl	4.87×10^{-2}	10.9
	isopropyl	7.97×10^{-4}	13.6

is similar to the value for pyridine, 3.9, and much smaller than the increment for triethylamine, 6.3 kcal mole^{-1}. Consequently, the results support the steric interpretation.

We had previously used the pyridine bases to investigate steric effects in molecular addition compounds. With their rigidity and relative simplicity of structure, they appeared to be ideal to investigate steric effects. Consequently, we undertook to examine the reaction of these bases with methyl, ethyl, and isopropyl iodides.[13] The results are summarized in Table VIII-4.

Indeed, the reaction rates decrease sharply from methyl to ethyl to isopropyl iodide, with the activation energies showing corresponding increases. Introduction of an alkyl group into the 3- or 4-position results in a small increase in rate, with no significant difference for a methyl, ethyl, isopropyl, or *t*-butyl derivative. In the 2-position these substituents bring about decreases in rate and increases in activation energies that become more and more pronounced with the increasing steric requirements of the alkyl group.

If the steric requirements of the base are maintained constant, the rate of reaction decreases and the energy of activation increases

[13] H. C. Brown and A. Cahn, *J. Amer. Chem. Soc.*, **77**, 1715 (1955).

Table VIII–4. Rate Data for the Reaction of Pyridine Bases with Alkyl Halides in Nitrobenzene Solution

Pyridine RC$_6$H$_4$N, R =	Methyl iodide		Ethyl iodide		Isopropyl iodide	
	$10^6 k_2^{25}$	E_{act}	$10^6 k_2^{25}$	E_{act}	$10^6 k_2^{25}$	E_{act}
Hydrogen	343	13.9	18.3	16.0	0.941	17.7
2-Methyl	162	14.0	4.27	16.5	0.0509	19.2
2-Ethyl	76.4	14.2	1.95	16.6		
2-Isopropyl	24.5	14.8	0.555	17.1		
2-t-Butyl	0.080	17.5				
3-Methyl	712	13.6	40.0	15.5	1.73	17.4
3-Ethyl	761		41.0		1.81	
3-Isopropyl	810		40.4		1.68	
3-t-Butyl	950		43.3		1.56	
4-Methyl	760	13.6	41.9	15.8	1.99	17.3
4-Ethyl	777		42.1		2.01	
4-Isopropyl	767		42.2		1.98	
4-t-Butyl	757	13.7	41.9		2.00	

with the increasing steric requirements of the alkyl halide (Figure VIII-1). Similarly, if the steric requirements of the alkyl halide are maintained constant, the rate decreases and the activation energy increases with the steric requirements of the pyridine base (Figure VIII-2). Simultaneous increases in the steric requirements of both the alkyl halide and the pyridine base cause cumulative changes in

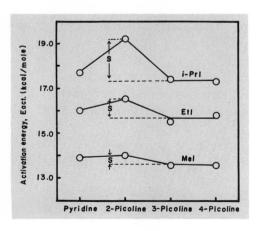

Figure VIII-1. Effect of steric strain (S) on activation energies in the picoline series.

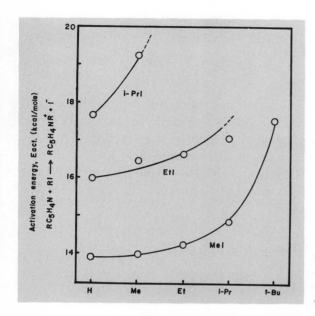

Figure VIII-2. Effect of increasing steric strain on activation energies in the reactions of 2-alkyl pyridines.

the reaction rates and activation energies. Clearly these results are in excellent accord with the operation of steric effects as the dominant factor.

One further experimental approach may be instructive. A useful procedure for separating polar and steric effects is to separate the group from the reaction center by means of an unsaturated system (aryl, vinyl, or ethynyl).[14] The introduction of methyl groups in the γ-position of allyl chloride should not alter the steric factor while relaying the polar effect of the methyl groups to the reaction center. Yet the exchange reaction with chloride ion in acetone shows an increase in rate accompanying the presence of the methyl groups.[15] In the α-position the methyl group sharply decreases the rate of the exchange reaction.

$$CH_2{=}CHCH_2Cl \qquad 1.00$$
$$CH_3CH{=}CHCH_2Cl \qquad 2.52$$
$$(CH_3)_2C{=}CHCH_2Cl \qquad 21.3$$
$$CH_2{=}CHC(CH_3)_2Cl \qquad 0.043$$

[14] P. D. Bartlett and L. J. Rosen, *J. Amer. Chem. Soc.*, **64**, 543 (1942).
[15] C. A. Vernon, *J. Chem. Soc.*, 4462 (1954).

Again, these results argue against the polar interpretation for the consistent decrease in reactivity exhibited for SN2 substitution reactions in the methyl—*t*-butyl series.

4. The F-, B-, I-Strains[16]

Up to this point the discussion has concerned itself with steric effects arising directly from the conflicting steric requirements of the two atoms or groups of atoms involved in the reaction. Thus the instabilities of triethylamine-trimethylborane (**8.3**) and of 2-picoline-trimethylborane (**8.4**) obviously arise from the large strains accompanying the steric compression of parts of the amine and trimethylborane molecules.

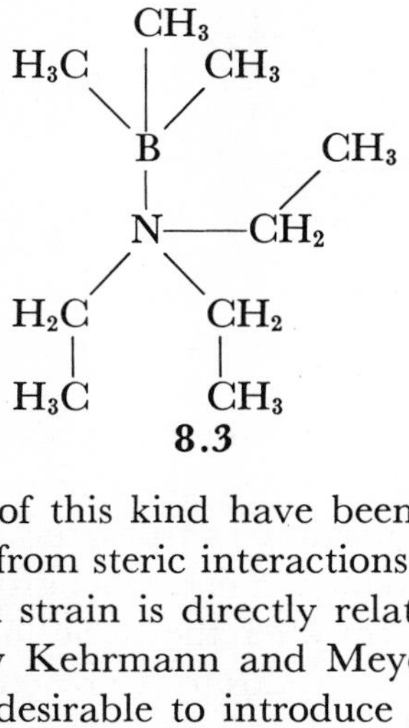

8.3 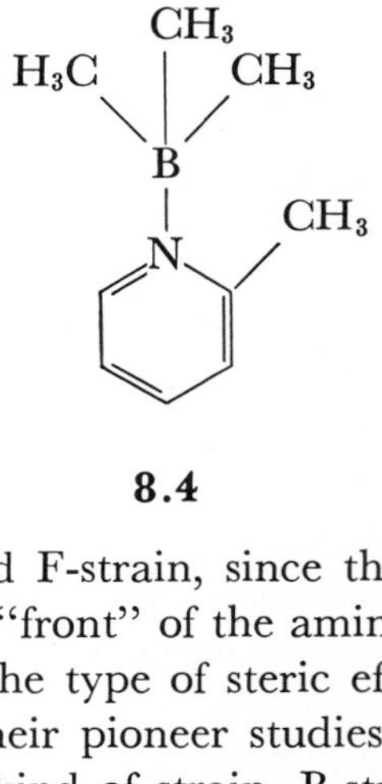   **8.4**

Strains of this kind have been termed F-strain, since they arise primarily from steric interactions at the "front" of the amine molecule. Such strain is directly related to the type of steric effect envisaged by Kehrmann and Meyer in their pioneer studies. It has appeared desirable to introduce a new kind of strain, B-strain, located at the "back" of the amine molecule, in order to account for the complete behavior of the aliphatic amines and other highly substituted systems.[17]

The $+I$ inductive effect of alkyl groups should result in an increase in the strength of aliphatic amines, $NH_3 < RNH_2 < R_2NH < R_3N$. On the other hand, the steric requirements of the aliphatic groups

[16] H. C. Brown, *Record Chem. Progress*, **14,** 83 (1953).

[17] H. C. Brown, H. Bartholomay, Jr., and M. D. Taylor, *J. Amer. Chem. Soc.*, **66,** 435 (1944).

should tend to reduce the apparent strength of the amines. In situations where the steric effect outweighs the polar effect, the opposite order should be observed: $NH_3 > RNH_2 > R_2NH > R_3N$.

By varying the steric requirements of the reference acid, it should be possible to vary the relative importance of the steric and the polar factor within wide limits. An increase in the bulk of the alkyl group should have the same effect. Consequently, by decreasing the strain systematically through an alteration in these two variables, we should be able to realize a systematic variation in the sequence from 1 to 7.

1. $R_3N < R_2NH < RNH_2 < NH_3$
2. $R_3N < R_2NH < NH_3 < RNH_2$
3. $R_3N < NH_3 < R_2NH < RNH_2$
4. $NH_3 < R_3N < R_2NH < RNH_2$
5. $NH_3 < R_3N < RNH_2 < R_2NH$
6. $NH_3 < RNH_2 < R_3N < R_2NH$
7. $NH_3 < RNH_2 < R_2NH < R_3N$

As pointed out above, 1 and 7 represent two possible extremes: dominant control by the polar effect of the alkyl substituents (sequence 7) and dominant control by the steric effect of the substituents (sequence 1).

We undertook to vary the steric factor systematically by varying the steric requirements of the alkyl substituent in the amines (t-Bu $>$ i-Pr $>$ Et $>$ Me) and by varying the steric requirements of the reference acid ("t-Bu$_3$B"[18] $> i$-Pr$_3$B $>$ Et$_3$B $>$ Me$_3$B $>$ BF$_3$ $>$ H$^+$). The observed results[19] are summarized in Table VIII-5.

It is apparent that decreases in the steric factor bring about systematic changes in the observed sequences from 1 to 2, to 3, and so on as predicted. However, a reduction of the steric factor to the minimum realizable, both by reducing the steric requirements of the alkyl groups in the amine and by reducing the steric requirements of the reference acid to boron trifluoride and to the proton, fails to

[18] We used the "tri-t-butylborane" described by E. Krause and P. Nobbe, *Ber.*, **64**, 2112 (1931). It was later reported to be largely isomerized material, probably t-butyldiisobutylborane: G. F. Hennion, P. A. McCusker, E. C. Ashby, and A. J. Rutkowski, *J. Amer. Chem. Soc.*, **79**, 5190 (1957).

[19] E. A. Lawton, Ph.D. Thesis, Purdue U., 1952.

Table VIII–5. Observed Sequences for the Alkylamines as Determined with Various Reference Acids

Amines	"t-Bu$_3$B"	i-Pr$_3$B	Et$_3$B	Me$_3$B	BF$_3$	H$^+$
t-Butyl[a]	1	1	1	1	1 or 2	—
Isopropyl[b]	1	1 or 2	1 or 2	2	2	5
Ethyl	1 or 2	2	2	3	3	5
Methyl	2 or 3	2 or 3	3	5	5	5

[a] Complete order deduced from the behavior of ammonia and mono-t-butylamine.
[b] Complete order deduced from the behavior of ammonia and mono- and di-isopropylamine.

achieve the theoretical limiting sequence for the complete absence of the steric factor. Instead, the sequences appear to be approaching 5 as a limit. It is apparent that the steric requirements of the proton must be quite small. As pointed out earlier, no steric effect toward the proton is indicated by 2,6-lutidine in spite of the large steric requirements that prevent it from combining with trimethylborane (Table V-3). Consequently we appear to be forced to the conclusion that even with reference acids of essentially negligible steric requirements the limiting sequence contains a contribution from the steric factor.

It was proposed that in the reaction of tertiary amines with a reference acid Z there is an increase in steric strain that is essentially independent of the steric requirements of the reference acid, Z (8.3).

$$
\begin{array}{cc}
\underset{\text{strained}}{\text{R}\!-\!\!\!\overset{\displaystyle \text{R}}{\underset{\displaystyle \text{R}}{\text{N}}}:\; +\; \text{Z}} & \rightleftharpoons \quad \underset{\text{more strained}}{\text{R}\!-\!\!\!\overset{\displaystyle \text{R}}{\underset{\displaystyle \text{R}}{\text{N}}}:\text{Z}}
\end{array}
\qquad 8.3
$$

This proposal implies that the strains present in the product can be divided into two factors—the strains arising from the interactions of the groups R and Z (F-strain), and strain that originates in the crowding of the three R groups (B-strain) accompanying formation of the donor-acceptor bond.

If the strain in trimethylamine is the result of the crowding of the three methyl groups around the small nitrogen atom, the effect should be greatly reduced or absent in the corresponding phosphorus derivatives with their much larger central atom. Indeed, the methylphosphines yield the sequence corresponding to order 7, $PH_3 <$ $MePH_2 < Me_2PH < Me_3P$, with H^+,[20] BMe_3,[20] and BF_3[21] as reference acids.

We obtained data on (1) ionization in aqueous solution, (2) dissociation of addition compounds of boron trifluoride, (3) dissociation of addition compounds of trimethylborane, (4) volatility of 'onium salts, (5) vapor pressures in systems with acetic and trifluoroacetic acids, and (6) dissociation of compounds with n-dodecanesulfonic acid.[20] In all cases the results indicated order 5 with the amines and order 7 with the phosphines.

The interpretation of base strengths of amines in aqueous solution is complicated by the fact that both the free amine and the protonated amine (the ammonium ion) are strongly solvated. This has resulted in proposals that order 5 for the methylamines may be the result of solvation effects rather than the operation of B-strain.[22] However, there appears to have been no attempt to account for all of the data, including the results not utilizing aqueous systems. The question could be resolved by determination of the proton affinities of ammonia and the methylamines in the gas phase. It is to be hoped that such data will be forthcoming to settle this question definitely.

Ring compounds exhibit an unusual chemistry, and I-strain was introduced to account for the interesting behavior of these derivatives. Data for the dissociation of the addition compounds of cyclic imines with trimethylborane are summarized in Table V-1. It will be noted that the ethylenimine derivative is highly dissociated. It was suggested that this could be the result of an increase in internal strain accompanying the change in coordination of the ring atom associated with the formation of the addition compound.

The I-strain concept proved to be especially valuable in accounting for the behavior of carbocyclic ring systems. For example, 1-

[20] S. Sujishi, Ph.D. Thesis, Purdue U., 1949.
[21] E. A. Fletcher, Ph.D. Thesis, Purdue U., 1952.
[22] F. E. Condon, *J. Amer. Chem. Soc.*, **87,** 4481, 4485, 4491, 4494 (1965).

105

methyl-1-chlorocyclopentane is unusually reactive in solvolysis (8.4), whereas 1-methyl-1-chlorocyclohexane is relatively slow in its hydrolysis[23] (8.5). On the other hand, cyclopentanone is relatively slow (8.6) and cyclohexanone relatively fast (8.7) in its reactions with reagents, such as sodium borohydride.[24]

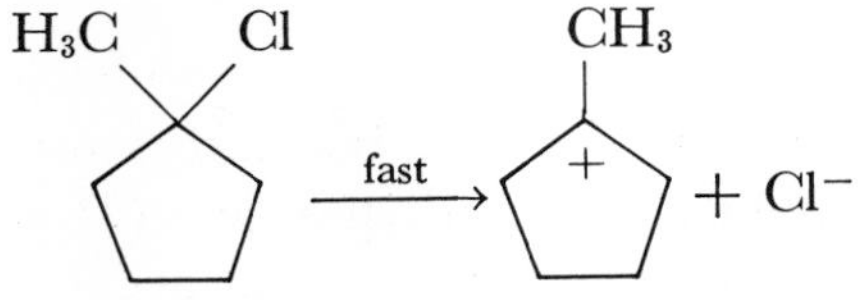

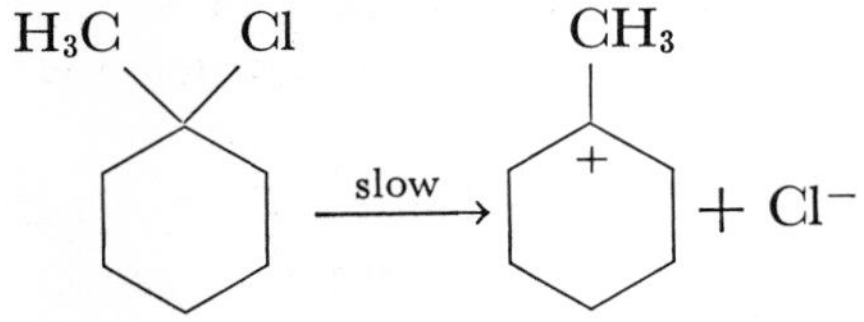

In other words, a change in coordination of a ring atom from four to three is favored in the 5-ring and disfavored in the 6-ring. Conversely, a change in coordination number of a ring atom from three to four is favored in the 6-ring and disfavored in the 5-ring. This behavior was attributed to the changes in internal strain accompanying the change in coordination of the ring atom.

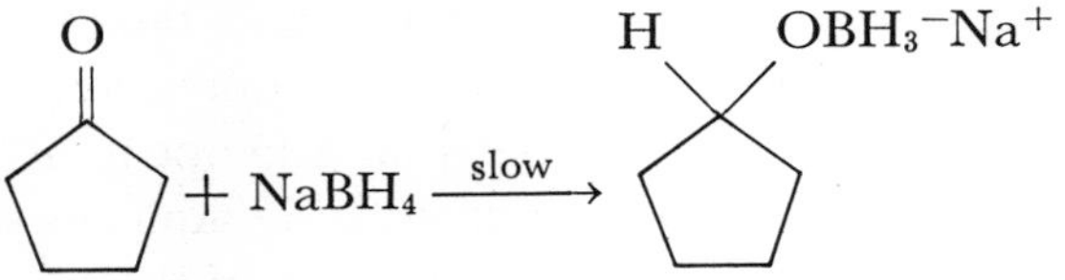

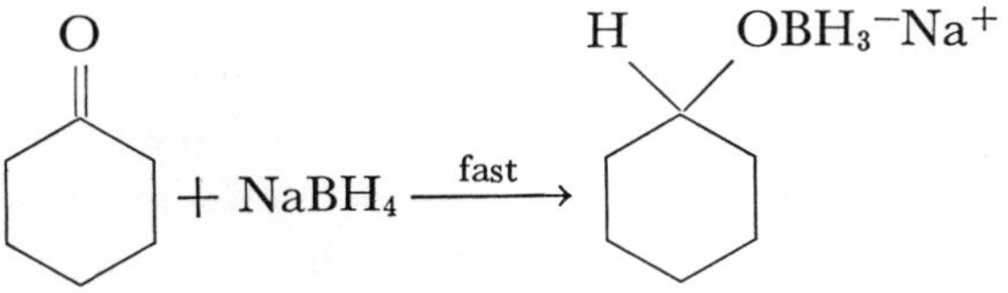

This may be made clear by the following simplified discussion of

[23] H. C. Brown and M. Borkowski, *J. Amer. Chem. Soc.*, **74**, 1894 (1952).
[24] H. C. Brown and K. Ichikawa, *Tetrahedron*, **1**, 221 (1957).

an idealized system. Consider the cyclopentane molecule (**8.5**). If fully planar, there would be ten bonds in opposition, leading to a predicted strain of approximately 10 kcal mole^{-1}. Loss of one bond would be accompanied by a decrease in bond oppositions to 6 (**8.6**) and therefore a decrease in internal strain to 6 kcal mole^{-1}. There-

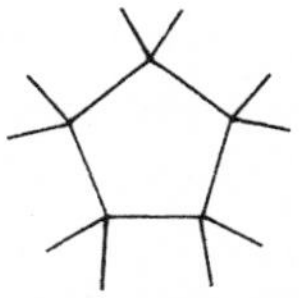

10 bond oppositions
10 kcal mole^{-1} strain
8.5

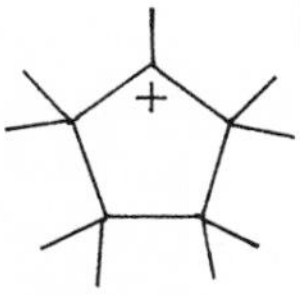

6 bond oppositions
6 kcal mole^{-1} strain
8.6

fore the loss of a bond on the conversion of a ring carbon atom from the tetrahedral configuration to the 3-covalent state will be strongly favored over the corresponding change in open-chain derivatives. It is known that the molecules do not have the planar configurations shown, but are puckered to minimize the strain. This will affect somewhat the estimate of the strain, but it should not alter the conclusion that a change in the coordination number of a ring atom from 4 to 3 should be strongly favored in cyclopentane derivatives. On this basis, the reverse change from coordination number 3 to 4, as in the reduction of cyclopentanone, should be accompanied by an increase in internal strain and should be relatively slow.

In cyclohexane all the bonds are nicely staggered, so there is no strain due to bond opposition forces. Loss of a bond to introduce a trigonal atom in the ring results in a much less symmetrical structure with some bond opposition. Cyclohexane derivatives therefore resist a change in the coordination number of a ring atom from 4 to 3, and strongly favor the reverse change.

These ideas can be extended to systematize the chemistry of a wide range of derivatives with rings of 5 and 6 members.[25] Extension to rings of other sizes will be discussed later (Section VIII-8).

[25] H. C. Brown, J. H. Brewster, and H. Schechter, *J. Amer. Chem. Soc.*, **76**, 467 (1954).

5. Steric Effects Involving the Proton [26]

It was previously pointed out that the relative base strengths of a series of pyridine derivatives vary in a simple and consistent manner with reference acids of increasing steric requirements. Thus there was observed a regular change in the observed order of base strengths in the series, pyridine, 2-picoline, and 2,6-lutidine, as the steric requirements of the reference acid were increased from H^+ to BH_3 to BF_3 and to BMe_3 (Figure V-1). The same kind of regular change was also observed in the series pyridine, 2-methyl, 2-ethyl, 2-isopropyl-, and 2-t-butylpyridine (Figure VI-1). These changes in the apparent base strengths have been attributed to the effect of increasing steric strain accompanying an increase in the steric requirements of the alkyl substituents R and the reference acid A (**8.7**).

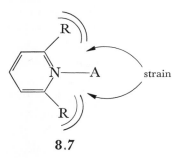

8.7

The regular order exhibited by the pK_a values of pyridine (5.17), 2-picoline (5.97), and 2,6-lutidine (6.75) (Table V-2) indicates that there is no significant steric effect involving the proton as reference acid. This is supported by the observation that the heats of reaction of these bases with methanesulfonic acid in nitrobenzene solution also increase linearly: $\Delta H = 17.1$, 18.3, and 19.5 kcal mole^{-1}, respectively (Table V-2). On the other hand, there is a hint that there may be some steric influence upon the addition of a proton to 2-alkylpyridines in the decreasing pK_a and ΔH values with increasing bulk of the alkyl substituent exhibited by these derivatives (Table VIII-6).

---

[26] H. C. Brown and B. Kanner, *J. Amer. Chem. Soc.*, **88**, 986 (1966).

Table VIII–6. pK_a Values and Heats of Reaction with Methanesulfonic Acid for the 2-Alkylpyridines

Base	pK_a (25°)	$-\Delta H$ (CH$_3$SO$_3$H), kcal/mole
2-Picoline	5.97	18.34
2-Ethylpyridine	5.92	18.22
2-Isopropylpyridine	5.83	18.10
2-t-Butylpyridine	5.76	18.02

Steric effects accompanying the addition of the proton have been suggested as a factor in the ionization constants of *ortho* substituted aromatic amines and phosphines.[27] For example, in these derivatives it is observed that the introduction of a methyl group into the *meta* and *para* position of aniline (pK_a 4.58) results in an increase in base strength, in accordance with the inductive influence of the methyl group, while introduction of a methyl group into the *ortho* position has the opposite effect (pK_a 4.39). Indeed, introduction of a second methyl substituent in the remaining *ortho* position has an even greater effect (pK_a 3.42). *o-t*-Butyl groups produce a much larger effect.[28]

The interpretation of the results for these aniline bases is complicated by the possibilities for rotation of the amino group and variable resonance of the amino group with the aromatic ring. It appeared of interest, therefore, to extend our earlier studies of pyridine bases to derivatives with very bulky substituents in the 2,6-positions (**8.7**) to see if we could increase the steric effects to the point where Lewis acids such as BF$_3$ could not add and steric hindrance even toward the addition of the proton might be realized. Consequently, we synthesized 2-methyl-6-*t*-butyl-, 2-ethyl-6-*t*-butyl-, 2-isopropyl-6-*t*-butyl-, 2,6-di-*t*-butyl-, and 2,6-diisopropylpyridine and examined their behavior. The base strengths are summarized in Table VIII-7.

As was observed earlier for water solutions, there was a regular increase in strength in the present mixed solvent (required by the low solubility of the higher bases in water) for pyridine, 2-picoline,

[27] H. C. Brown and A. Cahn, *J. Amer. Chem. Soc.*, **72,** 2939 (1950).

[28] B. M. Wepster, *Rec. trav. Chim.*, **76,** 357 (1957).

Table VIII–7. Dissociation Constants of Pyridine Bases in 50 Percent Ethanol-Water

Pyridine	pK_a	Pyridine	pK_a
Pyridine	4.38	2,6-Lutidine	5.77
2-Picoline	5.05	2-Methyl-6-*t*-butyl-	5.52
2-Ethyl-	4.93	2-Ethyl-6-*t*-butyl-	5.36
2-Isopropyl-	4.82	2-Isopropyl-6-*t*-butyl-	5.13
2-*t*-Butyl-	4.68	2,6-Di-*t*-butyl-	3.58
		2,6-Diisopropyl-	5.34

and 2,6-lutidine (Figure VIII-3). A similar linear increase in base strength is observed for the series pyridine, 2-isopropyl-, and 2,6-diisopropylpyridine. However, in the case of the *t*-butyl derivatives, following the usual increase for the first alkyl group there is observed a sharp decrease with the second (Figure VIII-3).

Had the usual linearity been followed, the predicted pK_a value for 2,6-di-*t*-butylpyridine would have been 4.98 (4.38 + 0.30 + 0.30). The observed value of 3.58 therefore represents a discrepancy of 1.4 pK_a units.

The effect apparently requires a minimum of two *t*-butyl groups; it is absent in 2-isopropyl-6-*t*-butylpyridine. In this compound the predicted pK_a value, assuming additivity, is 5.12 (4.38 + 0.44 +

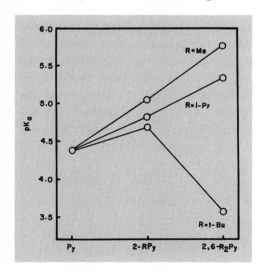

Figure VIII-3. Effect of increasing steric requirements of the alkyl groups in the 2,6-positions on the strengths of pyridine bases (in 50 percent ethanol).

0.30). The observed value is 5.13. The suddenness and magnitude of the effect is further illustrated by a comparison of the behavior of the 2-alkyl-pyridines with the 2-alkyl-6-*t*-butylpyridines (Figure VIII-4).

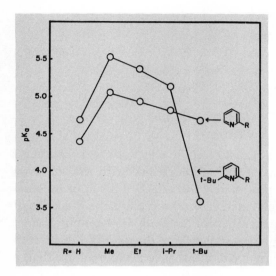

Figure VIII-4. Effect of increasing steric requirements of an alkyl group in the 2-position on the strengths of pyridine bases (in 50 percent ethanol) .

In the case of *ortho* substituted aniline bases we suggested that the steric requirements of the amino group, —NH$_2$, should be less than that of the ammonium group, —NH$_3$+.[26] Consequently, an increase in strain would accompany the addition of a proton to the lone pair. In effect, this postulates that the steric requirements of a lone pair on the nitrogen atom are less than the steric requirements of a lone pair bonding a proton to the nitrogen atom.

This concept can readily be extended to the 2,6-dialkylpyridine bases. An ethyl or isopropyl group is capable of rotating in such a manner as to minimize its steric interactions with the group associated with the nitrogen atom (Section VI-2). Consequently, the behavior of 2-ethyl- and 2-isopropyl-6-*t*-butylpyridine is not significantly different from that of 2-methyl-6-*t*-butylpyridine. The presence of two *t*-butyl groups in 2,6-di-*t*-butylpyridine, however, with their incapability for rotation to minimize the strain, results in steric interactions with the nitrogen-hydrogen bond and an enhanced tendency for ionization (8.8).

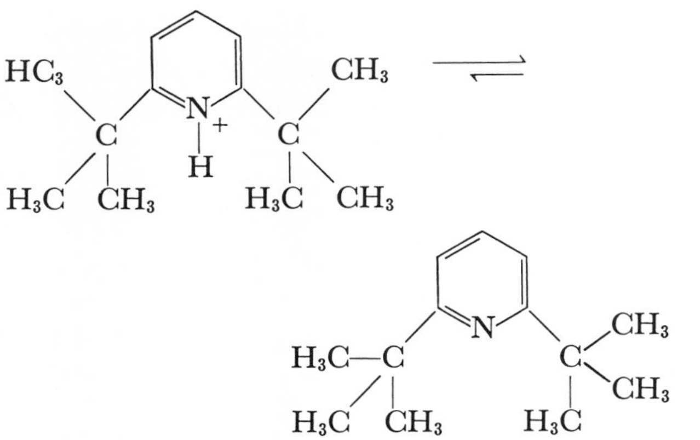

Alternatively, it has been argued that the effect of *ortho* substituents is predominantly the result of their interference with the solvation of the ionic charge in the protonated species.[28,29] This appears compatible with the small regular decreases observed in the 2-alkylpyridines and for the first three members of the 2-alkyl-6-*t*-butylpyridines (Figure VIII-4). However, the sudden large drop for 2,6-di-*t*-butylpyridine appears more suggestive of a direct steric interaction than a secondary steric influence on the general solvation of the ion.

Such a direct steric interaction could be the steric compression of the nitrogen-hydrogen bond (8.8), or it could involve steric interference with a water molecule, which is normally hydrogen-bonded to the protonated base[22] (**8.8**).

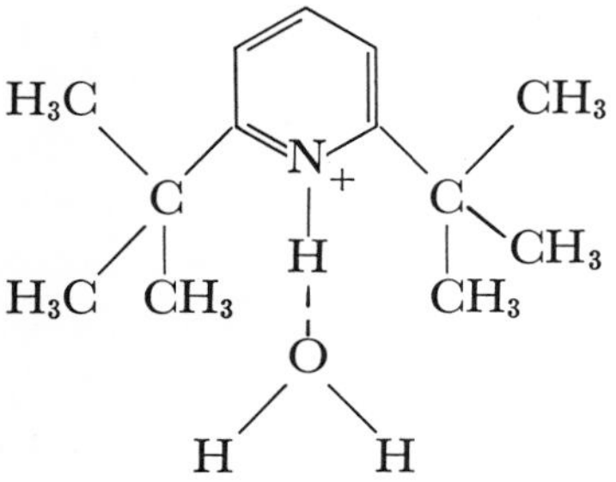

8.8

[29] P. D. Bartlett, M. Roha, and R. M. Stiles, *J. Amer. Chem. Soc.*, **76**, 2349 (1954).

(The importance of such hydrogen bonding is indicated by the results observed in the reaction of this base with hydrogen chloride under anhydrous conditions, as described later.) Unfortunately, no means appear available at this time to decide how each factor may contribute to the observed effect.

At one time we had hoped to apply the concepts of homomorphs to resolve the question. If the effect is really due to a direct steric interaction of the two neighboring *t*-butyl groups on the intervening nitrogen-hydrogen bond (**8.9**), we should expect to find similar strains in *m*-di-*t*-butylbenzene (**8.10**) and in 3-*t*-butylpyridine-trimethylborane (**8.11**), and the strains could be detected by the

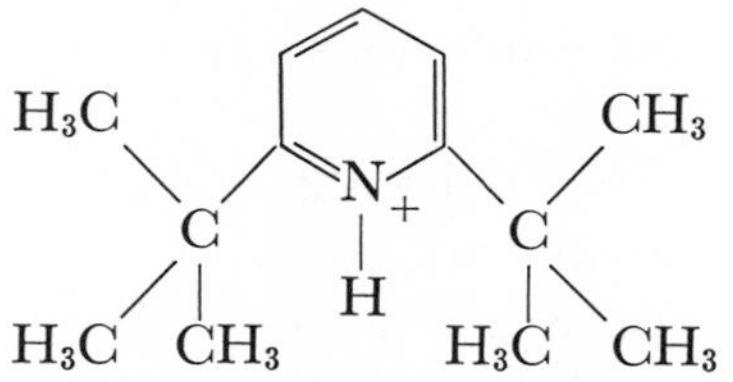

8.9

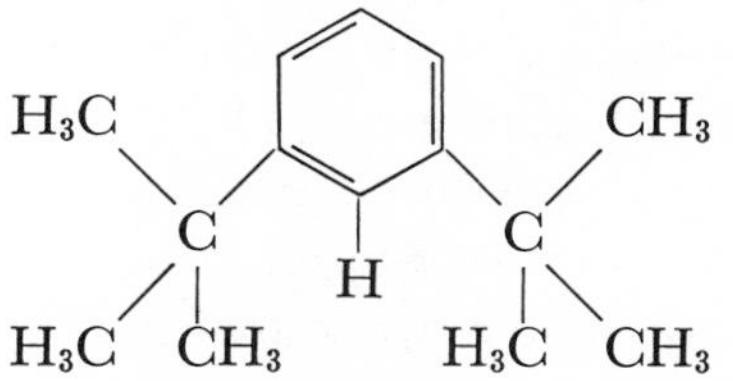

8.10

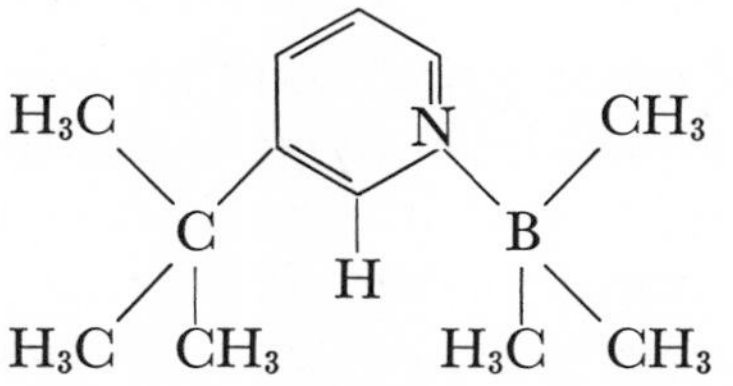

8.11

usual thermochemical approaches. Unfortunately, we have been distracted into other areas of research, so this test of our suggestion of a direct steric compression of the protonated center still awaits experimental examination. Of course, if it becomes possible to measure experimentally the proton affinities of amines, this and other interesting problems involving the strengths of amine bases could be answered.

The hindered base, 2,6-di-*t*-butylpyridine, exhibits highly unusual and interesting properties. Unlike other pyridine bases, it fails to form a simple hydrochloride. Only the dihydrochloride forms. Apparently hydrogen bonding involving the nitrogen-hydrogen bond and chloride ion, which stabilizes simple pyridinium chlorides, cannot occur in this derivative. A second molecule of hydrogen chloride is required to hydrogen-bond with the chloride ion to permit the transfer of the proton.

Boron trifluoride fails to react with this base. Consequently, it appears capable of distinguishing between protonic acids and Lewis acids.

Over a period of several months it fails to react with methyl iodide.[30] It should therefore be useful in situations where it is desired to have a base present that can neutralize acid, but is itself incapable of forming acid through elimination reactions.

Finally, in contrast to less hindered pyridine bases, it readily reacts with sulfur dioxide under aprotic conditions (liquid sulfur dioxide at $-10°$) to form the substitution product, 2,6-di-*t*-butylpyridine-3-sulfonic acid. Apparently, with coordination to the nitrogen atom effectively blocked by the two *t*-butyl groups, the electrophilic reagent readily attacks the heterocyclic nucleus.

6. Steric Effects in Highly Branched Carbon Compounds

If three alkyl groups attached to nitrogen constitute a center of strain, it follows from the similarity in atomic dimensions that three alkyl groups attached to carbon should also constitute a center of strain. Such strain should manifest itself in an increased reactivity of molecules in reactions where the transition state affords a decrease in strain.

Classically, steric effects have been considered synonymous with

---

[30] A reaction at very high pressures has been reported. Y. Okamoto and Y. Shimagawa, *Tetrahedron Letters*, 317 (1966). However, the product appears to be the hydroiodide. W. J. le Noble and Y. Ogo, *Tetrahedron*, **26**, 4119 (1970).

steric hindrance. The proposal that steric effects may serve to assist a reaction represented a sharp departure from tradition.[31]

The validity of the proposal was examined in the solvolysis of the tertiary alkyl chlorides. It was considered that the strain present in the initial tetrahedral molecule should be greatly reduced in the formation of the transition state leading to the planar carbonium ion (8.9).

$$
\begin{array}{ccc}
R & & R \\
\diagdown & & | \\
R\!-\!\!-\!\!C\!-\!Cl & \xrightarrow{k_1} & C^+ \;\;\; + \; Cl^- \\
\diagup & & \diagup \;\; \diagdown \\
R & & R \quad R
\end{array}
\qquad 8.9
$$

tetrahedral   planar
(strained)   (less strained)

If steric assistance were indeed a factor in this reaction, the introduction of one or more bulky R groups into a simple model tertiary halide, such as t-butyl chloride, should be accompanied by an enhanced rate of solvolysis. The replacement of one of the methyl groups in t-butyl chloride by a t-butyl group causes an increase in the first-order rate constant (80 percent ethanol) from 0.033 hr^{-1} to 0.040. A t-amyl group causes the rate to rise further to 0.188 hr^{-1}. These increases are in the direction predicted by the theory.[32] However, it is necessary to inquire whether they indeed arise from steric assistance, or might be the result of other factors.

A major factor in the rapid solvolysis of t-butyl chloride must be the stabilization of the carbonium ion through hyperconjugation (8.10).

$$
\begin{array}{ccc}
H \;\; CH_3 & & H \;\; CH_3 \\
| \;\;\;\; | & & | \;\;\;\; | \\
H\!-\!C\!-\!C^+ & \longleftrightarrow \; H^+ & C\!\!=\!\!C \\
| \;\;\;\; | & & | \;\;\;\; | \\
H \;\; CH_3 & & H \;\; CH_3
\end{array}
\qquad 8.10
$$

Carbon-carbon hyperconjugation is less effective than carbon-hydrogen hyperconjugation.[33] Consequently, the replacement of the methyl

[31] H. C. Brown, *Science*, **103,** 385 (1946).

[32] H. C. Brown and R. S. Fletcher, *J. Amer. Chem. Soc.,* **71,** 1845 (1949).

[33] H. C. Brown, J. D. Brady, M. Grayson, and W. H. Bonner, *J. Amer. Chem. Soc.,* **79,** 1897 (1957).

group by the *t*-butyl group or the *t*-amyl group would be expected to result in a decrease in the hyperconjugative stabilization and a decrease in the observed rate of solvolysis (8.11).

$$\underset{\underset{H_3C}{|}}{\overset{\overset{H_3C}{|}}{H_3C-C}}-\underset{\underset{CH_3}{|}}{\overset{\overset{CH_3}{|}}{C^+}} \quad \longleftrightarrow \quad H_3C^+ \quad \underset{\underset{H_3C}{|}}{\overset{\overset{H_3C}{|}}{C}}=\underset{\underset{CH_3}{|}}{\overset{\overset{CH_3}{|}}{C}} \qquad 8.11$$

Separation of the alkyl groups from the reaction center by a phenyl, ethenyl, or ethynyl linkage should permit transmission of the electronic influences of the alkyl substituents without the complicating intervention of steric effects. Application of this technique reveals that the *t*-butyl group, when it is remote from the reaction center, exerts the postulated effect, decreasing the rate of solvolysis (8.12, 8.13).

$$\underset{\underset{CH_3}{|}}{\overset{\overset{CH_3}{|}}{CH_3C\equiv CC}}-Cl \qquad\qquad (CH_3)_3CC\equiv \underset{\underset{CH_3}{|}}{\overset{\overset{CH_3}{|}}{CC}}-Cl \qquad 8.12$$

$$\qquad\quad 1.00 \qquad\qquad\qquad\qquad 0.55$$

$$CH_3-\!\!\!\bigcirc\!\!\!-\overset{\overset{CH_3}{|}}{\underset{\underset{CH_3}{|}}{C}}-Cl \qquad H_3C-\overset{\overset{CH_3}{|}}{\underset{\underset{CH_3}{|}}{C}}-\!\!\!\bigcirc\!\!\!-\overset{\overset{CH_3}{|}}{\underset{\underset{CH_3}{|}}{C}}-Cl \qquad 8.13$$

$$\qquad\quad 1.00 \qquad\qquad\qquad\qquad 0.55$$

Consideration of other contributing factors, such as the ease of solvation of the incipient carbonium ion and the ease of electrophilic attack of the solvent on the halogen, likewise leads to the prediction that the bulky *t*-butyl and *t*-amyl substituents should decrease the solvolysis rate. The observed increases may therefore be considered to support the postulated role of steric strain in assisting the ionization.

Similarly, the replacement of one of the methyl groups in *t*-butyl chloride by a neopentyl group, giving a homomorph of di-*t*-butylmethane (Section VII-2) results in a 21-fold increase in the reaction rate. Two neopentyl groups raise the rate to 580 times that of *t*-butyl chloride[34] (8.14).

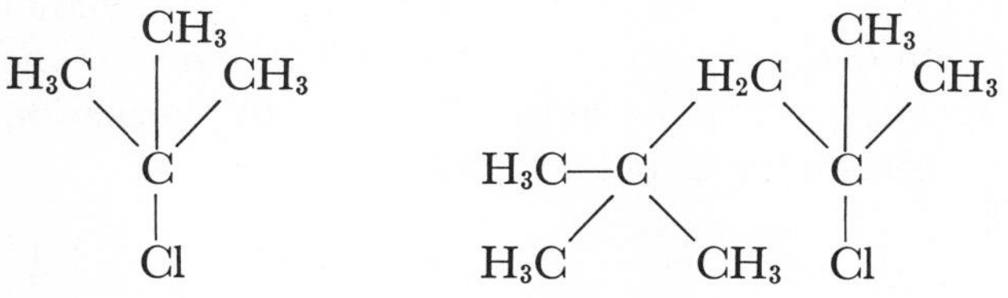

$$1.00 \qquad\qquad 21$$

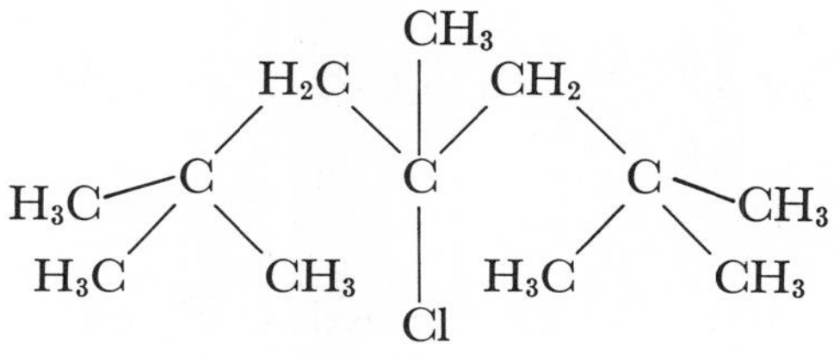

$$580$$

$$8.14$$

Much larger effects have been realized by Bartlett and his co-workers in the solvolysis of derivatives of highly hindered alcohols, such as tri-*t*-butylcarbinol.[34-37] Bridging by neighboring carbon was at one time considered a possible factor in these enhanced rates.[38] This led us into a study of the nonclassical ion problem (Chapters IX and X). However, Bartlett no longer considers carbon bridging to be significant in the enhanced rates exhibited by these highly branched derivatives.[37]

The carbonium ion, once formed, can undergo reaction by three different paths. It can react with solvent to regenerate the original tetrahedral system (k_s), it can lose a proton, forming an olefin (k_E), or it can undergo rearrangement to form a new carbonium ion. It is reasonable that a carbonium ion containing bulky substituents

[34] H. C. Brown and H. Berneis, *J. Amer. Chem. Soc.*, **75**, 10 (1953).

[35] P. D. Bartlett and M. S. Swain, *J. Amer. Chem. Soc.*, **77**, 2801 (1955).

[36] P. D. Bartlett and R. M. Stiles, *J. Amer. Chem. Soc.*, **77**, 2806 (1955).

[37] P. D. Bartlett and T. T. Tidwell, *J. Amer. Chem. Soc.*, **90**, 4421 (1968).

[38] P. D. Bartlett, *J. Chem. Educ.*, **30**, 22 (1953).

should react somewhat more slowly with the solvent. Consequently, such an ion should exhibit an enhanced tendency to undergo elimination and rearrangement. Both phenomena have been observed.

For example the solvolysis of *t*-butyl chloride in 80 percent ethanol results in the formation of 16 percent isobutene (8.15), whereas the solvolysis of *t*-butyldimethylcarbinyl chloride leads to the formation of 61 percent of the corresponding olefin[39] (8.16).

$$(CH_3)_3CCl \longrightarrow \quad (CH_3)_3C^+ \longrightarrow \quad (CH_3)_2C\!=\!CH_2 \qquad 8.15$$

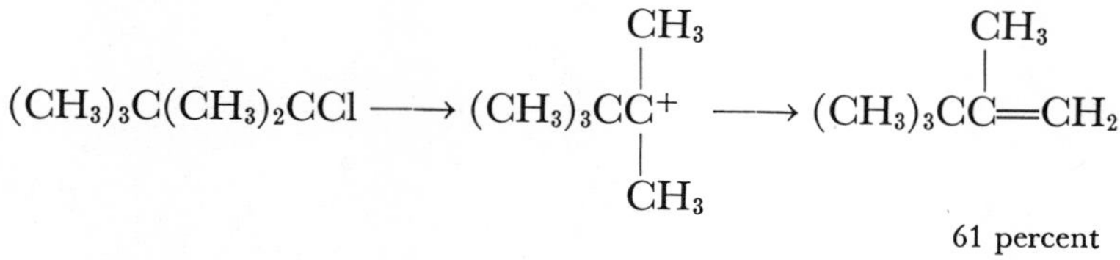

16 percent

$$(CH_3)_3C(CH_3)_2CCl \longrightarrow (CH_3)_3C\overset{\overset{\displaystyle CH_3}{|}}{\underset{\underset{\displaystyle CH_3}{|}}{C}}{}^+ \longrightarrow (CH_3)_3C\overset{\overset{\displaystyle CH_3}{|}}{C}\!=\!CH_2 \qquad 8.16$$

61 percent

The *t*-butyl cation has nine α-hydrogen atoms in position to be eliminated to form the olefin; the other cation has six. Thus on a statistical basis the amount of olefin produced from the triptyl cation (8.16) should be 10 percent, approximately two-thirds that from the *t*-butyl cation. The much higher yield of olefin, 61 percent, is presumably the result of a much lower rate of substitution of the much more hindered cation.

Moreover, we observed that the predominant product in the solvolysis of 2,4,4-trimethyl-2-chloropentane is the terminal olefin, not the internal olefin anticipated from operation of the Saytzeff rule (see the following section). Again we suggested that this unusual orientation must be the result of steric forces. The internal olefin, 2,4,4-trimethyl-2-pentene (**8.12**), is forced to have a methyl group

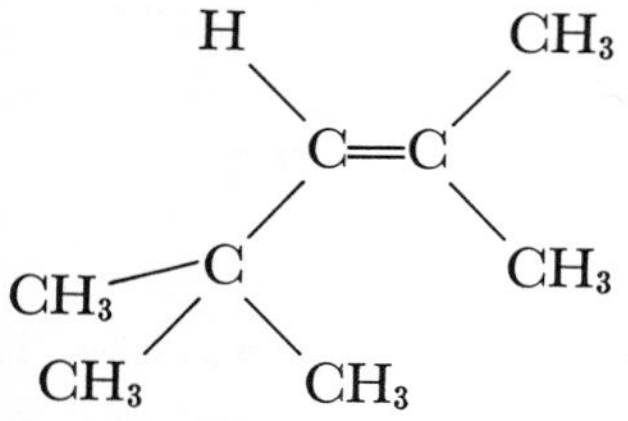

8.12

[39] H. C. Brown and R. S. Fletcher, *J. Amer. Chem. Soc.*, **72**, 1223 (1950).

and *t*-butyl group *cis* to each other, a strained relationship similar to that in homomorphs of *o-t*-butyltoluene.[39] Formation of the terminal olefin, 2,4,4-trimethyl-1-pentene, avoids this strain.

Again the degree of rearrangement observed in reactions of these highly branched derivatives can be correlated with the estimated steric hindrance involved in the capture of the nucleophile by the cation. For example, *t*-butyldiethylcarbinol rearranges on attempted formation of the tertiary chloride by treatment with hydrogen chloride at $0°$[40] (8.17).

$$
\begin{array}{cc}
\text{H}_3\text{C} & \text{C}_2\text{H}_5 \\
| & | \\
\text{H}_3\text{C}-\text{C}-\text{C}-\text{C}_2\text{H}_5 \\
| & | \\
\text{H}_3\text{C} & \text{OH}
\end{array}
\xrightarrow[0°]{\text{HCl}}
\begin{array}{cc}
\text{H}_3\text{C} & \text{CH}_3 \\
| & | \\
\text{H}_3\text{C}-\text{C}-\text{C}-\text{C}_2\text{H}_5, \text{ and so on.} \\
| & | \\
\text{Cl} & \text{C}_2\text{H}_5
\end{array}
\qquad 8.17
$$

On the other hand, the closely related structure, *t*-butylmethylethyl-carbinol is readily converted into the chloride without rearrangement[41] (8.18).

$$
\begin{array}{cc}
\text{H}_3\text{C} & \text{CH}_3 \\
| & | \\
\text{H}_3\text{C}-\text{C}-\text{C}-\text{C}_2\text{H}_5 \\
| & | \\
\text{H}_3\text{C} & \text{OH}
\end{array}
\xrightarrow[0°]{\text{HCl}}
\begin{array}{cc}
\text{H}_3\text{C} & \text{CH}_3 \\
| & | \\
\text{H}_3\text{C}-\text{C}-\text{C}-\text{C}_2\text{H}_5 \\
| & | \\
\text{H}_3\text{C} & \text{Cl}
\end{array}
\qquad 8.18
$$

This marked difference in the tendency toward rearrangement exhibited by these two structurally very similar alcohols could be rationalized in terms of the relative openness of the faces of the two carbonium ions for the capture of the nucleophile. The carbonium ion from *t*-butyldiethylcarbinol, the cation presumably formed initially, carries a bulky *t*-butyl group and two ethyl groups attached to the central carbonium ion. The ethyl groups cannot lie in the plane defined by the carbonium atom and its three bonds. With either of the two possible conformations maintaining a planar arrangement for the carbonium carbon and the two ethyl groups (**8.13, 8.14**),

[40] H. C. Brown and R. S. Fletcher, *J. Amer. Chem. Soc.*, **73**, 1317 (1951).
[41] H. C. Brown and R. B. Kornblum, *J. Amer. Chem. Soc.*, **76**, 4510 (1954).

$$H_3C-\overset{\overset{\displaystyle H_3C}{|}}{\underset{\underset{\displaystyle H_3C}{|}}{C}}-\overset{\overset{\displaystyle H_3C}{\diagdown}\quad\overset{\displaystyle CH_2}{\diagup}}{\underset{\underset{\displaystyle H_3C}{\diagup}}{\overset{\diagup}{C^+}}}{\underset{\displaystyle CH_2}{\diagdown}}$$

8.13

$$H_3C-\overset{\overset{\displaystyle H_3C}{|}}{\underset{\underset{\displaystyle H_3C}{|}}{C}}-\overset{\overset{\displaystyle CH_2}{\diagup}\overset{\displaystyle CH_3}{\diagdown}}{\underset{\underset{\displaystyle CH_2}{\diagdown}}{{}^+C}}\overset{\displaystyle CH_3}{\underset{\displaystyle CH_3}{}}$$

8.14

severe crowding would result in a high degree of steric strain. Such strain would be relieved by rotation of one ethyl group in one direction out of the plane of the carbonium carbon and rotation of the second ethyl group in the opposite direction (**8.15**). The molecular

$$H_3C-\overset{\overset{\displaystyle H_3C}{|}}{\underset{\underset{\displaystyle H_3C}{|}}{C}}-\overset{+}{C}\overset{\overset{\displaystyle CH_2}{\diagup}}{\underset{\underset{\displaystyle CH_2}{\diagdown}}{}} H_3C\overset{\displaystyle CH_3}{\diagup}\;CH_3$$

8.15

models shown in Figure VIII-5 may help to visualize the steric situation.

Although this last conformation results in a reduction of strain, it

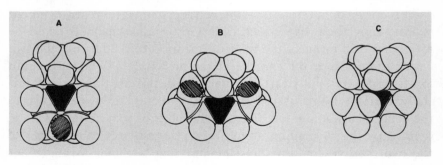

Figure VIII-5. Molecular models for the *t*-butyldiethylcarbonium ion showing three possible conformations: A, a planar conformation involving strain; B, the alternative planar conformation involving strain; C, a relatively strain-free conformation involving steric blocking of both faces of the carbonium carbon atom.

places the ethyl groups in such a position that they must block approach to the carbonium ion from both above and below the plane defined by the carbonium carbon atom. With approach to the carbonium carbon atom blocked in this way, reaction with chloride ion or with other molecular or ionic species to form the product should be severely hindered. This should result in a greatly increased half-life for the carbonium ion and give ample opportunity for facile rearrangement to structures that are both less strained and more favorable for the addition of chloride ion or other nucleophilic species.

In the carbonium ion derived from *t*-butylmethylethylcarbinol, this difficulty does not arise. With only one ethyl group attached to the central carbon atom (in addition to a *t*-butyl and a methyl group), the carbonium ion retains one face relatively free for the approach of the anion (**8.16**).

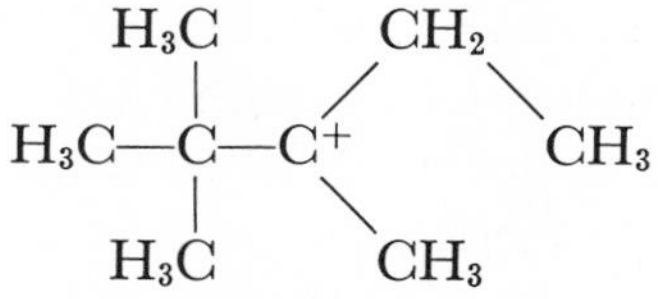

$$H_3C-\underset{\underset{H_3C}{|}}{\overset{\overset{H_3C}{|}}{C}}-C^+\overset{CH_2}{\underset{CH_3}{\diagdown}}CH_3$$

8.16

Thus simple steric considerations provide a simple explanation for the otherwise highly puzzling difference in behavior of these two remarkably similar structures. More generally, the concepts of steric assistance and steric hindrance provide unifying interpretations of the influence of structure upon the ionization, elimination, and rearrangement reactions of highly branched carbonium ions.

7. Steric Effects in Elimination Reactions [42, 43]

In 1948 a group of papers by Hughes and Ingold and co-workers described the first systematic attempt to provide a consistent inter-

[42] H. C. Brown and I. Moritani, *J. Amer. Chem. Soc.*, **77,** 3607 (1955) and accompanying papers.

[43] H. C. Brown, I. Moritani, and M. Nakagawa, *J. Amer. Chem. Soc.*, **78,** 2190 (1956) and accompanying papers.

pretation of elimination reactions.[44] These authors concluded that unimolecular eliminations proceed to give the most branched olefin. Such eliminations are considered to follow the Saytzeff rule of orientation. Bimolecular elimination of 'onium salts yields the least branched olefins. Eliminations of this kind follow the alternative Hofmann rule of orientation. They proposed that the Saytzeff rule represented control by the electromeric factor, whereas the Hofmann rule represented control by the polar factor rendered important by the positive charge.

This theory did not include steric effects as a significant factor. However, our studies of the solvolysis of highly branched tertiary chlorides had led us to conclude that steric strains assisted the ionization of such chlorides and influenced both the extent and direction of elimination. We so stated in our publication.[39]

In a paper entitled, "The Comparative Unimportance of Steric Strain in Unimolecular Elimination," Hughes, Ingold, and Shiner criticized this proposal and suggested possible ways of accounting for the results in electronic terms, without invoking the operation of steric factors.[45] As it happened, we had recently completed a detailed study of the role of the steric factor in E1 eliminations, so we counterattacked with a group of five papers, of which the final summary paper[42] was entitled, "The Importance of Steric Strains in the Extent and Direction of Unimolecular Elimination. The Role of Steric Strains in the Reactions of Highly Branched Carbonium Ions."

Unfortunately, space will not permit a detailed discussion of the alternative discussions. However, a reading of both papers could be highly instructive to the student interested in seeing how the same data can be subjected to alternative interpretations.[46]

Solvolysis of a series of closely related tertiary bromides in 85 percent aqueous *n*-butyl Cellosolve (to facilitate recovery of the olefin) revealed that the yield of olefin increased with increasing steric requirements of the group R in $RCH_2C(CH_3)_2Br$:

[44] E. D. Hughes, C. K. Ingold *et al.*, *J. Chem. Soc.*, 2093 (1948).

[45] E. D. Hughes, C. K. Ingold, and V. J. Shiner, Jr., *J. Chem. Soc.*, 3827 (1953).

[46] The same problem appears in the nonclassical carbonium ion problem (Chapters IX and X).

R = Me	27% olefin
Et	33%
i-Pr	46%
t-Bu	57%

Again, the ratio of 1-/2- olefin increased with the steric requirements of R, the increase being modest until the last member, t-butyl:

Me	1-/2- = 0.27
Et	0.41
i-Pr	0.70
t-Bu	4.3

As pointed out previously, the last case represents an E1 elimination in the Hofmann direction. We previously attributed it to the strains in the Saytzeff product arising from the steric interactions of the *cis* methyl and t-butyl groups (**8.12**).

Acetolysis of the related secondary derivatives, $RCH_2CH(OTs)CH_3$, does not show a similar anomaly:

Me	1-/2- = 0.11	*trans/cis* =	1.08
Et		.19	1.39
i-Pr		.25	1.94
t-Bu		.32	84

The difference in behavior of the secondary and tertiary derivatives containing R = t-Bu is instructive. The tertiary derivative cannot avoid placing a t-butyl and methyl group *cis* to each other in the 2-olefin. Consequently, it undergoes E1 elimination preferentially to the 1-olefin. The secondary derivative can avoid such strain by placing the t-butyl and methyl group *trans* to each other. Consequently, it undergoes E1 elimination preferentially to the *trans*-2-olefin.

These results provided convincing evidence as to the importance of the steric factor in E1 eliminations.

A detailed study was also made of E2 eliminations.[43] It was suggested that bimolecular elimination proceeds through the two transition states (**8.17, 8.18**).

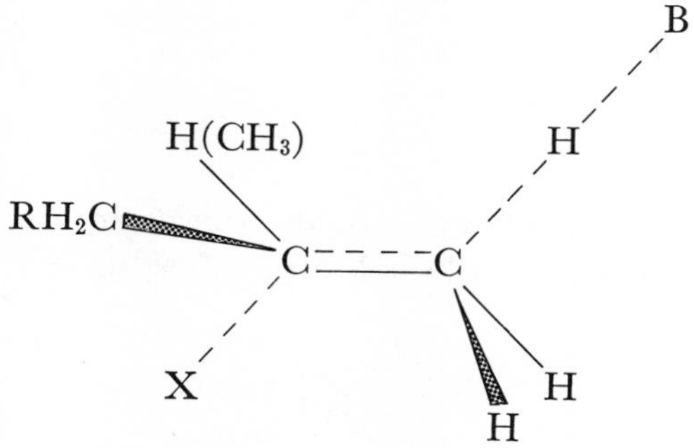

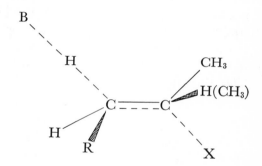

Transition state leading
to 1-olefin

8.17

Transition state leading
to 2-olefin

8.18

It was proposed that in the absence of steric strain the reaction will proceed preferentially through the transition state **8.18** to give the more alkylated of the two possible olefins (Saytzeff rule). With increasing steric requirements of either R or X (or both), steric strain will decrease the stability of this transition state. The alternative state **8.17** will be much less affected. Consequently there will occur an increased tendency for the reaction to proceed through transition state **8.17** to form the thermodynamically less stable 1-olefin. An increase in the steric requirements of the base B should also result in greater strain in transition state **8.18**, where a secondary hydrogen must be removed, than in transition state **8.17**, where a primary hydrogen is eliminated.

This interpretation accounts in a simple manner for the observation that a shift from the more alkylated to the less alkylated olefin can be achieved by (1) an increase in the steric requirements of the alkyl groups on the incipient double bond, (2) an increase in the steric requirements of the leaving group, and (3) an increase in the steric requirements of the attacking base.

For simplicity the present discussion has been restricted to the reactions of 2-substituted derivatives, whose elimination reactions yield 1- or 2-olefins. Generalization to more complex examples appears to offer no difficulty.

It therefore appeared to us that the available data supported the interpretation that elimination in accordance with the Saytzeff rule represents control by the hyperconjugative factor, whereas elimination in accordance with the Hofmann rule represents control by the steric factor.

Unfortunately, agreement as to the interpretation of E2 reactions has been far more difficult to achieve than for E1 reactions.[47,48] However, it now appears that a part of the difficulty may have been in the earlier generally accepted position that E2 eliminations exhibited a strong preference for *trans-anti* departure of the leaving group and β-hydrogen. Evidence has now been presented that *cis-syn* eliminations are not uncommon features of bimolecular eliminations.[49] With this new understanding of the reaction mechanism, there now appears to be a greater acceptance of the significance of steric forces in bimolecular eliminations.[49,50]

[47] J. F. Bunnett, *Angew. Chem. Int. Ed. Engl.*, **1**, 225 (1962).

[48] W. H. Saunders, Jr., in *The Chemistry of Alkenes*, S. Patai, ed. (Wiley-Interscience, London, 1964), Chap. 2.

[49] See D. S. Bailey and W. H. Saunders, Jr., *J. Amer. Chem. Soc.*, **92**, 6904 (1970) for leading references.

[50] I. N. Feit and W. H. Saunders, Jr., *J. Amer. Chem. Soc.*, **92**, 1630 (1970); D. S. Bailey, F. C. Montgomery, G. W. Chodak, and W. H. Saunders, Jr., *J. Amer. Chem. Soc.*, **92**, 6911 (1970).

8. Steric Effects in Ring Systems [24, 51]

In a masterful discussion of the chemistry of ring systems, Prelog revealed that chemical behavior varied in a fairly consistent manner with ring size.[51] It was of interest to see if the concepts that had been used with considerable success to rationalize the chemistry of 5- and

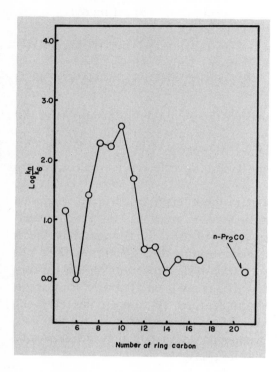

Figure VIII-6. The effect of ring size on the rate of acetolysis of the cycloalkyl *p*-toluenesulfonates at 70°.

6-membered ring systems[25] (Section VIII-4) could be extended to the larger and smaller rings.

We have examined the rates of solvolysis of the 1-methyl-1-chloro-cycloalkanes,[23] the acetolysis of the cycloalkyl tosylates,[52,53] and the rates of reaction of sodium borohydride with the cyclanones.

We previously pointed out that reactions involving a change in

[51] V. Prelog, *J. Chem. Soc.*, 420 (1950).

[52] H. C. Brown and G. Ham, *J. Amer. Chem. Soc.*, **78**, 2735 (1956).

[53] R. Heck and V. Prelog, *Helv. Chim. Acta*, **38**, 1541 (1955).

coordination number from four to three are favored in the 5-ring, presumably because the change is accompanied by a decrease in internal strain. This reaction is unfavorable in the 6-ring system. The 7-ring is strongly strained and reactions of this kind are favored. The strain increases to a maximum at the 10-ring.[51] We should expect to find the rate a maximum at this point, and the acetolysis of the cyclodecyl tosylates indeed reveals a maximum at this point (Figure VIII-6).

Contrariwise, for reactions involving a change in coordination number from three to four, as in the reduction of the ketones with sodium borohydride, the reverse should be true. Indeed, in this reaction cyclohexanone is relatively reactive and cyclopentanone is relatively inert (Section VIII-4). If this parallelism is followed by the larger ring systems, a minimum in reactivity should be observed for cyclodecanone. Indeed, this was observed[51] (Figure VIII-7).

Linear free-energy relationships involving aliphatic derivatives

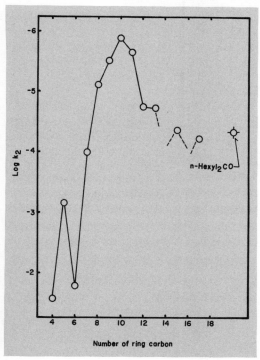

Figure VIII-7. The effect of ring size on the rate of reaction of the cyclanones with sodium borohydride in isopropyl alcohol at 0°.

generally are not observed. Consequently it is of interest that the ring compounds from 5- through 10-ring members provide a reasonably good linear free-energy relationship between the rates of acetolysis of the tosylates and the rates of reaction of the ketones with the borohydrides (Figure VIII-8).

It is of interest to point out that Heck and Prelog considered the possibility that the fast rate for cyclodecyl tosylate might be the result of transannular σ-participation, leading to a nonclassical cation,[53] rather than the result of a decrease in internal strain (I-strain). These increasingly numerous proposals in the literature accounting for enhanced rates in terms of stabilized σ-bridged transition states leading to σ-bridged intermediates led us to our explorations in the nonclassical ion area, discussed in Part Three.

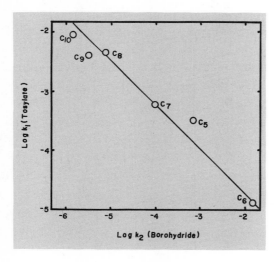

Figure VIII-8. Free-energy relationship between the rates of acetolysis of the cycloalkyl tosylates and the rates of reaction of the cyclanones with sodium borohydride.

THE NONCLASSICAL ION PROBLEM

IX. The Nonclassical Ion Problem[1,2]

1. Background

The study of the rearrangement of camphene hydrochloride into isobornyl chloride (9.1) has had major consequences for the development of carbonium ion theory.

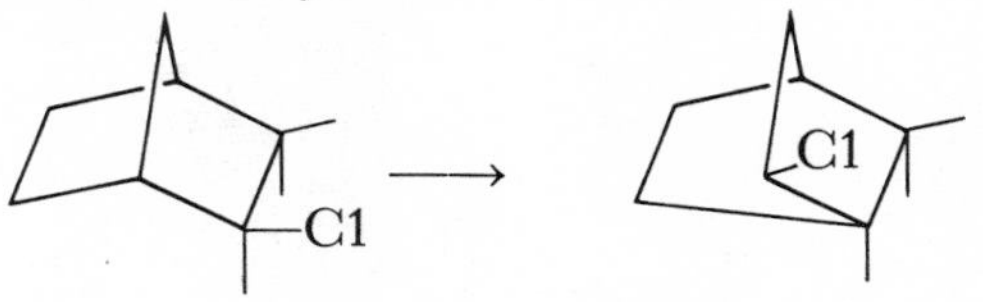

$$9.1$$

Meerwein's proposal in 1922 that this rearrangement involves the prior formation of a carbonium ion[3] appears to be the first application of carbonium ion intermediates to account for such a molecular transformation. In 1939 C. L. Wilson and his associates suggested that such a rapidly equilibrating pair of ions (9.2) might exist instead

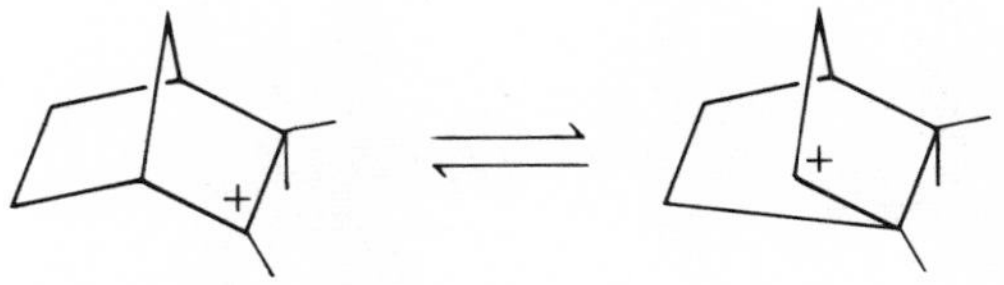

$$9.2$$

as the mesomeric species[4] **9.1**.

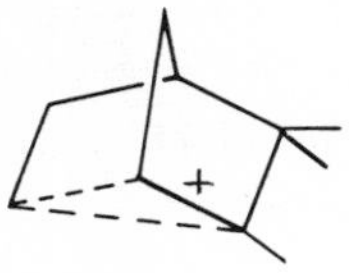

9.1

[1] H. C. Brown, "The Transition State," *Spec. Publs. Chem. Soc.*, 1962, pp. 140–158, 174–178.

[2] H. C. Brown, *Chem. Eng. News*, **45,** 86, Feb. 13, 1967.

[3] H. Meerwein and K. van Emster, *Ber.*, **55,** 2500 (1922).

[4] T. P. Nevell, E. de Salas, and C. L. Wilson, *J. Chem. Soc.*, 1188 (1939).

Wilson's proposal caught the fancy of physical organic chemists. The concept was widely adopted and used. Indeed, with the possible exception of the methyl cation, nonclassical structures apparently have been considered for nearly every known aliphatic, alicyclic, and bicyclic carbonium ion.[5] Representative systems for which nonclassical ions have been considered are shown in Figures IX-1 and IX-2.

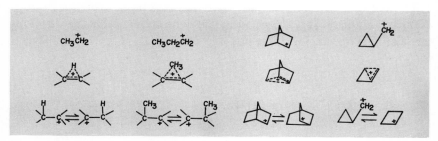

Figure IX-1. Carbonium ions for which electron-deficient nonclassical structures have been considered.

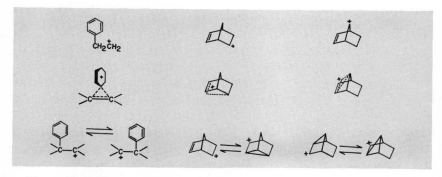

Figure IX-2. Carbonium ions for which normal or electron-sufficient nonclassical structures have been proposed.

2. The Problem: What Is the Basis for Rate Enhancements in Strained Systems?

In those cases for which nonclassical structures have been considered, it is also possible to represent the intermediate as an equilibrating pair of cations. On this basis the nonclassical structures

[5] P. D. Bartlett, *Nonclassical Ions* (W. A. Benjamin, New York, 1965).

would represent the transition states for the equilibrating systems. Chemists are thus faced with the problem of deciding, in individual systems, whether the structure is best represented in its classical form, or as an equilibrating pair (or set) of cations or ion-pairs, or as a resonance hybrid so stable that these prior alternative structures need not be considered.

By using isotopic tags, it is possible to demonstrate in a large number of systems that the carbonium ion can be generated and transformed into products without the equilibration of the carbon atoms required by the bridged structure. For example, solvolysis of the tagged 2,3,3-trimethyl-2-butyl 3,5-dinitrobenzoate in methanol produces the methyl ether product with negligible scrambling of the tag[6],[7] (9.3).

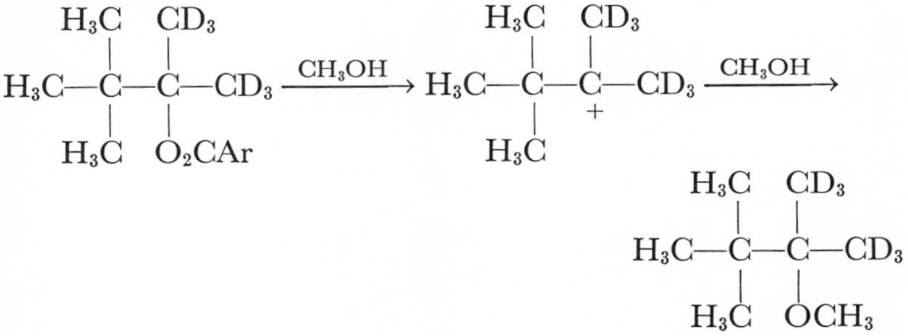

$$
\begin{array}{ccc}
\underset{\underset{H_3C\ \ \ O_2CAr}{|\quad\quad|}}{\overset{\overset{H_3C\quad CD_3}{|\quad\quad|}}{H_3C-C-C-CD_3}} & \xrightarrow{CH_3OH} & \underset{\underset{H_3C}{|}}{\overset{\overset{H_3C\quad CD_3}{|\quad\quad|}}{H_3C-C-C-CD_3}} \ + \\
\end{array}
\xrightarrow{CH_3OH}
$$

$$
\underset{\underset{H_3C\ \ \ OCH_3}{|\quad\quad|}}{\overset{\overset{H_3C\quad CD_3}{|\quad\quad|}}{H_3C-C-C-CD_3}} \quad 9.3
$$

Clearly the reaction cannot involve the symmetrical methyl bridged intermediate (**9.2**).

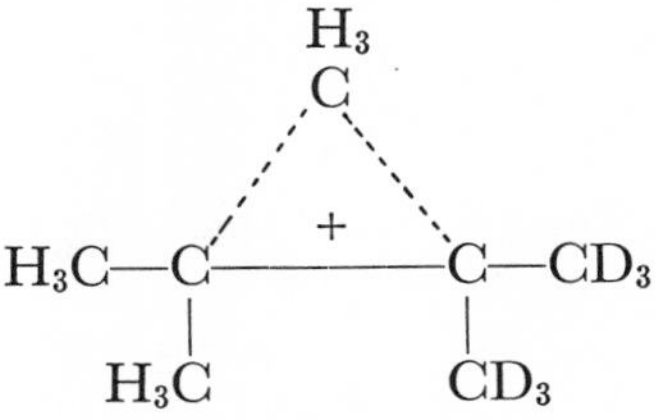

$$
\underset{\underset{H_3C}{|}}{H_3C-C} \overset{\overset{\overset{H_3}{C}}{\diagup \quad + \quad \diagdown}}{\longrightarrow} \underset{\underset{CD_3}{|}}{C-CD_3}
$$

9.2

[6] H. C. Brown and C. J. Kim, *J. Amer. Chem. Soc.*, **90**, 2082 (1968).

[7] For an earlier study using ^{14}C-labeled alcohol, see J. D. Roberts and J. A. Yancy, *J. Amer. Chem. Soc.*, **77**, 5558 (1955).

Formation of this symmetrical intermediate would have required 100 percent scrambling (50 percent rearrangement) of the tag in the product.

Similarly, 3-phenyl-2,3-dimethyl-2-butyl 3,5-dinitrobenzoate was subjected to methanolysis.[6] In this case the methyl ether product revealed 45 percent scrambling (22.5 percent rearrangement) (9.4).

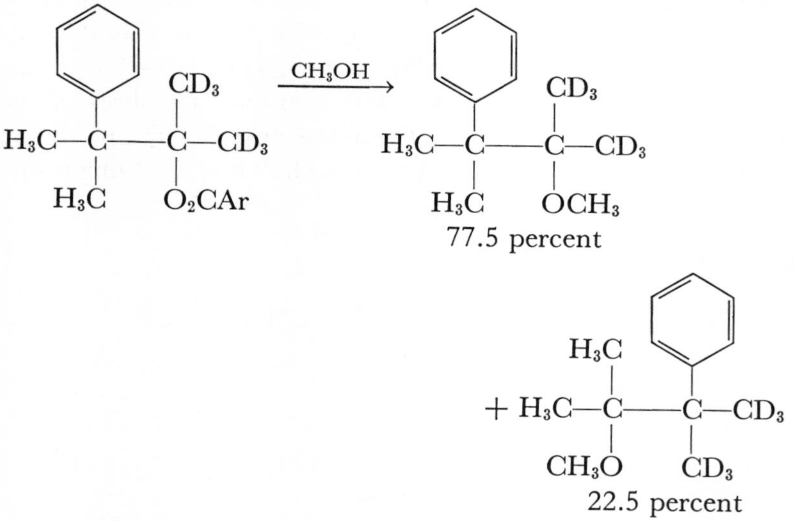

77.5 percent

$$+ \; H_3C-C-C-CD_3 \quad 9.4$$

22.5 percent

This result clearly demonstrates that the phenonium ion intermediate **9.3** is not the sole reaction path for the solvolysis.

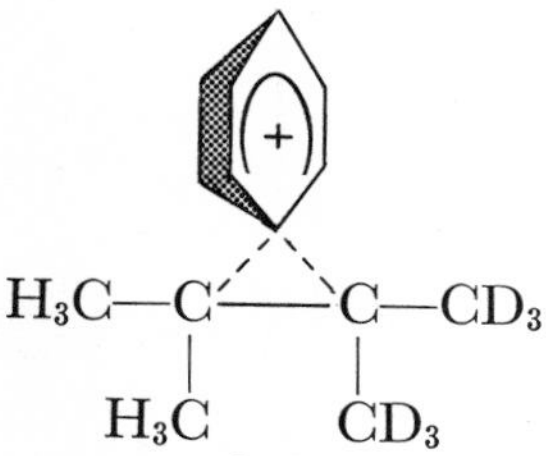

9.3

The solvolysis could be interpreted as proceeding 45 percent through the phenonium ion intermediate and 55 percent through the classical ion. However, this would require that the phenonium ion and the classical have equal energies. If so, there appears to be no reason to retain the phenonium ion in the interpretation of the reaction pathway. It is conceptually simpler to consider the solvolysis as proceeding through the classical ion, which equilibrates rapidly with its isomer (9.5).

$$
\underset{\substack{\mathrm{H_3C} \quad \mathrm{CD_3}}}{\mathrm{H_3C-C-\overset{+}{C}-CD_3}} \quad\rightleftharpoons\quad \underset{\substack{\mathrm{H_3C} \quad \mathrm{CD_3}}}{\mathrm{H_3C-\overset{+}{C}-C-CD_3}} \qquad 9.5
$$

The precise amount of scrambling that will occur in a particular reaction will then depend in a relatively simple manner upon the speed with which a particular nucleophile captures the cation.[6]

In cases such as the above, the use of isotopes provides a simple tool to rule out the involvement of bridged nonclassical cations in the solvolysis. However, in other cases, such as norbornyl, cyclopropylcarbinyl, and 3-phenyl-2-butyl, equilibration of the tag approaches that predicted by the nonclassical structure. Accordingly, these systems cannot be represented as involving static classical cations. Rather they must involve either bridged ions or their kinetic equivalent, a rapidly equilibrating pair (or set) of ions. The problem then becomes one of deciding between these two alternatives. This has proven to be quite difficult, as will be discussed in detail for norbornyl (Chapter X).

3. Steric Assistance vs. Nonclassical Resonance

It might be of interest to review how my own attention was attracted to the problem. In 1946 I suggested that the accumulation of bulky alkyl groups around a central carbon atom would constitute a center of strain that should assist unimolecular ionizations involving the formation of a planar, less strained carbonium ion (Section VIII-6) (9.6).

$$
\underset{\substack{\text{tetrahedral}\\\text{(strained)}}}{\underset{\mathrm{R}}{\overset{\mathrm{R}}{\mathrm{R}-C-Cl}}} \longrightarrow \underset{\substack{\text{planar}\\\text{(less strained)}}}{\underset{\mathrm{R}\quad\mathrm{R}}{\overset{\mathrm{R}}{\mathrm{C^+}}}} + \mathrm{Cl^-} \qquad 9.6
$$

A number of systems of this kind were investigated. The results, such as the following (9.7), appeared to support the proposal (Section VIII-6).

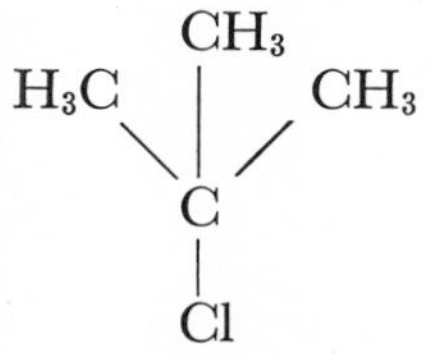

$$
\begin{array}{ccc}
\text{1.00} & \text{21} & \text{580}
\end{array}
$$

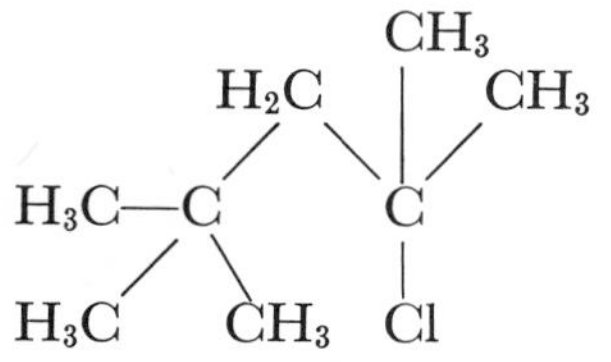

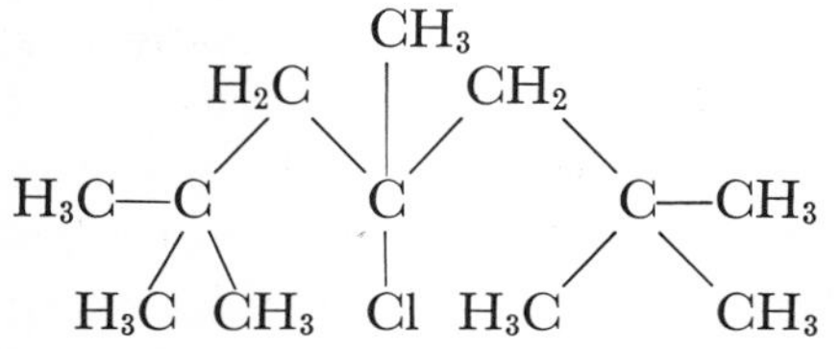

9.7

The concept appeared to be generally accepted. Accordingly, my associates and I abandoned work in this field for new areas.

At this time, however, a new concept appeared on the scene, and it was used with increasing frequency to account for rate acceleration that seemed to me to be adequately accounted for by relief of steric strain. Thus Ingold and his co-workers reported that camphene hydrochloride undergoes ethanolysis at a rate 6000 times greater than t-butyl chloride. This enhanced rate was attributed not to relief of steric strain accompanying the separation of chloride ion from its highly crowded environment, but to the driving force provided by the formation of a nonclassical (synartetic, mesomeric) cation[8] (9.8).

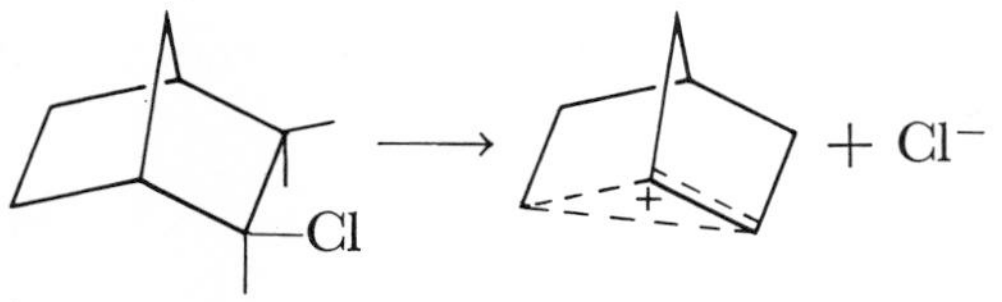

9.8

Similarly, Bartlett suggested that the formation of a methyl-bridged tri-t-butylcarbinyl cation might provide driving force for the

[8] F. Brown, E. D. Hughes, C. K. Ingold, and J. F. Smith, *Nature*, **168,** 65 (1951).

observed fast rates of solvolysis of tri-*t*-butylcarbinyl derivatives[9] (9.9).

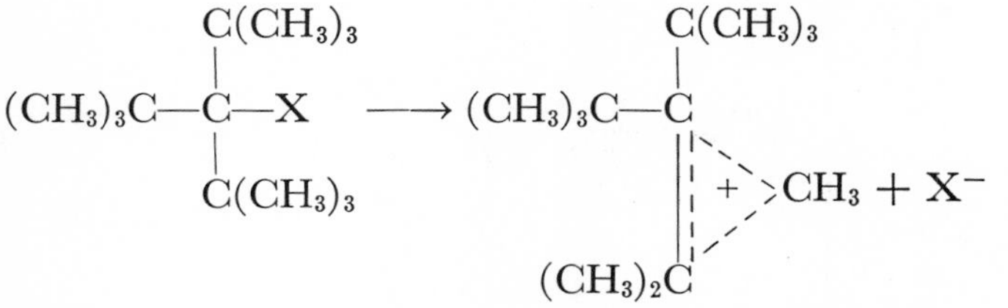

$$\text{(CH}_3)_3\text{C}\!-\!\overset{\overset{\displaystyle \text{C(CH}_3)_3}{|}}{\underset{\underset{\displaystyle \text{C(CH}_3)_3}{|}}{\text{C}}}\!-\!\text{X} \longrightarrow \text{(CH}_3)_3\text{C}\!-\!\text{C} \quad \overset{+}{>}\text{CH}_3 + \text{X}^- \qquad 9.9$$

Then Heck and Prelog suggested that the high rate of solvolysis of cyclodecyl tosylate might be due to the formation of a stabilized bridged cyclodecyl cation[10] (9.10), rather than to the decrease in internal strain as we had suggested previously.

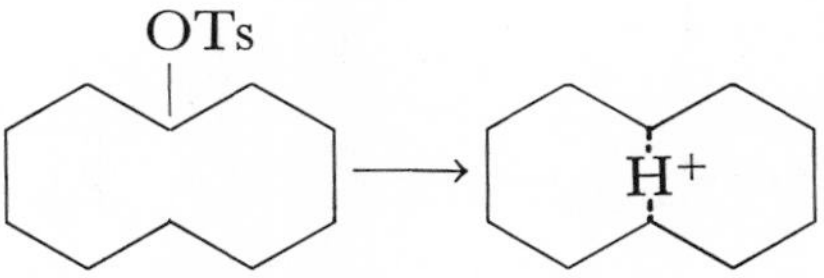

$$9.10$$

It might be desirable to clarify what these alternative proposals involve. *t*-Cumyl chloride undergoes solvolysis at a rate approximately 4600 times that of *t*-butyl chloride. The enhanced rate of solvolysis is attributed to resonance stabilization of the corresponding carbonium ion (9.11)—resonance that is partially present in the transition state and stabilizes that state.

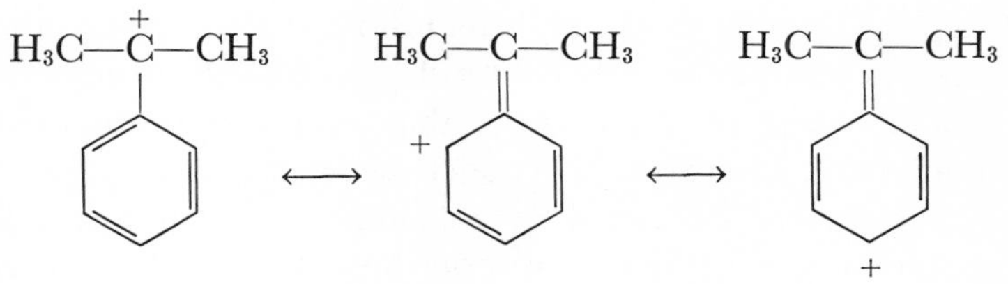

$$9.11$$

Consequently, the enhanced rate is primarily the result of the formation of a transition state that is lower in energy than the corre-

[9] P. D. Bartlett, *J. Chem. Educ.*, **30**, 22 (1953).

[10] R. Heck and V. Prelog, *Helv. Chim. Acta*, **38**, 1541 (1955).

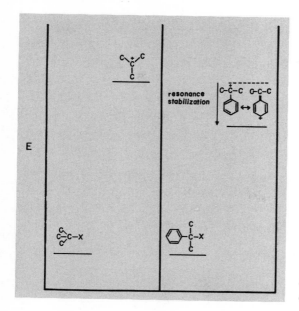

Figure IX-3. Effect of classical resonance in facilitating solvolysis.

sponding transition state in *t*-butyl chloride (Figure IX-3). (In these figures it is convenient to represent the transition states in terms of the fully formed cations. However, it should be recognized that this is a simplification; the leaving group would be partially bonded to the incipient carbonium ion in the actual transition state.)

Similarly, the enhanced rate of solvolysis of 2,4,4-trimethyl-2-chloropentane is attributed to an increase in the energy of the initial state (Figure IX-4). Conceivably, the fast rate of solvolysis of molecules such as camphene hydrochloride, tri-*t*-butylcarbinyl esters, and cyclodecyl tosylate could also be the result of steric crowding in the ground state. However, the nonclassical proposal is that these transition states are stabilized by σ-bridging and that such nonclassical resonance is primarily responsible for the enhanced rates.

Originally I had no reason to question those proposals. However, it was interesting that the phenomenon appeared to be significant only for structures where my co-workers and I had anticipated that relief of steric strain would be an important factor. Accordingly, I began to examine in great detail the available data for a number of systems in order to reach a decision as to how much of the observed

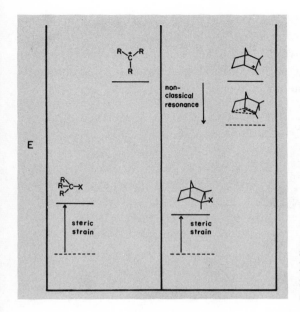

Figure IX-4. Effect of steric strain and postulated nonclassical resonance facilitating solvolysis.

rate enhancement might be due to relief of steric strain and how much to this new phenomenon of bridging by saturated carbon. We then undertook experimental studies of selected systems.

These studies forced us to conclude that in these systems we could find no convincing evidence for σ-bridging. We could account for all of the observed rate enhancements without including this new feature.

4. What to Do?

At this point I was faced with one of those agonizing decisions. The nonclassical concept had proved to be exceedingly popular. A large fraction of the physical organic chemists in the United States were engaged in research in this area. The subject had entered the textbooks, and thousands of graduate students were being taught and examined annually on the subject matter and approved interpretation.

In the meantime the theory was going on to even more elaborate structures, including Dewar's formulation as π-complexes.[11] This

[11] M. J. S. Dewar and A. P. Marchand, *Ann. Rev. Phys. Chem.*, **16,** 321 (1965).

might be considered to be the rococo period of carbonium ion structures, as shown by the examples presented in Figure IX-5. Obviously, to challenge the theory would lead to severe repercussions. However, to remain silent might subject future generations of students to needless studies of erroneous concepts.

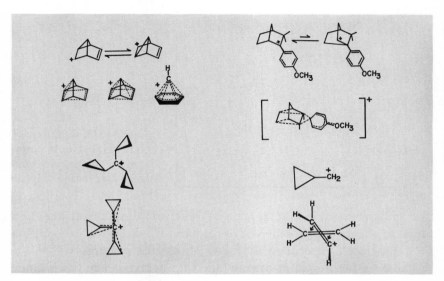

Figure IX-5. Examples of carbonium ions from the "rococo period" of carbonium ion structures.

There comes a time when one cannot remain on the sidelines. It becomes necessary to stand forth and declare, "But the Emperor is naked!"

Accordingly, in March 1961, at the Symposium on Carbonium Ions held at the St. Louis Meeting of the American Chemical Society, I pointed out that, in my opinion, many of the proposed nonclassical structures for carbonium ions rested on exceedingly fragile experimental foundations, and I suggested the desirability of further experimental work to test and reaffirm these foundations.

I had hoped that my cautious comments[1] would set in motion a critical objective reexamination of the field. Instead they were apprehended as a heresy, triggering what appeared to be a "holy war" to prove me wrong.

What could be done in this situation?

5. The Difficulties in Questioning an Accepted Theory

Many theoretical proposals are advanced and subjected to experimental test. The uncovering of unfavorable experimental data results in the ready revision of the proposal or its withdrawal.

The situation is quite different for a theory that has reached the stage of wide acceptance. Such a theory appears in the textbooks, is taught to a new generation of students, and becomes accepted as a fixed part of the fabric of established chemical theory. It becomes exceedingly difficult to question such an accepted concept.

In theory the various experiments that are done involving the theory will appear in print. An unsatisfactory theory will result in many experimental results that will not correlate with the theory. As these accumulate, there will be general recognition that the theory is unsatisfactory and revision will be instituted. In fact, under present conditions, this is not the case—there is an actual selection of experimental results to favor an established theory.

Let us consider an example. A young man receives a post in a university. He considers possible research problems. He has just been taught the fascinating new nonclassical ion theory and it appears an interesting area to explore. He makes a prediction based on the theory and subjects it to experimental test. The nonclassical theory is qualitative; consequently, even on a purely statistical basis, the chances are 50:50 that the results will support his prediction.

If this should be the case, he is happy as a lark. The experimental study is written up and submitted to a journal. The editor, reading what appears to be an interesting study supporting an accepted theory, sends it off to two experts for review. These will be supporters of the theory, since nonsupporters of an accepted theory will be rare (and under suspicion). These supporters of the nonclassical paper will read it, find it interesting, and return it with a recommendation that it be accepted. Shortly it appears in print.

However, suppose that the experimental result proves to be

contrary to the prediction. The young man will be puzzled. Here is a generally accepted theory that he has been taught, yet the results do not fit. The chances are that he will decide to put the study on the shelf for a while, to permit it to "mature," and often it will remain there forever.

Suppose that the young man is a rare individual who decides that this contrary result should be published. He submits the paper. The editor sees that a young unknown worker is proposing to publish results contrary to a well-established theory. He immediately sends it to two of the leading workers in the field, strong supporters of the theory. These referees ask: "Who is this young man who dares to attack such a well-established theory?" Clearly there must be something wrong with his experiments. They return the paper with the recommendation that the manuscript not be published until the author has done additional experiments, and they outline sufficient additional work to take years to complete.

Consequently there will be a consistent selection of papers to support an established theory.

How can one individual with one research group undertake to question such a theory? It is not easy. Obviously he cannot go back to reexamine each of the thousand papers in the literature purporting to support the theory. While he is busy demonstrating that a particular publication can be reinterpreted, another hundred papers supporting the theory will appear.

I decided that the only feasible strategy was to select what was generally believed to be the best possible example for a nonclassical ion. There appeared to be general agreement that this was norbornyl. Then I undertook to select what appeared to be the experimental foundations for the proposed nonclassical structure in this system. Next we subjected these to experimental test. Finally, I decided to report each test in Communication form, so that I could avoid engaging in unprofitable polemical discussions before I was prepared to reach a decision.

I believe we are now ready to reach a decision. The data are presented and discussed in the next lecture (Chapter X). However, before we undertake to consider the data in detail, it is desirable to consider a suitable definition for just what is meant by a nonclassical ion, to undertake a suitable definition for it, and to clarify the concepts through a consideration of the cyclopropylcarbinyl system.

6. Relationship to Neighboring Group Phenomena

The brilliant research work of Winstein and his co-workers in the forties[12,13] established that donor atoms in the β-position could greatly enhance the rates of solvolytic reactions and simultaneously control the mechanism of substitution (9.12).

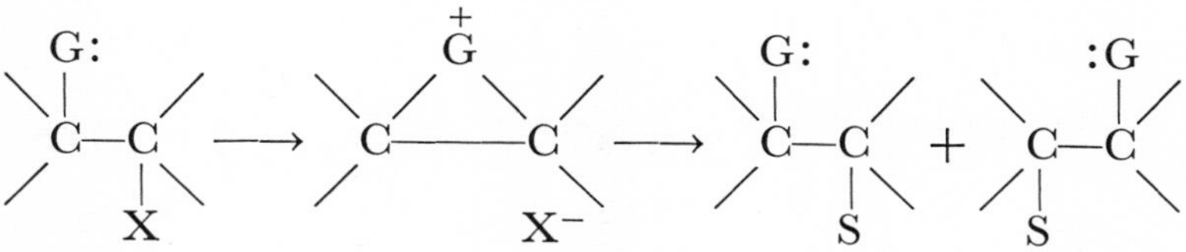

$$
\underset{X}{\overset{G:}{C-C}} \longrightarrow \underset{X^-}{\overset{\overset{+}{G}}{C \longrightarrow C}} \longrightarrow \underset{S}{\overset{G:}{C-C}} + \underset{S}{\overset{:G}{C-C}} \qquad 9.12
$$

Originally, those working in the field believed that a large rate enhancement was essential in order to postulate the formation of a stabilized bridged intermediate. For example, the fourfold rate enhancement observed for *trans*-2-chlorocyclohexyl brosylate, compared with the *cis* isomer, was not considered significant by Winstein and his co-workers, and the formation of a chloronium intermediate was not proposed[14] (9.13). But the large factors of 800 for the bromine

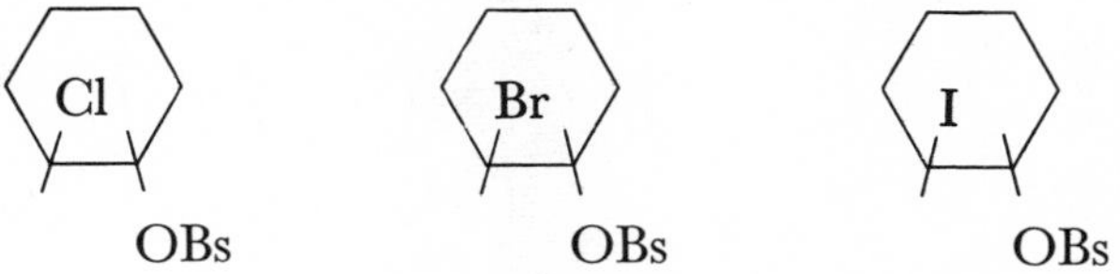

	Cl	Br	I	9.13
	OBs	OBs	OBs	
$\dfrac{trans}{cis}$	4	800	2,700,000	

and 10^6 for the iodine derivatives were believed to reflect the driving force accompanying the formation of the bridged intermediates. Consequently even with the excellent donors involved in the *n*-subdivision of neighboring groups (Table IX-1), we observe a varia-

[12] A. Streitwieser, Jr., *Solvolytic Displacement Reactions* (McGraw-Hill, New York, 1962).

[13] B. Capon, *Quart. Rev. (London)*, **18**, 45 (1964).

[14] S. Winstein and E. Grunwald, *J. Amer. Chem. Soc.*, **70**, 846 (1948); E. Grunwald, *J. Amer. Chem. Soc.*, **73**, 5458 (1951).

tion from groups such as thioalkoxy and iodo, which participate strongly, to groups such as methoxy and chloro, which participate very weakly or not at all.

Table IX–1. Classification of Neighboring Groups

$n-$	$\pi-$	$\sigma-$

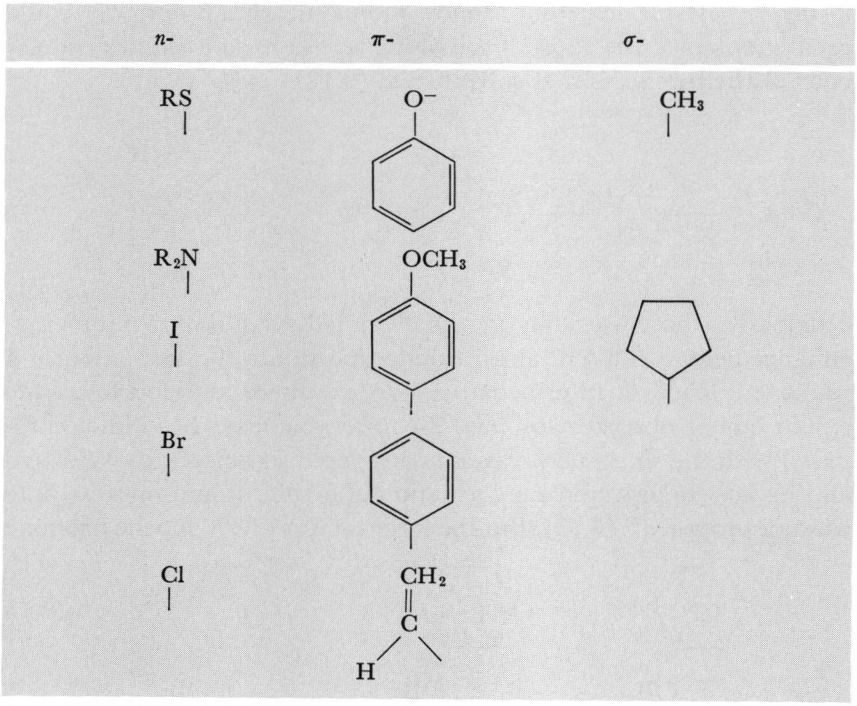

The π-subdivision of neighboring groups also exhibits a wide range of effectiveness[13] (9.14).

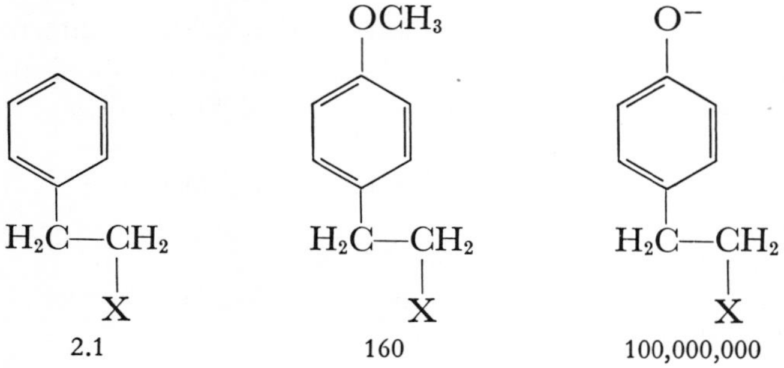

9.14

Consequently, within each of the donor classes there is a wide variation in the ability to participate, varying from strong participation to very weak or negligible participation. The question we face is whether a saturated carbon-carbon bond, the σ-classification, can serve as a donor to a developing carbonium ion center.

Strong Lewis acid acceptors, such as aluminum bromide and gallium chloride, form relatively stable addition compounds with donors of the n-class. Although the interaction is weaker, members of the π-class, such as benzene and mesitylene, also form complexes with aluminum bromide.[15] However, to my knowledge, similar donor-acceptor interaction has never been demonstrated for saturated alkanes or cycloalkanes, such as would be involved in the extension of participation to the proposed σ-class. Although this should make us cautious in postulating such participation by neighboring alkyl and cycloalkyl groups, it does not provide any basis for ruling out participation. The interaction of a neighboring group with a developing electron-deficient center may be subject to far different physical laws than the interaction of a donor molecule with a strong Lewis acid. The experimental data should be examined critically and objectively to see whether they require such participation by neighboring alkyl and cycloalkyl groups with its attendant formation of relatively stable intermediates containing carbon bonded to five atoms.

7. Just What Is Meant by the Term Nonclassical Ion?

Surprising as it may seem, despite the vast literature on the subject, apparently no attempt was made to define the term "nonclassical ion" until relatively recently.[16] Roberts seems to have been first to use the term when he proposed the tricyclobutonium structure (**9.5**) for the cyclopropylcarbinyl cation[17] (**9.4**).

[15] H. C. Brown, H. W. Pearsall, L. P. Eddy, W. J. Wallace, M. Grayson, and K. L. Nelson, *Ind. Eng. Chem.*, **45**, 1462 (1953).

[16] H. C. Brown, K. J. Morgan, and F. J. Chloupek, *J. Amer. Chem. Soc.*, **87**, 2137 (1965).

[17] J. D. Roberts and R. H. Mazur, *J. Amer. Chem. Soc.*, **73**, 3542 (1951).

9.4

9.5

The *structure* proposed (**9.5**) was clearly different from the classical structure (**9.4**) for the intermediate discussed. The second appearance of the term is apparently due to Winstein, who refers to the "nonclassical structures" of norbornyl, cholesteryl, and 3-phenyl-2-butyl cations.[18] The third such reference is apparently the statement of Roberts: "Recent interest in the structures of carbonium ions has led to speculation as to whether the ethyl cation is most appropriately formulated as a simple solvated electron-deficient entity, $CH_3CH_2^+$,

a "nonclassical" bridged ethyleneprotonium ion, $H_2C \underset{}{\overset{H}{\underset{+}{-----}}} CH_2$, or possibly as an equilibrium mixture of the two ions."[19]

It is clear from these references that the emphasis is on *structures* that differ markedly from the usual classical *structures*.

Consequently, we suggested the following definition:[16] a nonclassical ion is a carbonium ion in which the position in the structure of one or more atoms is markedly different from that predicted on the basis of classical structural principles.

We also proposed two subclasses. The first subclass consists of ions, such as β-phenylethyl in its bridged phenonium form, that possess sufficient electron-pairs for all of the required bonds. No extension of generally accepted bonding concepts is required to account for these structures. We suggested that such structures be termed "normal" or "electron-sufficient" nonclassical ions (Figure IX-2). The second subclass consists of ions, such as the norbornyl cation in its σ-bridged form, that lack sufficient electrons to provide a pair for all of the bonds required by the proposed structure. A new bonding concept

[18] S. Winstein and D. Trifan, *J. Amer. Chem. Soc.*, **74**, 1154 (1952).

[19] J. D. Roberts and J. A. Yancy, *J. Amer. Chem. Soc.*, **74**, 5943 (1952).

not yet established in carbon structures is required. We suggested that such structures be termed "electron-deficient" nonclassical ions (Figure IX-1).

More recently, Bartlett[5] and Sargent[20] have urged that the term nonclassical be restricted to ions of the electron-deficient group. After all, there is nothing nonclassical about a protonated aromatic[15] (**9.6**) or the heptamethylbenzene cation[21] (**9.7**).

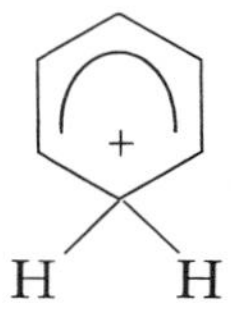

9.6 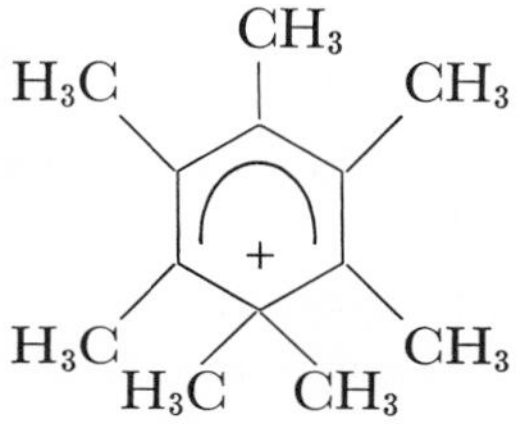   **9.7**

Actually there would be major advantages in restricting the term nonclassical ion to cations containing σ-bridges. Then a nonclassical ion could be defined as a cation that contains a carbon atom bonded to five atoms including the carbonium carbon. However, until there is some general agreement as to what is meant by the term nonclassical ion, it may be better to use instead the more definitive, less ambiguous term, σ-bridged cation.

8. The Cyclopropylcarbinyl Cation

The fast rate of solvolysis of cyclopropylcarbinyl tosylate was originally attributed to the stabilization of the transition state leading to the formation of the presumably highly stabilized symmetrical tricyclobutonium ion[17] (**9.5**). This species is σ-bridged. Note that it contains three carbon atoms bonded to five different atoms.

Later it was observed that the reactions of tagged cyclopropyl-carbinyl derivatives did not show the full equilibration of the tag required by the tricyclobutonium ion.[22] Consequently, it was pro-

---

[20] G. D. Sargent, *Quart. Rev.* (London), **20,** 301 (1966).

[21] W. von E. Doering, M. Saunders, H. G. Boyton, H. W. Earhart, E. F. Wadley, W. R. Edwards, and G. Laber, *Tetrahedron*, **4,** 178 (1958).

[22] R. H. Mazur, W. N. White, D. A. Semenov, C. C. Lee, M. S. Silver, and J. D. Roberts, *J. Amer. Chem. Soc.*, **81,** 4390 (1959).

posed that the cyclopropylcarbinyl cation exists as a rapidly equilibrating set of three equivalent bicyclobutonium ions (**9.8**). This is a

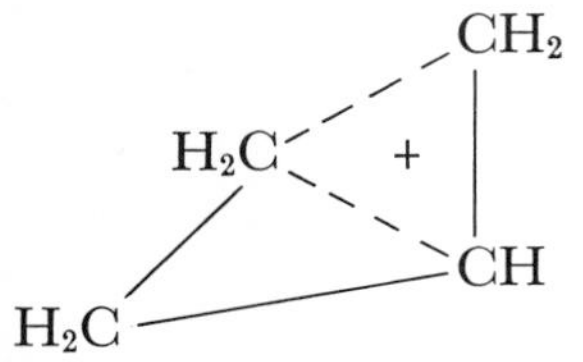

9.8

σ-bridged species with one carbon atom bonded to five nearest neighbors.

Hart and Sandri showed that a number of secondary and tertiary derivatives containing cyclopropyl groups undergo solvolysis with similar rate enhancements but without rearrangements.[23] Consequently, the cyclopropyl group is capable of providing electron density to stabilize a carbonium ion without rearrangement of the structure. Similarly, a study of the rates of reduction of ketones containing cyclopropyl groups established that these rates are quite low[24] (9.15).

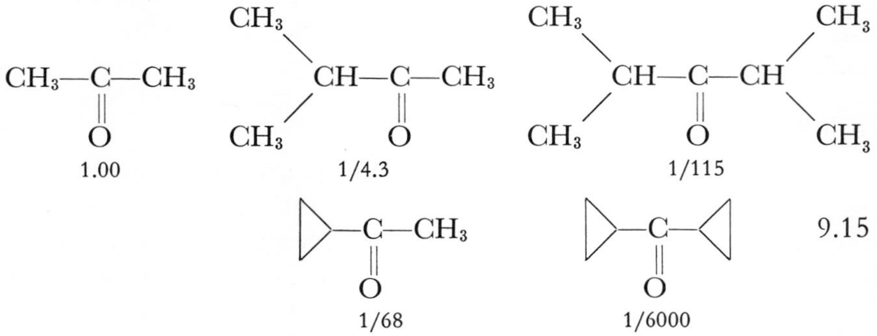

Consequently, the cyclopropyl group is capable of providing electron density to the carbonyl group in these ketones, as well as to the electron-deficient centers of carbonium ion.

The standard tool of the organic chemist in exploring electron deficiency in an organic system is to introduce substituents into the

[23] H. Hart and J. M. Sandri, *J. Amer. Chem. Soc.*, **81**, 320 (1959).

[24] R. H. Bernheimer, Ph.D. Thesis, Purdue U., 1960.

appropriate positions and ascertain the effect. For example, the proposed explanation for the stabilizing effect of the phenyl group in stabilizing the *t*-cumyl cation postulated delocalization of the positive charge from the carbonium carbon to the *ortho* and *para* positions of the aromatic ring (9.11). Introduction of methyl and methoxy substituents into the *para* position of the *t*-cumyl system should assist in satisfying this electron deficiency and result in an increase in the stability of the cation and an increase in the rate of solvolysis. This is observed[25] (9.16).

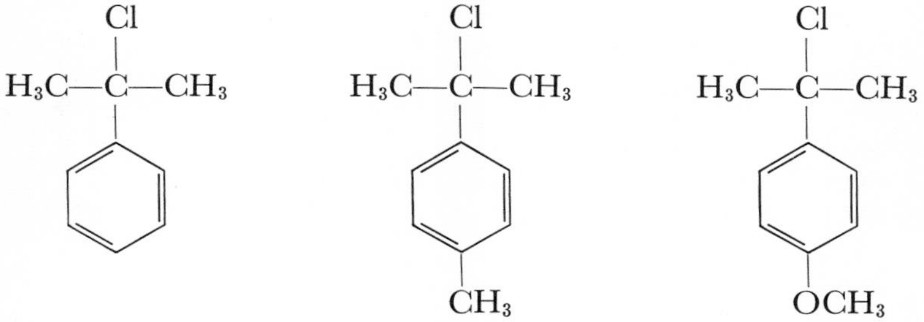

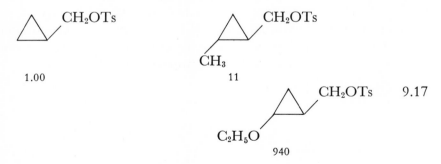

9.16

If the cyclopropyl group stabilizes a carbonium ion center to which it is attached, it should develop an electron deficiency in the ring. Indeed, the introduction of a methyl group or an ethoxy group into the ring results in large rate enhancements[26] (9.17). These rates give a linear plot against the σ^+ constants.[25]

9.17

[25] H. C. Brown, in *Steric Effects in Conjugated Systems*, G. W. Gray, ed. (Butterworth Scientific Publications, London, 1958), p. 100.

[26] P. von R. Schleyer and G. W. Van Dine, *J. Amer. Chem. Soc.*, **88**, 2321 (1966).

Table IX–2. Effect of Substituents on the Rates of Solvolysis of Substituted Cyclopropylcarbinyl 3,5-Dinitrobenzoates in 60 Percent Aqueous Acetone at 100°

Substituents[a]	$10^5 k$, sec^{-1}	Relative rates
H	0.0430	1.0
1-Me	0.213	5.0
1'-Me	43.7	1020
t-2-Me	0.475	11.0
c-2-Me	0.350	8.2
2,2-Me$_2$	3.97	92
t,t-2,3-Me$_2$	5.33	124
c,c-2,3-Me$_2$	3.53	82
c,t-2,3-Me$_2$	3.45	80
t-2,3,3-Me$_3$	21.2	490
2,2,3,3-Me$_4$	67.5	1570
t-2-EtO	40.3	940

a

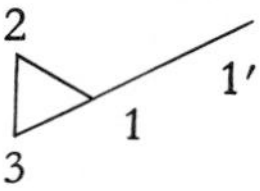

Moreover, the authors made a detailed study of the effect of cumulative methyl groups[26] (Table IX-2). They noted an unusually good additivity for each successive methyl substituent. They concluded that electron supply from the cyclopropane ring must involve a symmetrical contribution, not compatible with the bicyclobutonium ion formulation. Indeed, the bicyclobutonium formulation for the tetramethyl derivative would require a σ-bridge to a fully substituted carbon atom (**9.9**). This appears highly improbable.

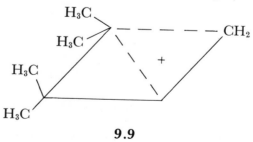

9.9

We explored the electronic contributions of the cyclopropyl substituent in the *para* position of the *t*-cumyl system.[27] We also concluded that the electronic contributions from the cyclopropyl substituent cannot involve σ-bridging through space.

A *p*-isopropyl group increases the rate of solvolysis of *t*-cumyl chloride in 90 percent aqueous acetone at 25° by a factor of 18. On the other hand, a *p*-cyclopropyl group is much more effective—it increases the rate by a factor of 157. A single *o*-methyl substituent, as in 3-methyl-4-cyclopropyl-*t*-cumyl chloride, increases the relative rate to 172. Correcting for the contribution of the *m*-methyl substituent, a factor of 2, reveals only a modest decrease in the effect of the cyclopropyl group accompanying the introduction of the single methyl substituent. On the other hand, the observed relative rate for 3,5-dimethyl-4-cyclopropyl-*t*-cumyl chloride is 37. Correcting this for the contribution of two *m*-methyl substituents, a factor of 4, reveals a sharp drop in the contribution of the cyclopropyl substituent to the rate, to a factor of only 9. Thus, with two *o*-methyl substituents, the contribution of the *p*-cyclopropyl substituent to the rate drops from its original high value of 157 down to a low value of 9, even lower than the effect of a simple alkyl substituent, such as 18 for isopropyl.

There is increasing evidence that the maximum interaction between a cyclopropane group and an adjacent electron-deficient center is achieved with the bisected conformation.[28] Such a bisected arrangement for the cyclopropyl substituent in the *t*-cumyl system readily accounts for the observations, as is apparent from the examination of the structures **9.10–9.12.**

The bisected arrangement of the cyclopropyl group, shown in **9.10,** would not be seriously affected by the introduction of a single methyl group, as shown in **9.11.** However, two methyl substituents effec-

[27] H. C. Brown and J. D. Cleveland, *J. Amer. Chem. Soc.*, **88,** 2051 (1966).

[28] A. D. Walsh, *Trans. Faraday Soc.*, **45,** 179 (1949); M. Y. Lunika, *Russ. Chem. Rev.*, **31,** 419 (1962); C. U. Pittman, Jr., and G. A. Olah, *J. Amer. Chem. Soc.*, **87,** 5123 (1965); G. L. Closs and H. B. Klinger, *J. Amer. Chem. Soc.*, **87,** 3265 (1965); N. C. Deno, H. G. Richey, Jr., J. S. Liu, D. N. Lincoln, and J. O. Turner, *J. Amer. Chem. Soc.*, **87,** 4533 (1965).

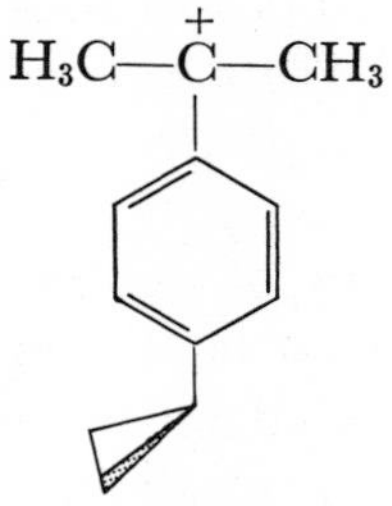

$$H_3C-\overset{+}{C}-CH_3 \qquad\qquad H_3C-\overset{+}{C}-CH_3 \qquad\qquad H_3C-\overset{+}{C}-CH_3$$

9.10 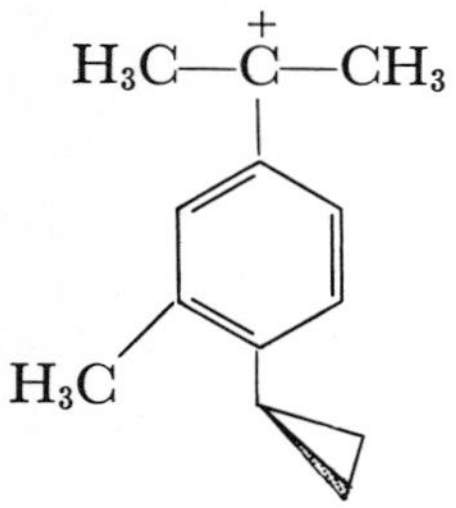   **9.11** 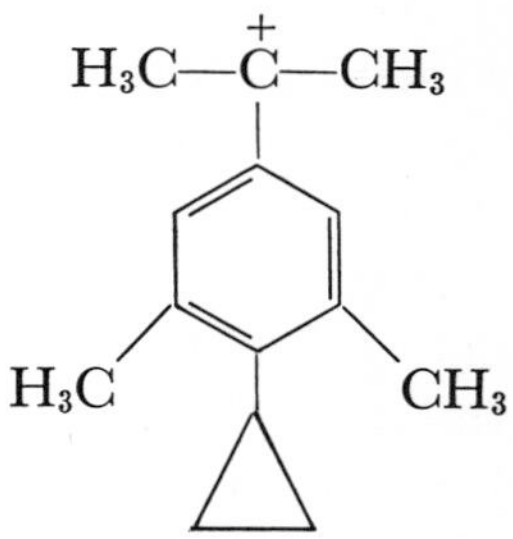   **9.12**

tively block this conformation (**9.12**), greatly reducing the electronic contributions from the cyclopropyl substituents.

Finally, solvolysis of optically active cyclopropylmethylcarbinyl derivatives does not yield optically active products.[29] A σ-bridged intermediate would have been expected to retain asymmetry.

Clearly, the enhanced rate of solvolysis of cyclopropylcarbinyl derivatives is not the result of σ-bridging through space of the carbonium ion center with one or both of the more distant carbon atoms of the ring.

[29] M. Vogel and J. D. Roberts, *J. Amer. Chem. Soc.*, **88,** 2262 (1966).

X. The Question of σ-Participation in Norbornyl Systems[1]

1. Beginnings

The origin of our interest in the question of σ-participation as a factor in the rates of solvolysis of highly branched or otherwise strained systems was discussed in the preceding chapter. In brief, in 1946 I proposed that the solvolysis of highly branched tertiary alkyl derivatives should be facilitated by the relief of steric strain.[2] A number of systems were examined, and the results appeared to support the proposal.[3] The concept seemed to receive general acceptance.[4] Consequently, we left this area for other problems.

At this point there began to appear in the literature examples of fast solvolysis rates for highly strained systems, whose enhanced rates were attributed not to relief of steric strains, but to a new phenomenon, σ-participation.[5] Typical examples were camphene hydrochloride,[6] tri-t-butylcarbinyl esters,[7] and cyclodecyl tosylate.[8]

It seemed a most remarkable coincidence that such cases of enhanced rates of solvolysis attributed to σ-participation all involved strained molecules, where relief of steric strain might have been considered a major contributory factor. A detailed examination of the evidence advanced for σ-participation proved unconvincing. Consequently, a detailed experimental program was undertaken to test this proposal.

[1] H. C. Brown, *Chem. Brit.*, **2**, 199 (1966).

[2] H. C. Brown, *Science*, **103**, 385 (1946).

[3] H. C. Brown and R. S. Fletcher, *J. Amer. Chem. Soc.*, **71**, 1845 (1949); H. C. Brown and H. L. Berneis, *J. Amer. Chem. Soc.*, **75**, 10 (1953).

[4] E. L. Eliel, in *Steric Effects in Organic Chemistry*, M. S. Newman, ed. (John Wiley, New York, 1956), Chap. 2.

[5] S. Winstein and D. S. Trifan, *J. Amer. Chem. Soc.*, **74**, 1147, 1154 (1952).

[6] F. Brown, E. D. Hughes, C. K. Ingold, and J. F. Smith, *Nature*, **168**, 65 (1951).

[7] P. D. Bartlett, *J. Chem. Educ.*, **30**, 22 (1953).

[8] R. Heck and V. Prelog, *Helv. Chim. Acta*, **38**, 1541 (1955).

2. The Experimental Approach

The norbornyl system was generally regarded as the best, most clear-cut example of σ-participation[9] (10.1). Accordingly, this system

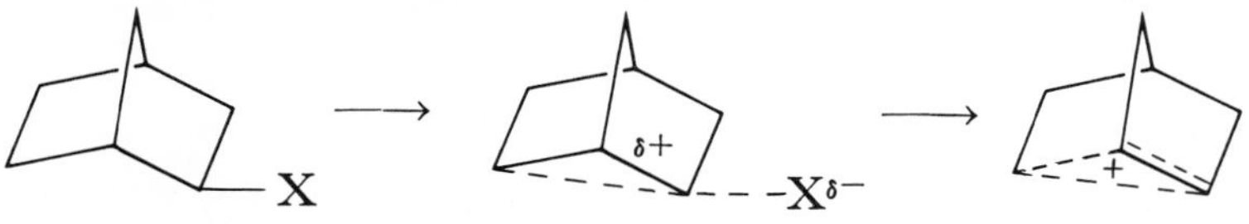

10.1

was selected for intensive study.

A detailed examination of the literature revealed three major foundations for the proposal of σ-participation in the solvolysis of 2-norbornyl derivatives:

1. Unusually fast rates of solvolysis for the *exo* derivatives, as illustrated by a rate for camphene hydrochloride 6000 times greater than that for *t*-butyl chloride.[6]

2. High *exo*:*endo* rate ratios, as indicated by a value of 350 for the acetolysis of the 2-norbornyl brosylates.[5]

3. Almost exclusive formation of *exo* acetates in the solvolysis of both *exo* and *endo* derivatives,[5] especially for compounds containing 7,7-dimethyl substituents.[9]

Each of these indicated foundations was subjected in turn to intensive examination.

3. Are Exo Rates Unusually Fast?

The magnitude of the enhanced rate of solvolysis exhibited by camphene hydrochloride (6000) as compared to *t*-butyl chloride (1.00) was considered to be incompatible with relief of steric strain.[6] Consequently, the enhanced rate was attributed to the driving force associated with the formation of a stabilized mesomeric cation[6] (10.2).

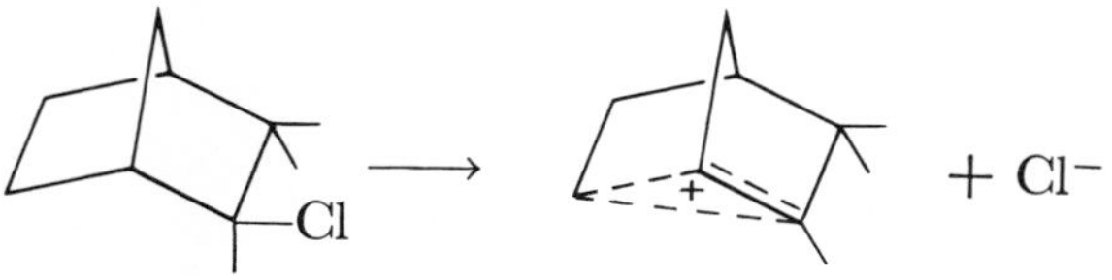

$+ \ Cl^- \quad 10.2$

[9] J. A. Berson, in *Molecular Rearrangements*, P. de Mayo, ed., Vol. 1 (Interscience Publishers, New York, 1963), Chap. 3.

In such analyses it is always important to select suitable models.[10] The question may be raised whether *t*-butyl chloride is a suitable model for the highly congested molecule, camphene hydrochloride. A more suitable molecule would doubtless be the pentamethylcyclopentyl chloride (**10.5**) obtained by opening the 5,6-ethylene bridge of camphene hydrochloride.

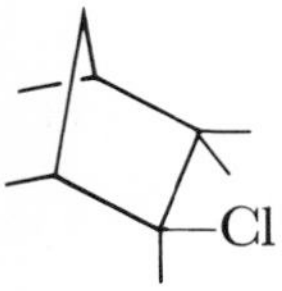

10.5

This structure has not yet been synthesized. However, many of the other methyl substituted 1-methylcyclopentyl chlorides have

[10] For example, a comparison of the rates of **10.1** and **10.2** led to the conclusion that the ion from **10.2,** containing a cyclopropylcarbinyl moiety, does not exhibit stabilization, presumably because of steric difficulties. J. D. Roberts, W. Bennett, and R. Armstrong, *J. Amer. Chem. Soc.,* **72,** 3329 (1950).

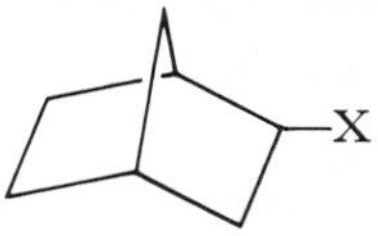

1.00	0.07
10.1	**10.2**

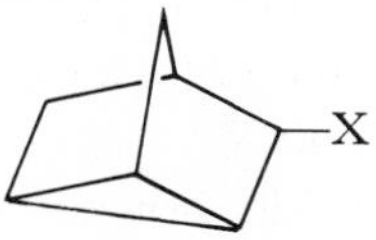

On the other hand, comparison of **10.3** and **10.4** reveals a major acceleration in the latter.

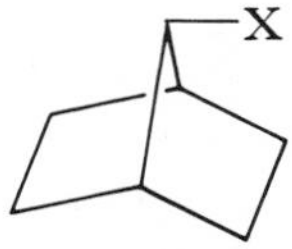

1.00	10^{10}
10.3	**10.4**

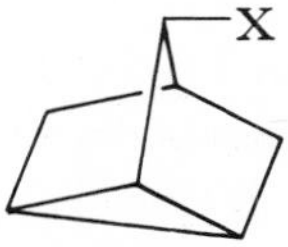

The enormous acceleration of rate in the latter derivative is presumably the result of a major contribution of the cyclopropane ring to the carbonium center.

The fact is that **10.2** and **10.4** are identical. Yet we have come to opposite conclusions as to the effectiveness of the cyclopropane ring in stabilizing the carbonium ion center merely through a selection of the model.

been prepared and their rates of ethanolysis at 25° determined.[11] The data do not support the conclusion that the rates for the norbornyl derivative are exceptional (10.3).

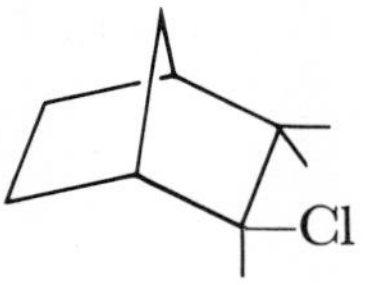

| 13,600 | 1.00 | 2,380 |

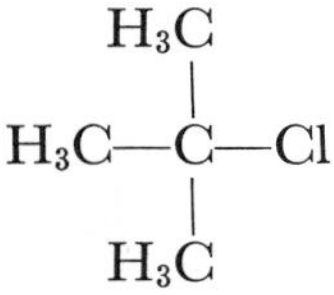

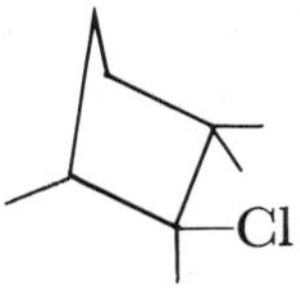

10.3

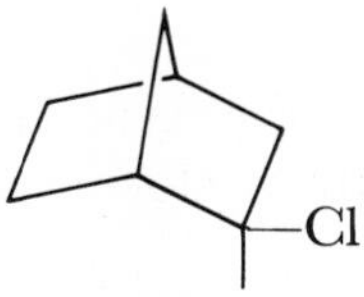

| 355 | 66 |

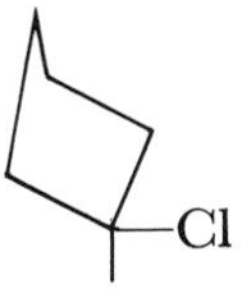

Quite clearly the results are in accord with the postulated effect of increasing steric strain in enhancing the rates of solvolysis of highly branched tertiary chlorides.[2,3] Similar results have been realized in other comparisons.[12]

These results do not disprove the formation of mesomeric nor-bornyl cations in the solvolysis of 2-norbornyl derivatives. However, they do eliminate the original argument that the rates of solvolysis of camphene hydrochloride and other *exo*-norbornyl derivatives are too fast to be explicable on any basis other than the formation of mesomerically stabilized cations.[13]

[11] H. C. Brown and F. J. Chloupek, *J. Amer. Chem. Soc.*, **85,** 2322 (1963). (A slightly higher rate was obtained for camphene hydrochloride.)

[12] H. C. Brown, F. J. Chloupek, and M.-H. Rei, *J. Amer. Chem. Soc.*, **86,** 1246, 1247, 1248 (1964).

[13] In line with the special objectives of this volume (Section I–1), it may be appropriate to record a very helpful comment from a referee. The Communication dealing with the rates of hydrolysis of camphene hydrochloride and the cyclopentane derivatives[11] came back from a referee with the usual recommendation in this field that the manuscript should not be published. The referee took the position that camphene hydrochloride was a tertiary chloride and that anyone knowledgeable in the area of nonclassical ions would immediately understand that tertiary

4. Does the High Exo:Endo Rate Ratio Require a σ-Bridged Cation?

The acetolysis of *exo*-norbornyl brosylate occurs at a rate 350 times that of the *endo* isomer.[5,14] This high *exo*:*endo* rate ratio was assigned to a fast rate for the *exo* isomer, attributed to σ-participation (10.1), with a normal rate for the *endo* isomer, attributed to the unfavorable stereoelectronic situation for similar σ-participation.

At a time when little was known about the stereochemical properties of the norbornyl system, it was quite reasonable to consider this high *exo*:*endo* rate ratio to be exceptional and to propose a unique explanation for the phenomenon. With the years, however, our knowledge of the stereochemical properties of the norbornyl system has grown. It is now evident that all reactions reveal a greater or lesser preference for the *exo* face of the norbornane structure over the *endo*.[15]

This subject will be discussed in greater detail in the next chapter. However, two examples may be given here. Hydroboration of norbornene (10.4) proceeds to give 99.5 percent *exo*, only 0.5 percent *endo*, a ratio of 200:1.[16]

$$10.4$$

99.5% *exo*

---

norbornyl cations should be too stabilized to force σ-participation from the 1,6 electron-pair (although many leading supporters of the nonclassical ion concept have proposed nonclassical structures even for stabilized tertiary derivatives). Fortunately I was able to persuade the editor to accept the manuscript. However, the comment started me along a line of thought. Why not prepare tertiary norbornyl cations of increasing stability and examine their properties? If such ions did indeed approach classical ions in character, the properties that had been considered to be associated with σ-participation would vanish. This proved to be a highly fruitful line of approach.

[14] If we correct for internal return, the ratio becomes 1600.

[15] P. D. Bartlett, G. N. Fickes, F. C. Haupt, and R. Helgeson, *Accounts Chem. Res.*, 3, 177 (1970).

[16] H. C. Brown and J. H. Kawakami, *J. Amer. Chem. Soc.*, **92,** 1990 (1970).

Similarly, the base-catalyzed deuterium exchange of norbornanone proceeds to give an *exo*:*endo* rate ratio of 715[17] (**10.6**).

10.6

Consequently, there is nothing unique about the high *exo*:*endo* rate ratio observed in the solvolysis of norbornyl brosylate. Equally high *exo*:*endo* ratios are now known for a wide variety of reactions not involving carbonium ions. For reactions such as the hydroboration of norbornene and the deuterium exchange of norbornanone the high *exo*:*endo* ratios are believed to arise as a result of the more difficult attack of the reagent upon the hindered *endo* position. It is therefore appropriate to consider whether the high *exo*:*endo* rate ratio observed for the solvolysis may not be the result of a normal *exo* rate, but a retarded *endo* rate. This question will be considered in detail in the next chapter.

One approach to the problem was to examine the *exo*:*endo* rate ratios in highly stabilized tertiary norbornyl derivatives.[13] It appears clear that the more stable the carbonium ion center, the less demand that center will make upon neighboring groups for stabilization. In-deed, a careful pmr examination of the 2-phenylnorbornyl cation revealed no evidence for charge delocalization from the 2- to the 1- and 6-positions.[18] If the 2-phenylnorbornyl cation is classical, there can be no question about the 2-anisyl derivatives.

The position that a cationic center stabilized by a *p*-anisyl group should make little demand upon neighboring groups for participation was recently tested experimentally by Gassman and his co-workers.[19] They observed that the 10^{11} rate enhancement arising from participation of the double bond in the solvolysis of *anti*-7-dehydronorbornyl derivatives (**10.7**) essentially vanishes in the corresponding 7-*p*-anisyl derivatives (**10.8**).

---

[17] T. T. Tidwell, *J. Amer. Chem. Soc.*, **92**, 1448 (1970).

[18] D. G. Farnum and G. Mehta, *J. Amer. Chem. Soc.*, **91**, 3256 (1969).

[19] P. G. Gassman, J. Zeller, and J. T. Lumb, *Chem. Comm.*, 69 (1968).

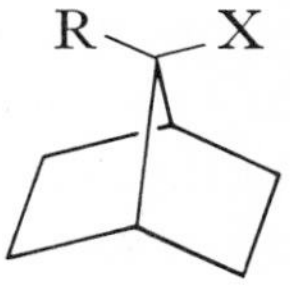

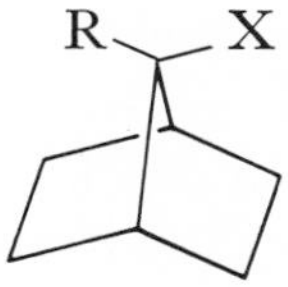

R = H	1.00	10^{11}
R = p-anisyl	1.00	2.4
	10.7	**10.8**

If the *p*-anisyl group can cause the truly enormous participation ($\times 10^{11}$) observed in *anti*-7-dehydronorbornyl to vanish, surely it should also cause the much smaller *σ*-participation ($\times 350$) proposed for *exo*-norbornyl to vanish. However, the *exo:endo* rate ratio in the solvolysis of 2-*p*-anisylnorbornyl *p*-nitrobenzoate is comparable, 284 (10.5).[20]

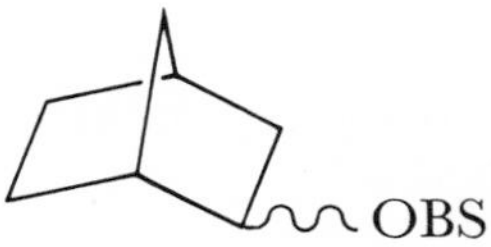

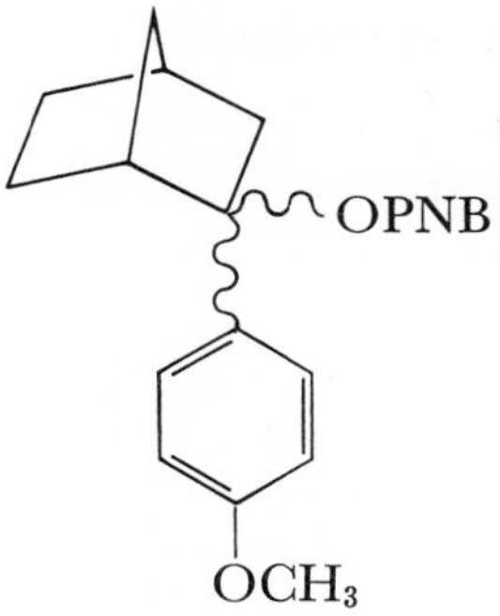

exo:endo   350   284

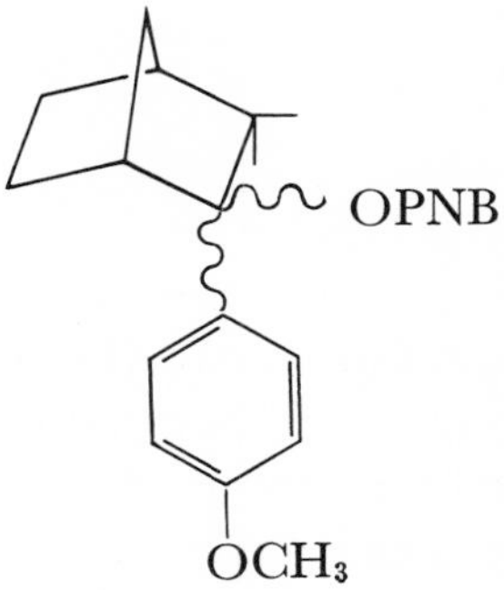

                                                                10.5

                                44,000

[20] H. C. Brown and K. Takeuchi, *J. Amer. Chem. Soc.*, **90**, 2691 (1968).

Moreover, the introduction of methyl groups into the 3-position increases the observed *exo:endo* rate ratio to 44,000 (10.5), largely as a result of a marked decrease in the rate of the *endo* isomer.[21]

Consequently, high *exo:endo* rate ratios occur even in stabilized norbornyl systems where σ-participation cannot be a factor. Therefore the high *exo:endo* rate ratios observed in the solvolysis of norbornyl derivatives do not *require* a σ-bridged cation. However, there still remains the possibility that high *exo:endo* rate ratios may have different origins in secondary and tertiary norbornyl derivatives. Is it possible that the high *exo:endo* rate ratio has its origin in steric influences in tertiary norbornyl derivatives, but in σ-participation in the parent norbornyl system? To find an answer, we turned to examine the effect of substituents on the rates of solvolysis of norbornyl derivatives.

5. The Effect of Substituents

The nonclassical norbornyl cation has been represented as a resonance hybrid of three structures[5] (10.6).

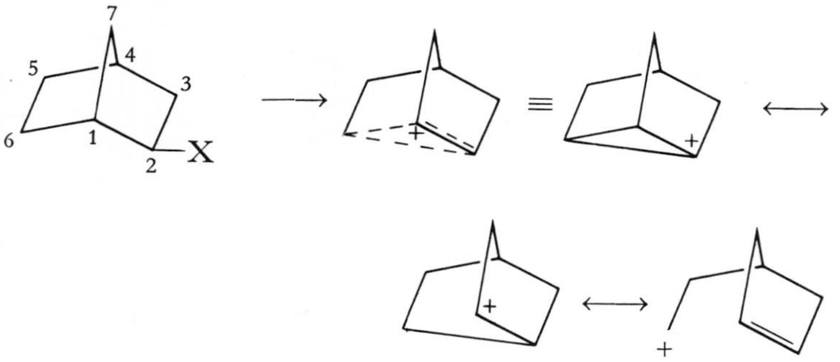

According to this formulation, charge is delocalized from the 2- to the 1- and 6-positions. If the lower energy of the *exo* transition state is the result of the fact that the stabilized σ-bridged norbornyl cation is partially formed in that transition state, it should be possible to obtain evidence for such charge delocalization by using the classical tool of the organic chemist, the effect of substituents in appropriate positions.

[21] *Ibid.*, 5268.

This tool was applied earlier to the cyclopropylcarbinyl problem (Section IX-8). Introduction of a methyl group into the *para* position of the *t*-cumyl chloride molecule increases the rate of solvolysis by a factor of 26. A methoxy substituent is much more effective, resulting in a rate enhancement by a factor of 3360. This is in agreement with the postulated delocalization of charge from the carbonium carbon atom of the *t*-cumyl cation to the *para* position of the aromatic ring.

Introduction of a methyl or an ethoxyl substituent into the cyclopropyl ring of a cyclopropylcarbinyl derivative results in rate increases of 11 and 940, respectively. Again this is in agreement with the postulated delocalization of charge from the carbonium carbon atom into the cyclopropyl system.

If charge were similarly delocalized to the 6-position of the norbornyl system in the transition state, then the introduction of methyl and methoxy groups into that position should result in similar rate increases. However, such rate increases are not observed (10.7).

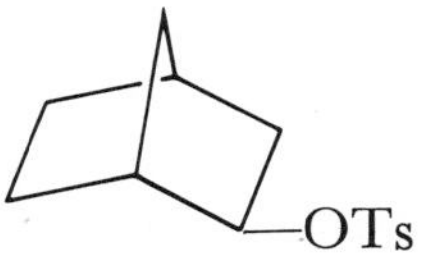

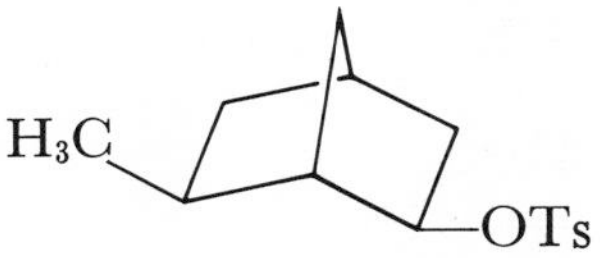

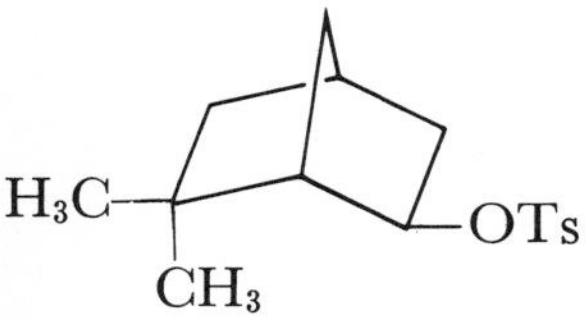

10.7

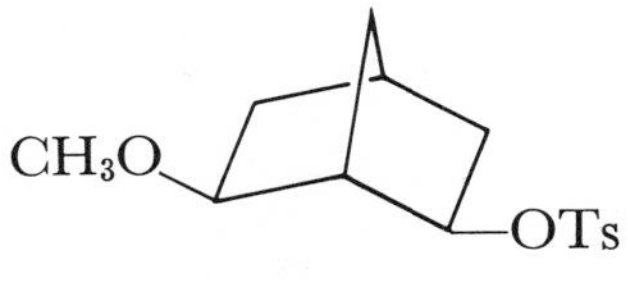

These results are clearly not in accord with the postulated acetolysis of *exo*-norbornyl derivatives with σ-participation of the 1,6 bonding-pair. Such σ-participation should be accompanied by

[22] The value for 6-methyl is a tentative value: private communication from P. von R. Schleyer.

[23] P. von R. Schleyer, M. M. Donaldson, and W. E. Watts, *J. Amer. Chem. Soc.*, **87**, 375 (1965).

[24] Corrected from the value reported: P. J. Stang and P. von R. Schleyer, *Abstracts of the 155th Meeting of the American Chemical Society* (San Francisco, 1968), P. 192.

charge delocalization to the 1- and 6-positions. The effect of methyl and methoxy substituents in the 6-positions does not support such charge delocalization.

It has been suggested that "there are good reasons to expect carbon bridging to lag behind C—X ionization at the transition state."[25] However, this does not appear to offer an escape from the difficulty. If σ-bridging is the factor responsible for the *exo:endo* rate ratio of 350, such σ-bridging must be present in the transition state to produce the observed rates. If such σ-bridging occurs in the transition state, why are the usual effects of methyl and methoxy substituents absent?

If the norbornyl cation is really a resonance-stabilized species, the effect of substituents in the 2-position should be greatly decreased over their effect in a typical aliphatic or alicyclic system. However, we observe identical effects[26] (10.8).

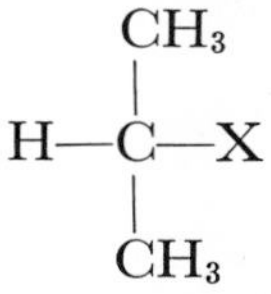

$$
\begin{array}{c}
CH_3 \\
| \\
H-C-X \\
| \\
CH_3
\end{array}
\qquad 1.00
$$

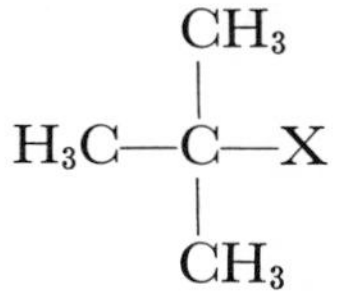

$$
\begin{array}{c}
CH_3 \\
| \\
H_3C-C-X \\
| \\
CH_3
\end{array}
\qquad 55{,}000
$$

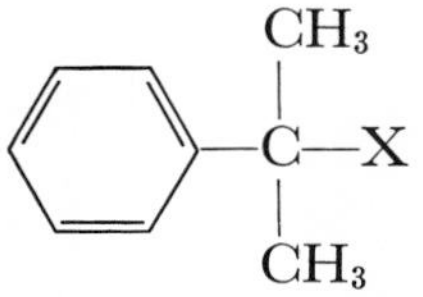

$$
\begin{array}{c}
CH_3 \\
| \\
C-X \\
| \\
CH_3
\end{array}
\qquad
\begin{array}{l}
\times\ 4600 \\
(2.6 \times 10^8)
\end{array}
$$

$$
CH_3O-
\begin{array}{c}
CH_3 \\
| \\
C-X \\
| \\
CH_3
\end{array}
\qquad
\begin{array}{l}
\times\ 3400 \\
(8.5 \times 10^{11})
\end{array}
$$

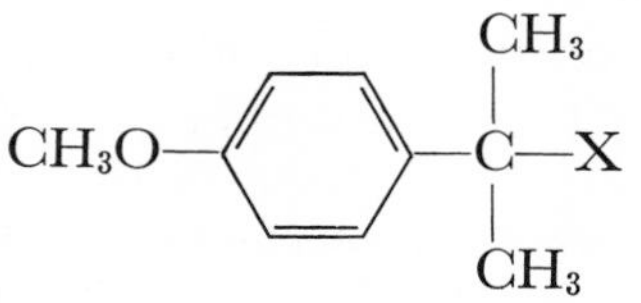

[25] S. Winstein, *J. Amer. Chem. Soc.*, **87**, 381 (1965).

[26] H. C. Brown and M.-H. Rei, *J. Amer. Chem. Soc.*, **86**, 5008 (1964); H. C. Brown and K. Takeuchi, *J. Amer. Chem. Soc.*, **88**, 5336 (1966).

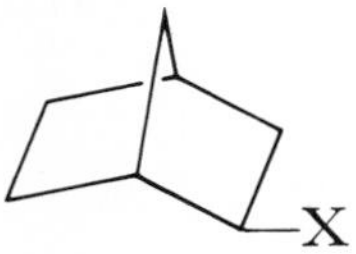

1.00

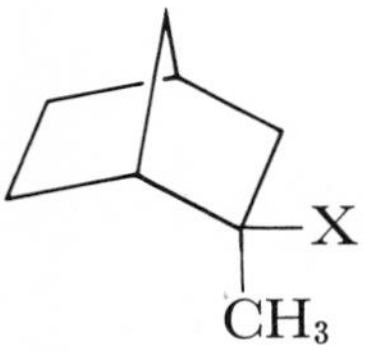

55,000

10.8

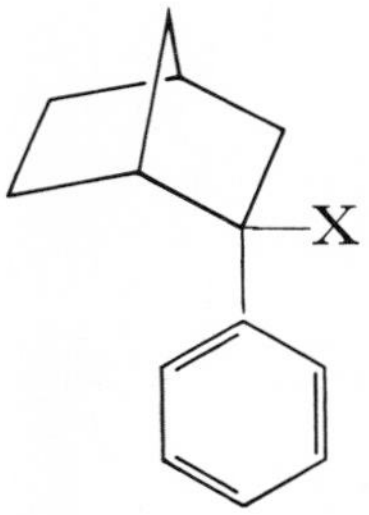

× 5300
(2.9 × 10^8)

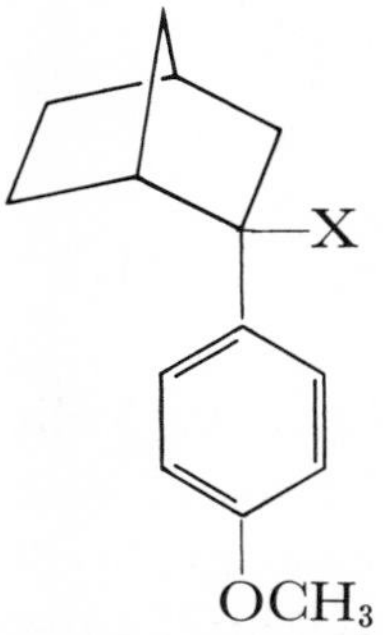

× 1700
(5.0 × 10^{11})

There is a difficulty in introducing substituents into the 1-position as a test for charge delocalization to that position. Norbornyl derivatives containing alkyl or aryl substituents in the 1-position solvolyze with rearrangement to tertiary norbornyl cations. This should provide a driving force that would enhance the effect of the 1-substituent and decrease the observed difference between the effects of the same substituent in the 1- and 2-positions[26,27] (10.9).

[27] D. C. Kleinfelter and P. von R. Schleyer, quoted by Berson, footnote 9, p. 154.

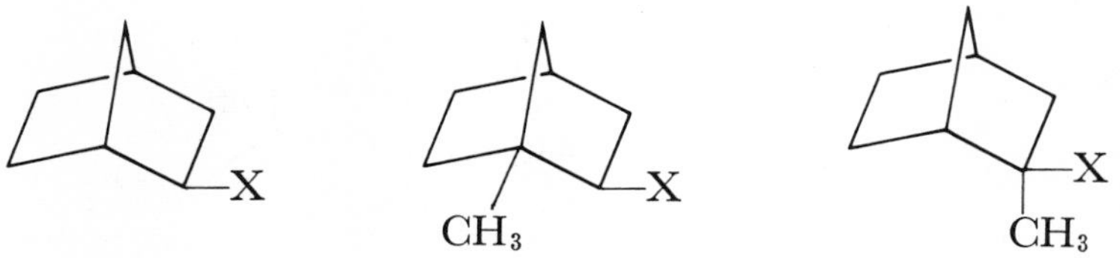

1.00   50   55,000

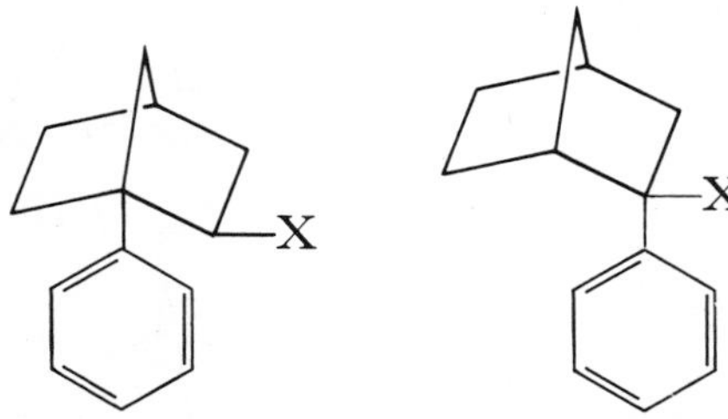

CH$_3$ CH$_3$

10.9

3.9   290,000,000

OCH$_3$ OCH$_3$
7.8   500,000,000,000

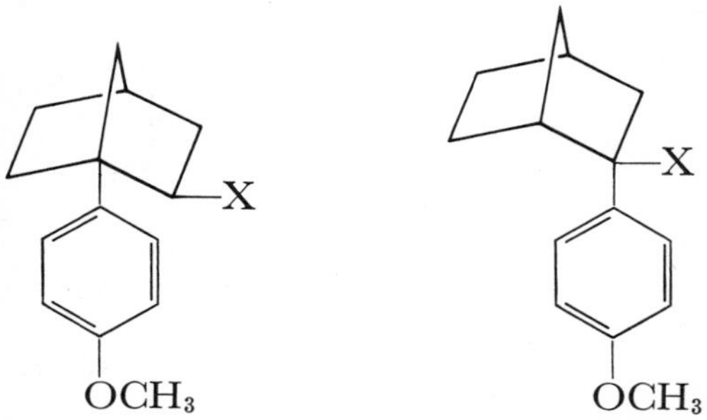

Finally, the study of symmetrically substituted 1,2-norbornyl derivatives is highly informative.

It was originally pointed out by Bunton that the great difference in stability between secondary and tertiary cations would appear to preclude resonance between structures of such great energies[28] (10.10).

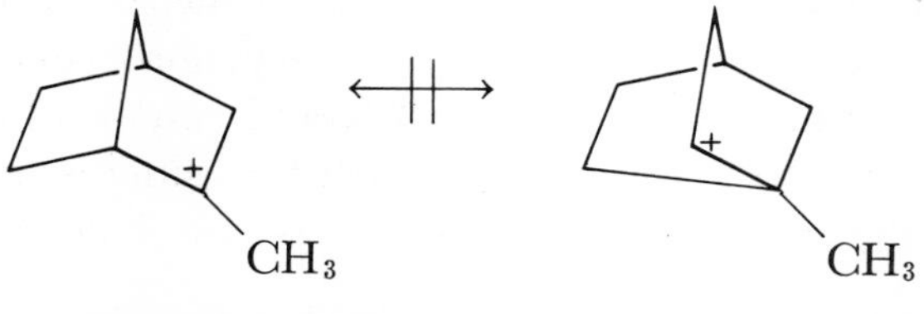

10.10

CH$_3$ CH$_3$

[28] C. A. Bunton, *Nucleophilic Substitution at a Saturated Carbon Atom* (Elsevier, New York, 1963).

Consequently, it was suggested that a tertiary norbornyl cation, such as 2-methylnorbornyl, should be essentially classical in nature.

As indicated earlier[26] (10.8), a methyl group stabilizes the transition state for the limiting solvolysis of an isopropyl derivative by 6.6 kcal/mole, a phenyl group by 11.7, and an anisyl group by 16.6. Presumably, these values can be considered to be lower limits for the differences in energies between the corresponding secondary and tertiary canonical structures (10.11).

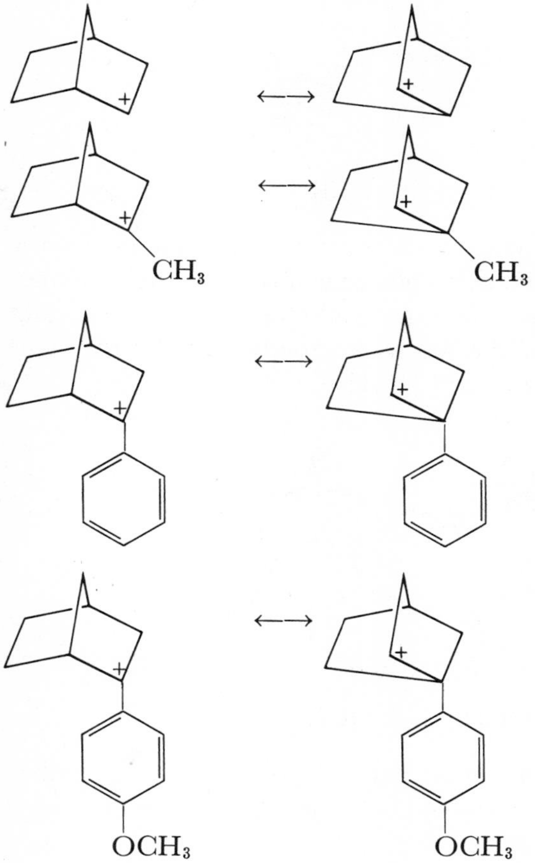

ΔE 0 kcal/mole

ΔE 6.6

ΔE 11.7   10.11

ΔE 16.6

Clearly, with the exception of the secondary-secondary case, resonance between structures that differ so greatly in energy should not be significant. However, by introducing an identical substituent in the 1-position, we again produce symmetrical resonance structures. All of the resonance energy lost in the secondary-tertiary systems should now return in the tertiary-tertiary systems (10.12).

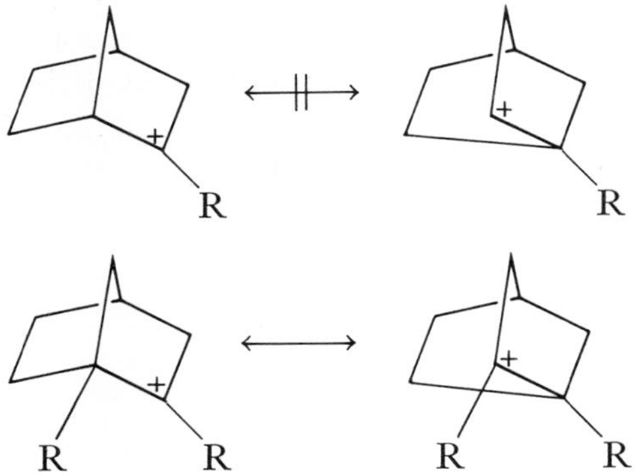

10.12

The 1,2-dianisylnorbornyl cation was subjected to intensive study by Schleyer and his co-workers.[29] They concluded that their thermochemical, chemical reactivity, and ultraviolet and nmr spectral data supported the existence of the ion as a rapidly equilibrating pair rather than as a resonance hybrid (10.13).

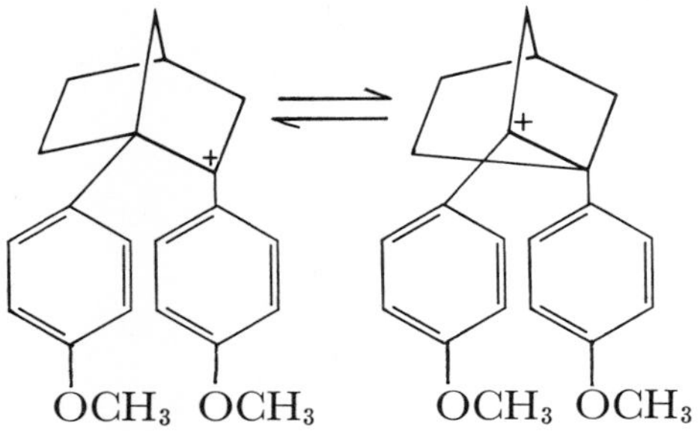

10.13

Similarly, a phenyl substituent in the 1-position retards, rather than enhances, the rate of solvolysis of 2-phenylnorbornyl chloride[30] (10.14).

[29] P. von R. Schleyer, D. C. Kleinfelter, and H. G. Richey, Jr., *J. Amer. Chem. Soc.,* **85,** 479 (1963).

[30] Unpublished research with M.-H. Rei.

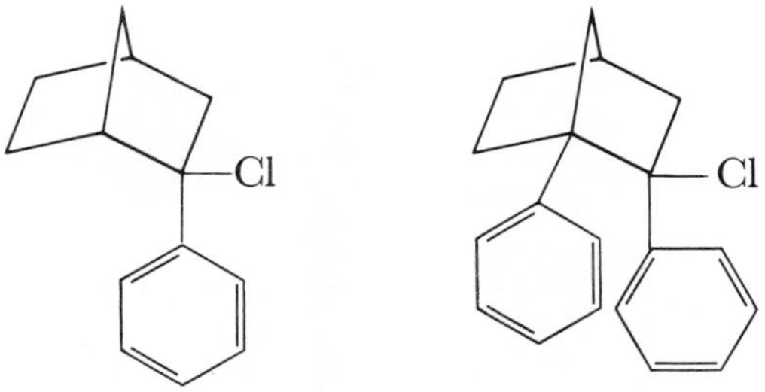

1.00 $\quad\quad\quad\quad\quad\quad\quad\quad$ ⅟₁₆

A possible difficulty with this experimental approach lies in the argument that it may be impossible, for steric reasons, to place both aryl groups into the orientation required to permit them to participate in resonance-stabilization of the ion. This difficulty does not arise in the 1,2-dimethylnorbornyl derivatives.[31] Here the introduction of a methyl group into the 2-position of 1-methylcyclopentyl *p*-nitrobenzoate results in a rate enhancement of 3.4, presumably in the combined result of the steric and inductive effect of the methyl substituent. Introduction of a methyl substituent into the 1-position of 2-methyl-*endo*-norbornyl *p*-nitrobenzoate results in a similar rate enhancement of 4.8 over the parent compound. The effect is obviously similar to that observed in the cyclopentyl derivative. The critical case is the *exo* isomer. Here also, however, the introduction of a methyl substituent of 2-methyl-*exo*-norbornyl produces a rate enhancement of the same magnitude, 4.3 (10.15).

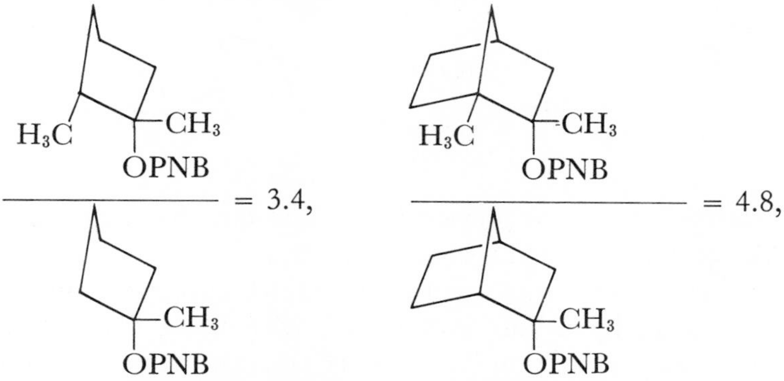

[31] H. C. Brown and M.-H. Rei, *J. Amer. Chem. Soc.*, **86**, 5004 (1964).

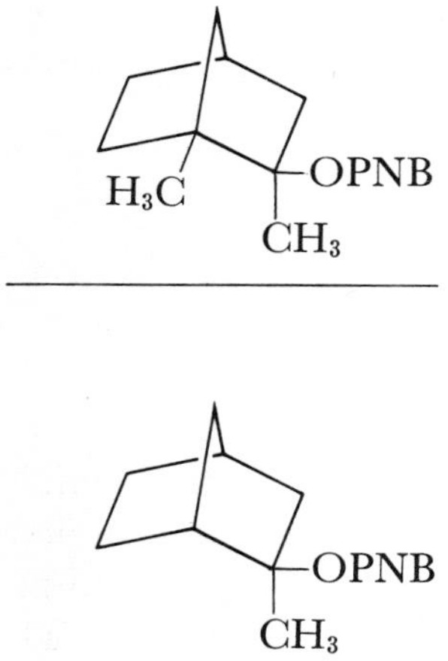

$$\frac{\text{H}_3\text{C} \quad \text{OPNB} \atop \text{CH}_3}{\text{OPNB} \atop \text{CH}_3} = 4.3 \qquad\qquad 10.15$$

Consequently, there is no evidence for σ-participation even in these symmetrically substituted 1,2-norbornyl derivatives. Furthermore, all of these different approaches point to the absence of significant σ-participation even in the parent system.

6. Does the High Exo:Endo Product Ratio Require a σ-Bridged Cation?

The acetolysis of *exo*-norbornyl brosylate produces the *exo* acetate in high stereochemical purity.[5] It was argued that this high stereospecificity supported the proposed σ-bridged intermediate (10.16).

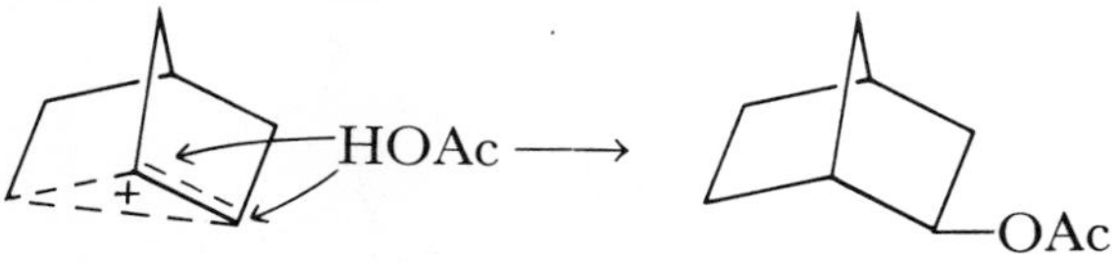

$$\text{HOAc} \longrightarrow \text{OAc} \quad + \text{H}^+ \quad 10.16$$

However, we have previously pointed out that reaction at the *exo* face of the norbornyl system (10.4, **10.6**) is not an unusual characteristic of the norbornyl system. Indeed, the classical 2-anisylnorbornyl is captured by sodium borohydride with essentially exclusive attack from the *exo* position.[20] Moreover, the *exo* and *endo* isomers of 2-

anisylcamphenilyl *p*-nitrobenzoates undergo solvolysis with the formation of $\geq$ 99.5 percent *exo* alcohol.[32]

A detailed study of the acetolysis of the parent norbornyl derivative revealed the formation of 99.98 percent *exo*-norbornyl acetate and 0.02 percent *endo*-norbornyl acetate.[33] It has been argued that such an extraordinarily high stereoselectivity is inexplicable without the intermediacy of a σ-bridged ion.[34] Fortunately, a study of the addition of acetic acid and trifluoroacetic acid to norbornene refutes the position.[35]

The addition of acetic acid to norbornene proceeds slowly at 100° to give 2-norbornyl acetate with a remarkably high *exo* stereoselectivity: 99.97 percent *exo* in the absence of sodium acetate and 99.99 percent *exo* in its presence. This stereoselectivity is comparable to (actually slightly higher than) the value realized, 99.95 percent *exo* in the acetolysis of *exo*-norbornyl tosylate at 100° in the presence of sodium acetate.

As has been mentioned, such an extraordinarily high stereoselectivity has been considered previously to require the intermediacy of a σ-bridged cation. On this basis the addition of deuterioacetic acid to norbornene should proceed with an equal distribution of the deuterium tag between the *exo*-3 and *syn*-7 positions (10.17).

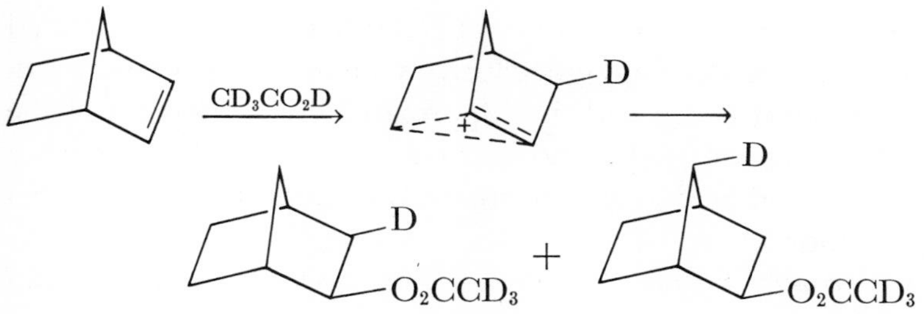

$$ 10.17 $$

[32] K. Takeuchi and H. C. Brown, *J. Amer. Chem. Soc.*, **90**, 5270 (1968).

[33] H. L. Goering and C. B. Schewene, *J. Amer. Chem. Soc.*, **87**, 3516 (1965).

[34] S. Winstein, E. Clippinger, R. Howe, and E. Vogelfanger, *J. Amer. Chem. Soc.*, **87**, 376 (1965).

[35] H. C. Brown, J. H. Kawakami, and K.-T. Liu, *J. Amer. Chem. Soc.*, **92**, 5536 (1970).

However, the product contains 69 percent of the deuterium at the *exo*-3 position, 19 percent at the *syn*-7, with 12 percent of hydride shifted material.

In contrast to the slow addition of acetic acid to norbornene at 100°, the addition of trifluoroacetic acid to this olefin is exceedingly rapid at 0°, the reaction being complete in 1 to 2 minutes. However, here also the product is highly stereoselective, consisting of 99.98 percent of *exo*- and 0.02 percent of *endo*-norbornyl trifluoroacetate.

The addition of deuteriotrifluoroacetic acid also exhibits the same high stereoselectivity. Even more significant, the deuterium tag is not equally distributed between the *exo*-3 and *syn*-7 positions (37 percent *exo*-3-*d*, 26 percent *syn*-7-*d*, and the remainder hydride-shifted products), clearly indicating an unsymmetrical structure of the carbonium ion in this typical carbonium ion reaction.

These results do not appear explicable in terms of a mechanism proceeding through the symmetrical nonclassical norbornyl cation as sole intermediate. They can be rationalized in terms of a rapidly equilibrating pair of classical cationic intermediates that are captured before they have been fully equilibrated.

Alternatively, it is possible to interpret the results in terms of two distinct, competing mechanisms—an ionic addition involving a non-classical cation and yielding product with equal distribution of the deuterium between the *exo*-3 and *syn*-7 positions, and a concerted *cis* addition that places the tag exclusively at the *exo*-3-position. On this basis, it might be argued that approximately 50 percent of the addition of the acetic acid to norbornene proceeds through such a concerted process and 50 percent proceeds through the symmetrical norbornyl cation.

This interpretation requires that the exo stereoselectivity of the postulated concerted addition process be comparable to that realized in the carbonium ion pathway. Adoption of this position would negate the argument that the observed unusually high stereoselectivity, 99.98 percent *exo*, is explicable only in terms of the unique stereochemical requirements of a σ-bridged intermediate.

There is a third possibility to consider. It could be assumed that the transfer of the proton from these acids to norbornene proceeds to the initial formation of the classical norbornyl cation. This cation then undergoes a closure to the nonclassical cation at a rate that is competitive with the capture of the nucleophile.[36] Thus, in the case of acetic acid, 40 percent of the product (not involving hydride shift) would be formed via the σ-bridged intermediate and 50 percent via the classical norbornyl cation captured prior to the conversion to the nonclassical ion. This mechanism would account for the isomer distribution. However, it requires that capture of the nucleophile by the classical norbornyl cation must proceed with stereoselectivity of essentially 99.98 percent *exo*. This again would be incompatible with the argument that such a high stereoselectivity is inexplicable save in terms of the unique steric requirements of a σ-bridged cation.[33,34]

These considerations suggest that the simplest, most consistent interpretation of these addition reactions[37] is that they proceed through rapidly equilibrating classical norbornyl cations and that such cations react with nucleophile with a high preference for *exo* capture. It follows that the high *exo:endo* product ratios realized in the solvolysis of 2-norbornyl derivatives do not require a σ-bridged cation.

7. Do the Results With 7,7-Dimethylnorbornyl Derivatives Require a σ-Bridged Cation?

Let us next turn our attention to the argument that the almost exclusive *exo* substitution observed in the solvolysis of 7,7-dimethylnorbornyl brosylate requires a bridged ion.[38,39]

In the reaction of sodium borohydride with norcamphor and apocamphor the presence of the 7,7-dimethyl substituents changes the direction of preferential attack from the *exo* direction for norcamphor

[36] Such a mechanism has been proposed to account for the formation of *exo*-norbornanol with 10 percent retention of optical activity in the deamination of optically active *exo*-norbornylamine: E. J. Corey, J. Casanova, Jr., P. A. Vatakencherry, and R. Winter, *J. Amer. Chem. Soc.*, **85,** 169 (1963).

[37] Including the addition of deuterium chloride: H. C. Brown and K.-T. Liu, *J. Amer. Chem. Soc.*, **89,** 466, 3898, 3900 (1967).

[38] A. Colter, E. C. Friedrich, N. J. Holness, and S. Winstein, *J. Amer. Chem. Soc.*, **87,** 379 (1965).

[39] R. Howe, E. C. Friedrich, and S. Winstein, *J. Amer. Chem. Soc.*, **87,** 381 (1965).

(**10.9**) to the *endo* direction for apocamphor[40] (**10.10**). It is argued

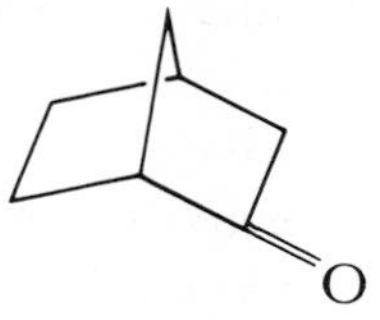

86% *exo* attack
10.9

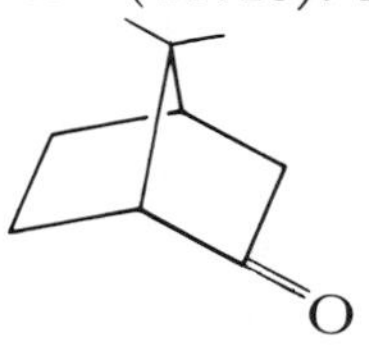

20% *exo* attack
10.10

that a similar reversal would be anticipated for the capture of solvent by the cationic intermediate in the absence of the special directive requirements introduced by σ-bridging.[38,39]

It is true that the direction of hydroboration[41] and epoxidation[42] of norbornene (10.18) is inverted in apobornene (10.19).

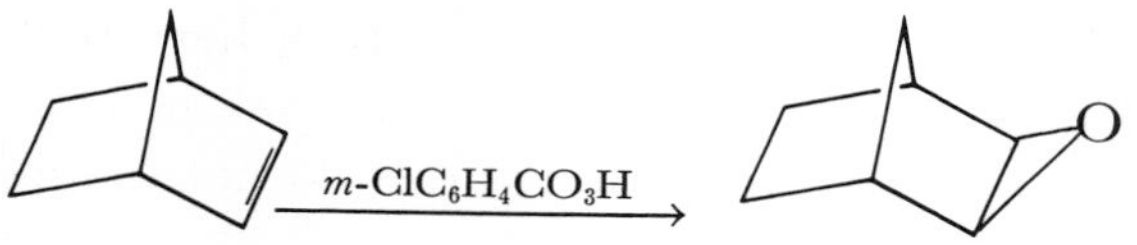

m-ClC$_6$H$_4$CO$_3$H

10.18

99.5% *exo*

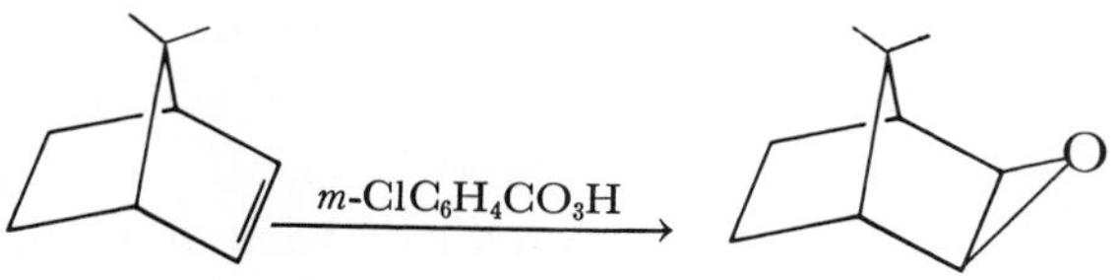

m-ClC$_6$H$_4$CO$_3$H

10.19

10% *exo*

However, these reactions proceed through a cyclic addition mechanism that places the adding moiety directly under the *syn*-7-methyl substituent. The situation proved to be quite different for reactions that proceed through two-stage noncyclic additions.[43]

[40] H. C. Brown and J. Muzzio, *J. Amer. Chem. Soc.*, **88**, 2811 (1966).
[41] H. C. Brown and J. H. Kawakami, *J. Amer. Chem. Soc.*, **92**, 1990 (1970).
[42] H. C. Brown, J. H. Kawakami, and S. Ikegami, *J. Amer. Chem. Soc.*, **92**, 6914 (1970).
[43] H. C. Brown and J. H. Kawakami, *J. Amer. Chem. Soc.*, **92**, 201 (1970).

Thus the free-radical addition of thiophenol to norbornene and apobornene gives the *exo* product preferentially in both cases (10.20).

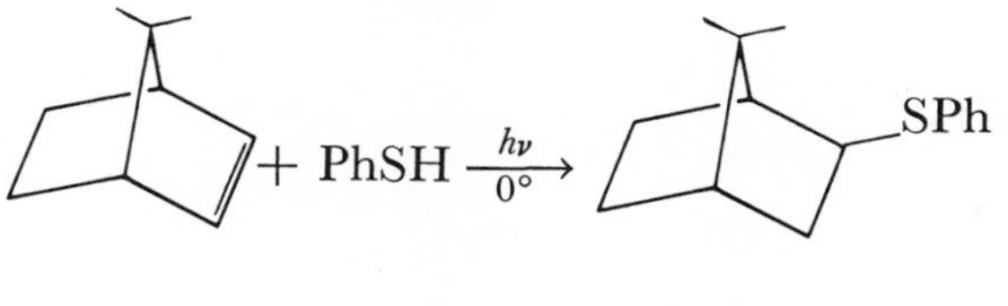

10.20

95%

This reaction involves free-radical intermediates that form bonds individually at the corners of the apobornane system (10.21).

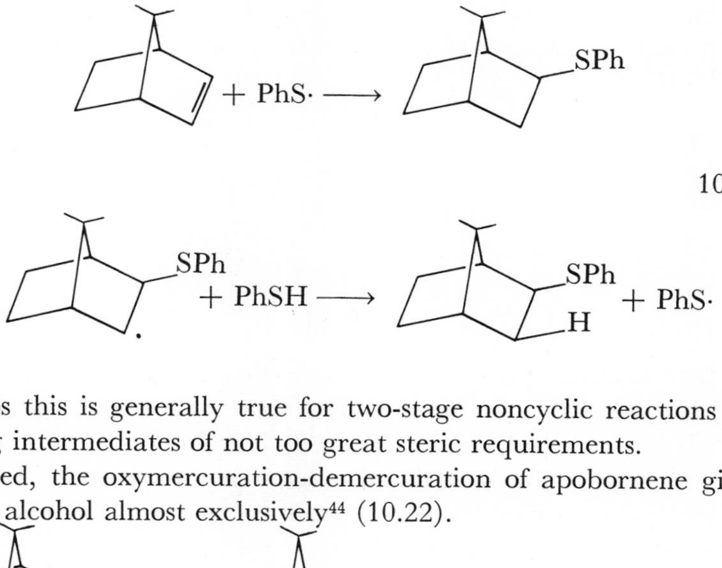

10.21

Perhaps this is generally true for two-stage noncyclic reactions involving intermediates of not too great steric requirements.

Indeed, the oxymercuration-demercuration of apobornene gives the *exo* alcohol almost exclusively[44] (10.22).

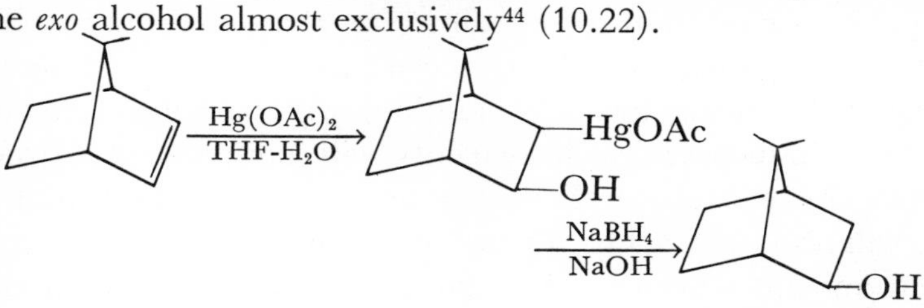

10.22

> 99.8% *exo*

[44] H. C. Brown, J. H. Kawakami, and S. Ikegami, *J. Amer. Chem. Soc.*, **89**, 1525 (1967).

Moreover, the addition of mercuric trifluoroacetate to norbornene and apobornene in benzene solution goes with remarkable rapidity directly to the *exo-cis* adducts[45] (**10.11, 10.12**).

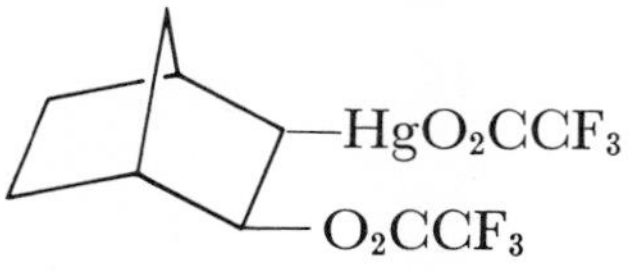

10.11 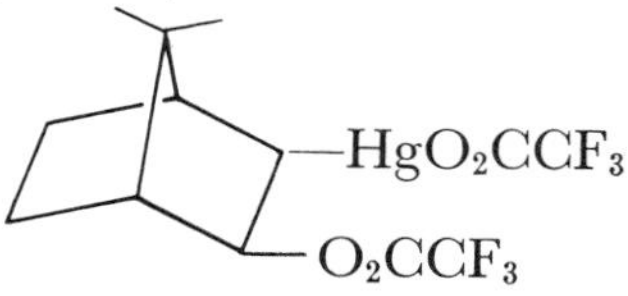   **10.12**

Free-radical substitution of the saturated hydrocarbons also shows preferential attack from the *exo* direction even in the presence of 7,7-dimethyl substituents.[15]

Finally, base-catalyzed deuteration of camphor also goes predominantly *exo* in spite of the presence of the 7,7-dimethyl substituents[17] (**10.13**).

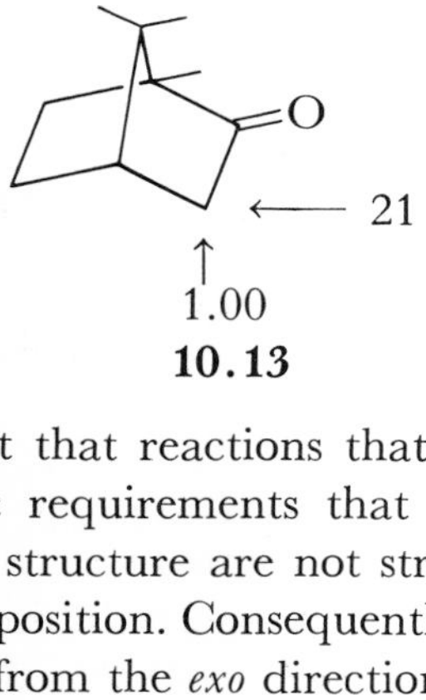

10.13

These results suggest that reactions that proceed through intermediates of low steric requirements that attack in stages at the corners of the bicyclic structure are not strongly influenced by the methyl groups at the 7-position. Consequently, such attacks evidently proceed preferentially from the *exo* direction. Reagents that involve cyclic addition, causing the adding species to come directly under the methyl substituents in the 7-position, will be subject to far greater steric influences. Such reagents will normally add *endo*. Finally, reagents of large steric requirements, such as the complex hydrides, will also be strongly affected sterically by the *syn*-7-methyl substituent and influenced to react preferentially from the *endo* direction.[44]

Finally, it is significant that the trapping of tertiary apobornyl

[45] H. C. Brown, M.-H. Rei, and K.-T. Liu, *J. Amer. Chem. Soc.*, **92**, 1760 (1970).

cations also takes place preferentially from the *exo* direction[46] (10.23, 10.24).

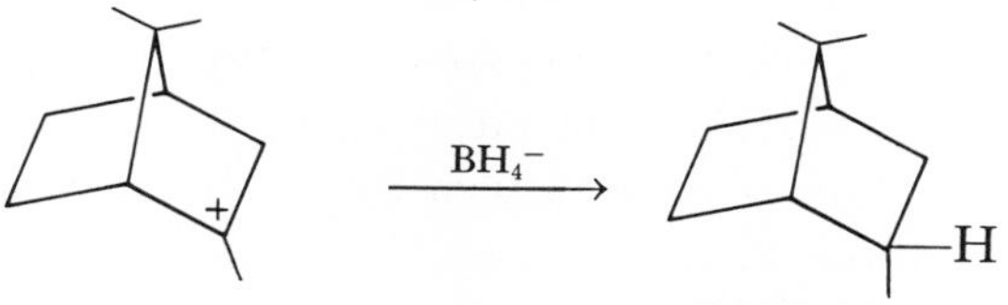

$$BH_4^- \longrightarrow$$

100% *exo*-2-H

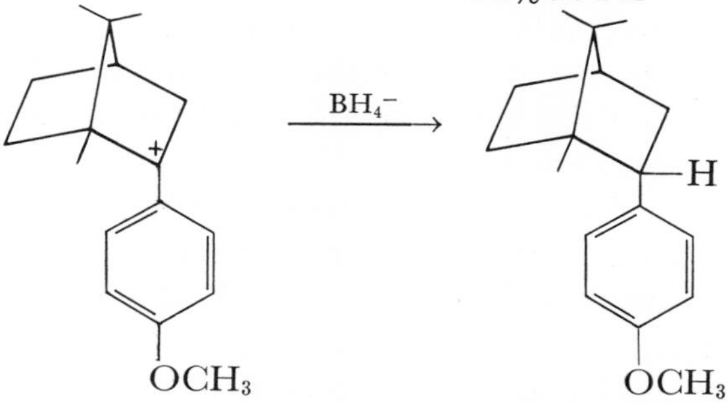

10.23

$$BH_4^- \longrightarrow$$

OCH₃

OCH₃

87% *exo*-2-H

10.24

The original argument supporting the nonclassical interpretation postulated that all reactions of the 7,7-dimethylnorbornyl system would be expected to go preferentially *endo* in the absence of the σ-bridging.[9] This argument can no longer be accepted. It is possible to fall back on the quantitative difference in the stereoselectivities exhibited by the parent secondary apobornyl system (99.95 percent *exo*) and the corresponding tertiary systems (10.24).[47] This is similar to the argument for the parent norbornyl system considered in the preceding section. However, at this time we do not have a clear picture of how solvent capture by the tight ion-pairs of secondary systems compares in stereochemical characteristics to capture of nucleophiles by stabilized tertiary ions. Consequently, it appears dangerous to rely upon this argument in the face of all of the contrary evidences.

In conclusion, it appears that the results obtained with 7,7-dimethylnorbornyl derivatives do not *require* a σ-bridged cation.

---

[46] H. C. Brown and H. M. Bell, *J. Amer. Chem. Soc.*, **86,** 5006, 5007 (1964).

[47] J. A. Berson, J. H. Hammons, A. W. McRowe, R. G. Bergman, A. Remanick, and D. Houston, *J. Amer. Chem. Soc.*, **89,** 2590 (1967).

8. Other Studies

In this chapter I have described our efforts to obtain experimental evidence for the oft-postulated σ-participation in the solvolysis of norbornyl derivatives. The topic has been a popular one, and numerous other studies have appeared bearing on the question of whether the norbornyl cation exists as a classical ion or possesses the σ-bridged structure of the nonclassical formulation. We cannot consider here all of the available studies. However, many of these were advanced as evidence in support of the nonclassical structure, and it may be instructive to point out how time and developments have answered many of the arguments.

At one time the Foote-Schleyer correlation was utilized to support the σ-bridged formulation for the norbornyl cation.[48] As discussed in detail in the next chapter, it proved incapable in its present form of handling steric hindrance to ionization, and it was concluded that the correlation could not be used to resolve the problem.[49] It now appears to be in even more serious difficulties in view of the recent evidence that the acetolysis of secondary alkyl arenesulfonates must involve a large solvent contribution.[50] The correlation does not include such a term.

Early calculations favored the σ-bridged structure for the norbornyl cation.[51] A recent, presumably more precise approach favors the classical structure.[52]

Gassman originally argued that the low *exo*:*endo* rate ratio for 7-ketonorbornyl supported a classical structure for this species and therefore a nonclassical structure for norbornyl itself.[53] More re-

[48] P. von R. Schleyer, *J. Amer. Chem. Soc.*, **86**, 1854, 1856 (1964).

[49] H. C. Brown, I. Rothberg, P. von R. Schleyer, M. M. Donaldson, and J. J. Harper, *Proc. Nat. Acad. Sci.*, **56**, 1653 (1966).

[50] J. L. Fry, C. J. Lancelot, L. K. M. Lam, J. M. Harris, R. C. Bingham, D. J. Raber, R. E. Hall, and P. von R. Schleyer, *J. Amer. Chem. Soc.*, **92**, 2538 (1970), fn. 21.

[51] G. Klopman, *J. Amer. Chem. Soc.*, **91**, 89 (1969).

[52] L. C. Allen and D. Goetz, *Abstracts of the Organic Chemistry Division of the ACS-CIC Joint Conference at Toronto, May 24–29, 1970.* See report in *Chem. Eng. News*, **48**, 76, June 8, 1970.

[53] P. G. Gassman and J. L. Marshall, *J. Amer. Chem. Soc.*, **88**, 2822 (1966).

cently he has concluded that in this derivative the rate of the *endo* isomer is enhanced by participation by the 7-keto group,[54] so the results have no bearing on the problem in norbornyl itself.

Bartlett and his co-workers observed that 2-(Δ^3-cyclopentenyl)-ethyl arenesulfonate solvolyzes in acetic acid 74 times faster than its saturated analog. The product is *exo*-norbornyl acetate. Clearly the reaction proceeds with participation of the double bond. A methyl group at the double bond increases the rate by a factor of 7; a second methyl group increases the rate by a factor of 5.5. It was considered that the almost equal increments in rates produced by the methyl groups required that both the transition states and the intermediates produced in the solvolysis, such as the norbornyl and 1,2-dimethyl-norbornyl cations, must be symmetrical and therefore nonclassical.[55] However, Goering and Humski have recently succeeded in trapping the 1,2-dimethylnorbornyl cation in optically active form.[56] They conclude, therefore, that the 1,2-dimethylnorbornyl cation must be classical. Consequently, the argument above is rendered doubtful.

Observations of the norbornyl cation in strong acids at low temperatures have been recently interpreted to favor the σ-bridged structure under these conditions.[57] The interpretation is not unambiguous. Moreover, interesting as these studies are, they are not pertinent to the question of what is responsible for the high *exo*:*endo* rate and product ratios in the solvolysis of norbornyl derivatives.

Numerous studies have been reported on the secondary isotope effect for norbornyl derivatives containing deuterium labels.[58] Unfortunately, general agreement as to the precise manner in which secondary isotopes exert their effects is not yet at hand (Section VI-5). Consequently, it would appear unsafe to rely on this uncertain tool to resolve a problem of such complexity.

[54] P. G. Gassman, J. L. Marshall, J. G. McMillan, and J. M. Hornback, *J. Amer. Chem. Soc.*, **91,** 4282 (1969).

[55] P. D. Bartlett *et al.*, *J. Amer. Chem. Soc.*, **87,** 1288, 1297 (1965).

[56] H. Goering and K. Humski, *J. Amer. Chem. Soc.*, **90,** 6213 (1968).

[57] G. A. Olah, A. M. White, J. R. DeMember, A. Commeyras, and C. Y. Lui, *J. Amer. Chem. Soc.*, **92,** 4627 (1970).

[58] For discussion and leading references, see B. L. Murr and J. A. Conkling, *J. Amer. Chem. Soc.*, **92,** 3462, 3464 (1970).

It has been suggested that the most favorable remaining evidence for σ-participation as a factor in the *exo*:*endo* rate and product ratios in the secondary norbornyl derivatives, in spite of the absence of such σ-participation in the tertiary compounds, is provided by the trimethylene derivatives.[59] The *exo*:*endo* rate ratio for the acetolysis of 5,6-trimethylene-2-norbornyl derivatives is small, approximately 10.[60] A similar factor is observed for the corresponding 4,5-trimethylene derivatives.[61] It has been argued that the strain engendered by the trimethylene bridge resists formation of the σ-bridged cation and that this is responsible for the low *exo*:*endo* rate ratio (10.25).

10.25

Relative rate   10   1.00

This is a reasonable argument. However, it may be desirable to point out how qualitative the nonclassical theory is. Had the authors observed a larger than normal *exo*:*endo* ratio, they could have accounted for it equally well by postulating that the strain introduced by the trimethylene bridge had caused the 1,6-bonding electrons to become more polarizable and better donors.

Another possible explanation for the low *exo*:*endo* rate ratio is that the trimethylene bridge distorts the norbornyl structure. Substituents can apparently introduce surprisingly large distortions of the norbornane skeleton.[62] Consequently, it is desirable to reserve judgment until these systems have been subjected to detailed study.

Finally, a major argument for the absence of significant σ-participation in the solvolysis of norbornyl derivatives is provided by the observation that the *exo*:*endo* rate ratio for norbornyl tosylate shows no major change in trifluoroacetic acid over that observed in acetic

[59] Private communication from Professor Paul von R. Schleyer.
[60] K. Takeuchi, T. Oshika, and Y. Koga, *Bull. Chem. Soc. Japan*, **38**, 1318 (1965).
[61] E. J. Corey and R. S. Glass, *J. Amer. Chem. Soc.*, **89**, 2600 (1967).
[62] C. Altona and M. Sundaralingam, *J. Amer. Chem. Soc.*, **92**, 1995 (1970).

acid.[63] Systems such as β-phenylethyl reveal enormous increases in the k_Δ/k_s ratios in trifluoroacetic acid.[64] The absence of such an increase in the norbornyl system argues strongly for the absence of σ-participation as the factor responsible for the *exo:endo* rate ratio.

9. Conclusion

Approximately ten years have passed (at the time of this writing) since we first began a systematic search for σ-participation in camphene hydrochloride and related tertiary norbornyl derivatives. There now appears to be considerable agreement that such tertiary norbornyl cations are largely, if not entirely, classical. Yet they exhibit *exo:endo* rate and product ratios comparable to the corresponding ratios exhibited by the parent secondary systems. Is it reasonable to continue to interpret such similar phenomena in terms of totally different physical bases? I believe that it is not and that it should not be done until some definitive piece of experimental evidence is made available to support σ-bridging in the solvolysis of secondary norbornyl derivative.

As pointed out earlier, our interest in the question of σ-participation was initiated by its proposal as an alternative explanation to relief of steric strain for certain enhanced rates: (1) camphene hydrochloride,[6] (2) tri-*t*-butylcarbinyl *p*-nitrobenzoate,[7] and (3) cyclodecyl tosylate.[8] It is of interest to observe what has happened to these three original cases in the intervening years. In the case of cyclodecyl tosylate, it was observed that the introduction of deuterium into the transannular positions failed to affect the rate significantly. Consequently, the transannular bridged formulation (9.10) was withdrawn.[65] A detailed study of the products from the solvolysis of tri-*t*-butylcarbinyl *p*-nitrobenzoate failed to support the σ-bridged formulation (9.9) and it has been withdrawn.[66] Finally, if one can

---

[63] J. E. Nordlander, S. P. Jindal, and W. J. Kelly, *Abstracts of the 155th Meeting of the American Chemical Society* (San Francisco, 1968), P. 191.

[64] J. E. Nordlander and W. G. Deadman, *J. Amer. Chem. Soc.*, **90,** 1590 (1968).

[65] V. Prelog, *Rec. Chem. Prog.*, **18,** 247 (1957). See also V. Prelog and J. G. Traynham, *Molecular Rearrangements*, P. de Mayo, ed. Vol. 1 (Interscience Publishers, New York, 1963), Chap. 9.

[66] P. D. Bartlett and T. T. Tidwell, *J. Amer. Chem. Soc.*, **90,** 4421 (1968).

extrapolate the conclusions of Goering and Humski[56] about the essentially classical nature of the 1,2-dimethylnorbornyl cation to the 2,3,3-trimethylnorbornyl (hydrocamphenilyl) cation, then this σ-bridged cation (9.8) is no longer with us.

It would appear that only the parent structure is still with us as a system in which σ-participation may contribute in a modest way to the *exo:endo* rate and product ratios. However, we cannot point to any really definitive experimental results that require the presence of such σ-participation.

I should like to close this chapter with a quotation from an earlier lecture.

"It is evident from the literature that this inquiry into the structure of the norbornyl cation and other cations for which nonclassical structures have been proposed has aroused considerable emotion and controversy. This is indeed unfortunate. Any concept in science should be subject to examination and re-examination. We can have no 'sacred cows.' I hope that this summary of my evidence and views will convince the chemical public interested in the structure of carbonium ions that there remains a problem to be solved. I am confident that if this is explored with the reason and ingenuity of which the human mind is capable, the problem will soon be resolved. Whichever way the decision goes, organic chemistry will be enriched by a new understanding of the factors influencing the behavior of carbonium ions."

XI. Steric Hindrance
to Ionization

1. The Steric Pattern of Reactions of the Norbornyl System

The norbornane molecule possesses a rigid structure with uncommon steric characteristics (**11.1**). Carbon atoms 1–6 constitute a

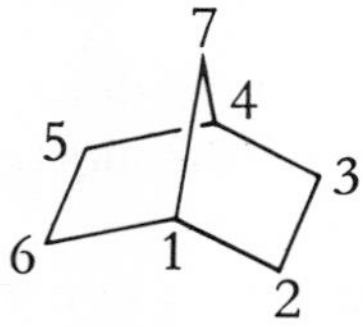

11.1

cyclohexane structure in the higher-energy boat conformation. Moreover, the 7-methylene group not only locks this ring system into a rigid boat conformation, but the constraint so produced accentuates the steric crowding within the boat structure (Figure XI-1).

It is unfortunate that so much emphasis was placed in the past on the high *exo*:*endo* rate ratio of 350 in the acetolysis of the 2-norbornyl brosylates[1] (11.1).

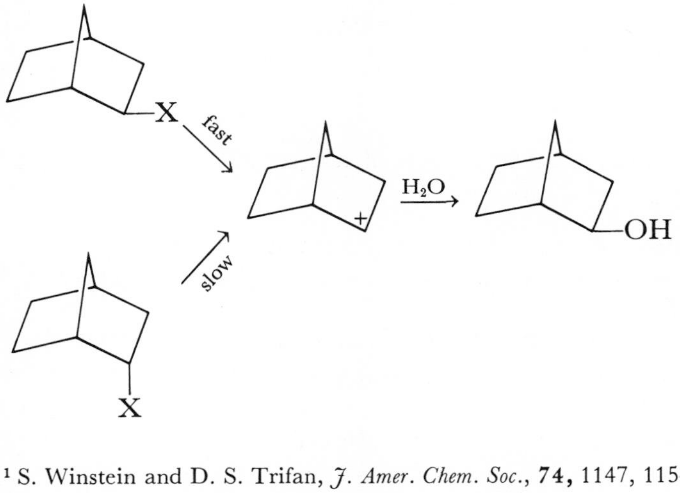

11.1

[1] S. Winstein and D. S. Trifan, *J. Amer. Chem. Soc.*, **74,** 1147, 1154 (1952).

As a result, it was largely overlooked that many non-carbonium ion reactions of the norbornyl system exhibit a comparable stereospecificity. For example, Kooyman reported that the free-radical chlorination of norcamphane with sulfuryl chloride yields approximately 99 percent *exo*-norbornyl chloride[2,3] (11.2).

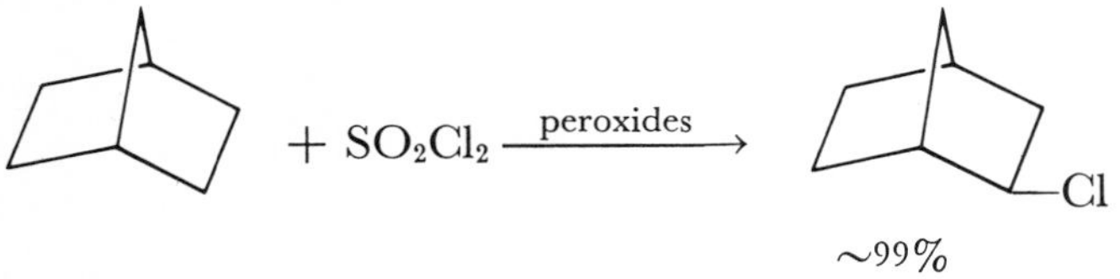

$$+ \; SO_2Cl_2 \xrightarrow{\text{peroxides}} \qquad 11.2$$

~99%

Accordingly, I decided to initiate a detailed study of the stereochemical properties of reactions of the norbornyl system not involving carbonium ions.

Hydroboration-oxidation of norbornene yields 99.5 percent *exo* product[4] (11.3).

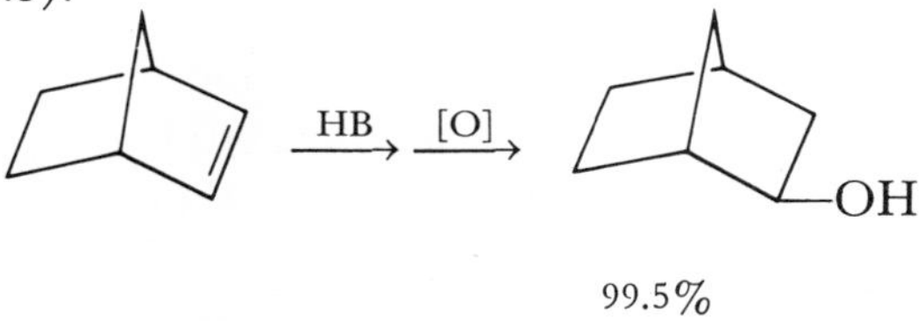

$$\xrightarrow{HB} \xrightarrow{[O]} \qquad 11.3$$

99.5%

Epoxidation of norbornene gives 99.5 percent *exo* epoxide[5] (11.4).

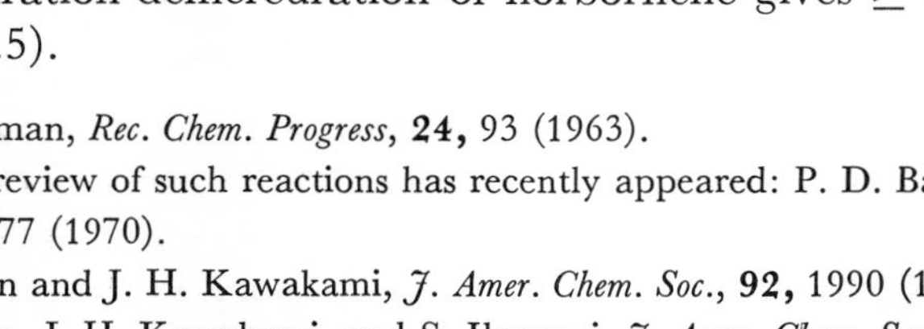

$$\xrightarrow{m\text{-}ClC_6H_4CO_3H} \qquad 11.4$$

99.5%

Oxymercuration-demercuration of norbornene gives ≥ 99.8 percent *exo*[6] (11.5).

---

[2] E. C. Kooyman, *Rec. Chem. Progress*, **24,** 93 (1963).

[3] A detailed review of such reactions has recently appeared: P. D. Bartlett, *Acc. Chem. Res.*, **3,** 177 (1970).

[4] H. C. Brown and J. H. Kawakami, *J. Amer. Chem. Soc.*, **92,** 1990 (1970).

[5] H. C. Brown, J. H. Kawakami, and S. Ikegami, *J. Amer. Chem. Soc.*, **92,** 6914 (1970).

[6] H. C. Brown, J. H. Kawakami, and S. Ikegami, *J. Amer. Chem. Soc.*, **89,** 1525 (1967).

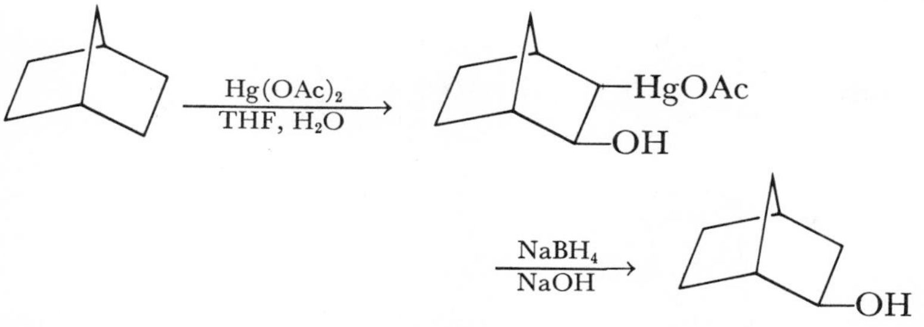

$$\geq 99.8\%$$

Free-radical addition of thiophenol at $0°$ gives 99.5 percent *exo*[7] (11.6).

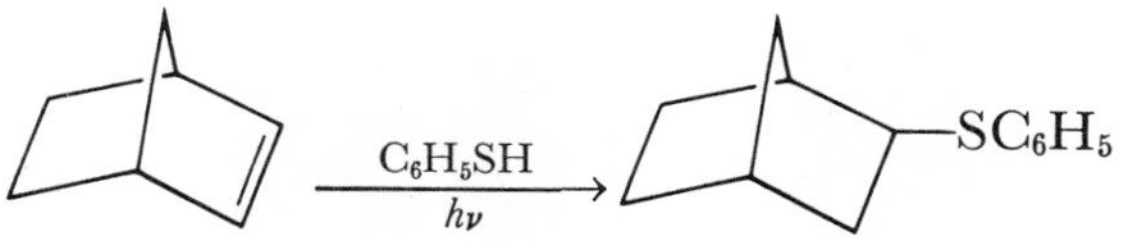

$$99.5\%$$

Finally, the base-catalyzed deuterium exchange of norcamphor yields an *exo:endo* ratio of 715[8] (11.2).

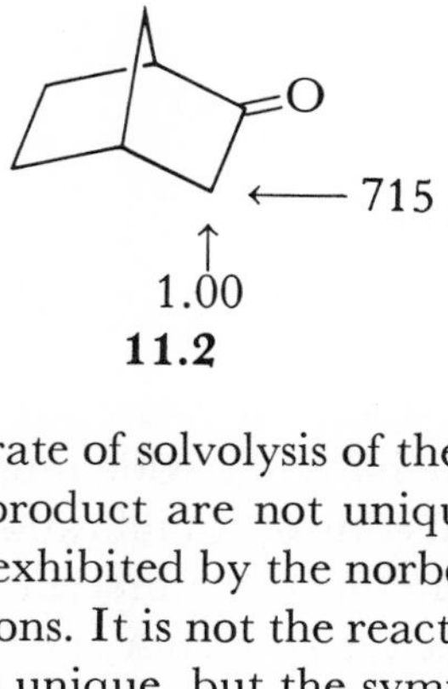

Therefore, the faster rate of solvolysis of the *exo* isomer and the *exo* stereochemistry of the product are not unique, but conform to the same reactivity pattern exhibited by the norbornyl system in varying degree in all of its reactions. It is not the reactivity and stereochemistry of solvolysis that are unique, but the symmetry properties of the cation[1] (11.3). The nonclassical structure (11.4) possesses such a

---

[7] H. C. Brown and J. H. Kawakami, *J. Amer. Chem. Soc.*, **92**, 201 (1970).

[8] T. T. Tidwell, *J. Amer. Chem. Soc.*, **92**, 1448 (1970).

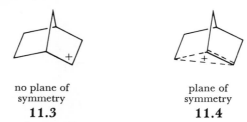

no plane of symmetry	plane of symmetry
11.3	**11.4**

plane of symmetry. However, this characteristic would also be duplicated by a pair of classical ions or ion-pairs that equilibrated rapidly compared to the rate of reaction with solvent (11.7).

$$\text{11.7}$$

Is it reasonable to postulate such a rapid rate of equilibration?

1,2-Dimethylnorbornyl has been assigned a classical structure.[9] Yet it undergoes 91 percent racemization during solvolysis. Consequently equilibration of the ion or ion-pair must proceed at a rate some ten times its capture by solvent (11.8).

$$\text{11.8}$$

Similarly the 1,2-di-*p*-anisylnorbornyl cation has been assigned a classical structure on the basis of a wide range of properties.[10] Yet at ordinary temperatures its rate of equilibration is so fast that pmr cannot distinguish between the two different anisyl groups. Only at lower temperatures does the spectrum exhibit changes suggestive of an equilibrating pair of ions (11.9).

[9] H. Goering and K. Humski, *J. Amer. Chem. Soc.*, **90,** 6213 (1968).

[10] P. von R. Schleyer, D. C. Kleinfelter, and H. G. Richey, Jr., *J. Amer. Chem. Soc.*, **85,** 479 (1963).

11.9

The problem then reduces to that of distinguishing between a resonance σ-bridged cation and a rapidly equilibrating pair of classical ions. The σ-bridged structure should reveal charge delocalization to the 1- and 6-positions. However, all attempts to obtain independent evidence for such charge delocalization have failed (Section X-5). Consequently, we undertook to search for a reasonable explanation not involving σ-participation for the high *exo*:*endo* rate ratio.

2. The Possibility of Steric Hindrance to Ionization in Endo-Norbornyl Derivatives

We have pointed out that non-carbonium ion reactions also show high *exo*:*endo* ratios. These high ratios are presumably the result of decreased rates of reaction in the sterically hindered *endo* direction of the U-shaped norbornyl structure (Figure XI-1). Consequently, it

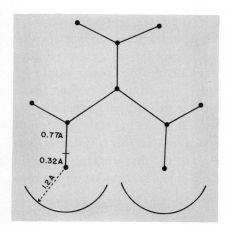

Figure XI-1. End-on view of the norbornane molecule.

185

appeared appropriate to consider the possibility that the high *exo : endo* rate ratio in the solvolysis of 2-norbornyl derivatives was actually the result of a normal *exo* rate combined with a very slow *endo* rate.

As discussed earlier, the concept of steric assistance to ionization was introduced in 1946. It was proposed that a tertiary derivative carrying bulky substituents would undergo solvolysis with relief of steric strain and would therefore exhibit enhanced rates (11.10).

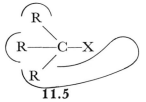

$$11.10$$

strained   less strained

This concept received ready acceptance (Section VIII-6).

The proposal of steric hindrance to ionization[11] has encountered much more resistance. Yet it appears quite reasonable that the structure of the compound undergoing solvolysis may be so shaped that the departure of the leaving group may be hindered, as indicated in **11.5**.

11.5

Indeed, *endo*-norbornyl derivatives would appear to possess this feature. Consider the *endo*-norbornyl chloride structure (Figure XI-2). In the ionization process the chlorine substituent would be expected to move along a curved path away from the carbon atom at the 2-position, maintaining the chlorine substituent perpendicular to the face of the developing carbonium ion so as to retain maximum overlap of the orbitals undergoing separation. In this way the system should pass through the transition state to the first intermediate, the idealized ion-pair shown in Figure XI-3. Clearly there would be ma-

---

[11] H. C. Brown, F. J. Chloupek, and M.-H. Rei, *J. Amer. Chem. Soc.*, **86**, 1248 (1964).

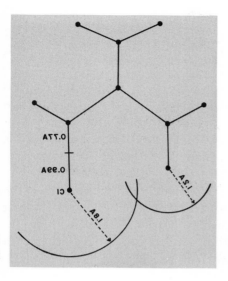

Figure XI-2. End-on view of the *endo*-norbornyl chloride molecule.

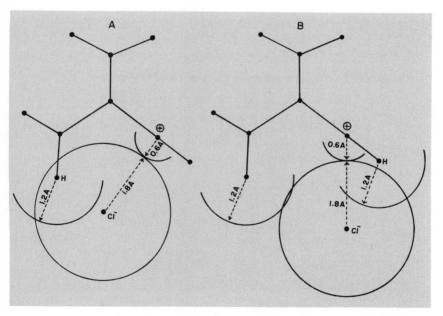

Figure XI-3. Molecular models for the hypothetical intimate ion-pairs from the two postulated reaction paths.

187

jor steric overlap of the chlorine substituent with the *endo*-6-hydrogen. Moreover, the group undergoing ionization should be strongly solvated by the medium, yet the U-shaped structure obviously makes such solvation difficult.

An alternate model for ionization has been suggested.[12] In this model the departing group would move initially along the direction of the C—X bond, leading to the first intermediate, the idealized intimate ion-pair shown in Figure XI-3. This path does not avoid the steric difficulty, although in this model it is largely transferred to the 2-hydrogen atom.

The large steric interactions of both models will presumably cause some other path, providing decreased steric interactions at the cost of poorer overlap, to be selected as a compromise, resulting in an increase in the energy of the transition state as compared to that for a derivative without this structural feature. This proposal of steric hindrance to ionization of the *endo*-norbornyl derivatives appeared to provide a reasonable alternative to the older interpretation involving σ-participation of the *exo*-norbornyl derivatives as a basis for the observed high *exo*:*endo* rate ratios. Consequently, we undertook to explore this possibility.

3. The Goering-Schewene Diagram for Norbornyl Derivatives

In 1965 Goering and Schewene introduced a means of looking at the norbornyl system that is particularly helpful in seeing the relationship between the *exo*:*endo* rate ratio and the *exo*:*endo* product ratio.[13]

The rate of acetolysis of *exo*-norbornyl brosylate is 350 times that for the corresponding *endo* derivative. If the data are corrected for internal return in the *exo*, then the *exo*:*endo* rate ratio becomes 1600. The relative rate of 1600 corresponds to a difference in the free energy of activation of 4.5 kcal/mole. The strain in *endo*-norbornyl arenesulfonates is estimated to be 1.3 kcal/mole.[14] This leads to a

---

[12] P. von R. Schleyer, M. M. Donaldson, and W. E. Watts, *J. Amer. Chem. Soc.*, **87**, 375 (1965).

[13] H. L. Goering and C. B. Schewene, *J. Amer. Chem. Soc.*, **87**, 3516 (1965).

[14] P. von R. Schleyer, *J. Amer. Chem. Soc.*, **86**, 1854, 1856 (1964).

difference in the energies of the two transition states of 5.8 kcal/mole (Figure XI-4).

The problem is clearly that of defining what factor or factors are responsible for the difference in energies of the two transition states. Is the transition state for the *exo* isomer stabilized by σ-participation, or is the *endo* transition state destabilized by steric hindrance to ioni-

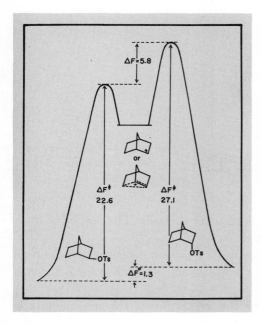

Figure XI-4. Free-energy diagram for the acetolysis of *exo*- and *endo*-norbornyl tosylate.

zation? Once ionization occurs, the ion or ion-pair will be in the central well. Recombination will obviously occur much more readily to produce the *exo* ester as compared to the *endo* ester.

What we would like to know is not the relative rates at which the cationic intermediate reacts with the anion to reform the original esters, but the relative rates at which the intermediate captures the solvent acetic acid to form the two products, *exo*- and *endo*-norbornyl acetate. Fortunately, the data are available. Goering and Schewene[13] measured the rates of racemization, the rates of exchange, and the

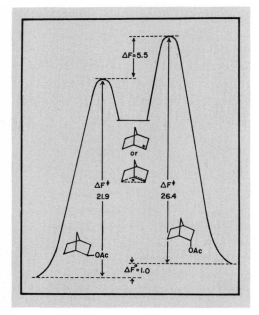

Figure XI-5. Free-energy diagram for the acetolysis of *exo-* and *endo*-norbornyl acetates.

equilibration of *exo-* and *endo*-norbornyl acetates. The results of their study are shown in Figure XI-5.[15]

We are now in a position to consider what happens to the cationic intermediate in reacting with the acetic acid solvent to form the two acetates.

The energy of activation for the ionization of the *endo* acetate is 4.5 kcal/mole higher than for the *exo*. If we correct for the higher ground-state energy of the *endo* isomer, 1.0 kcal/mole, the difference in the energies of the two transition states becomes 5.5 kcal/mole.

The two reactants pass over these two transition states to form the same intermediate, the norbornyl cation (or ion-pair). The principle

[15] The authors formulated their diagram in terms of enthalpies (H) rather than free energies (F). We have found it more convenient to use free energies because they are more precise, being directly related to the measured rate and equilibrium constants, and because the *exo:endo* product ratio is directly related to ΔF for the two transition states.

of microscopic reversibility requires that in the symmetrical system (the subject of their study) the norbornyl intermediate, in reacting with the solvent, must pass over the same two transition states that are involved in the solvolysis of the reactants. The authors established that the product from the solvolysis of norbornyl brosylate consisted of 99.98 percent *exo-* and 0.02 percent *endo*-norbornyl acetate. This is almost precisely the distribution predicted for a difference of 5.5 kcal/mole in the energies of the two transition states.

It follows that the factor responsible for the difference in energy between the *exo* and *endo* transition states must be largely responsible for the difference in the *exo* and *endo* rates of solvolysis. (The difference in ground-state energies also contributes, but this difference is relatively small in the present cases compared to the much larger difference in the energies of the transition states.) It also follows that the factor responsible for the difference in energy between the *exo* and *endo* transition states must be responsible for the stereoselectivity leading to the almost exclusive formation of the *exo* product. Unfortunately, the fact that we recognize this point does not aid us in understanding just what the factor may be.

The remarkable stereoselectivity exhibited in the solvolysis of norbornyl derivatives has been considered to require bridging in the norbornyl cation.[1] However, the diagram makes it clear that the amount of bridging that may or may not be present in the free ion is not directly involved in the stereoselectivity of the substitution. *It is the amount of bridging in the exo transition state, or whatever the factor responsible for the difference in stability of the two transition states, that will control the distribution of the norbornyl cation between exo and endo product.*

It is instructive to compare the Goering-Schewene diagram for 2-anisylcamphenilyl (Figure XI-6) with the diagram for norbornyl (Figure XI-4). The similarity is apparent. It was previously pointed out that σ-bridging cannot be a factor in norbornyl cations stabilized at the 2-position by a phenyl or *p*-anisyl group (Section X-4). Consequently, the difference in energy of the two transition states in norbornyl itself can be due to some factor other than σ-participation.

Let us consider what might be the factor or factors responsible for

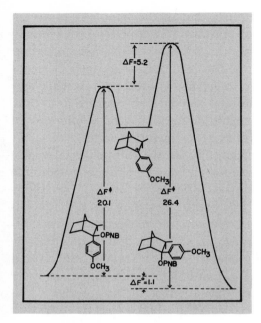

Figure XI-6. Free-energy diagram for the solvolysis of the 2-*p*-anisylcamphenilyl *p*-nitrobenzoates in 80 percent aqueous acetone at 25°.

the difference in energy between the *exo* and *endo* transition states. Four possibilities may be pointed out: (1) the *exo* transition state is stabilized by σ-participation, with the *endo* being normal; (2) the *endo* transition state is destabilized by steric strain, with the *exo* transition state being normal; (3) a combination of (1) and (2); or (4) some new factor not now recognized by current theory.[16]

At one time the Foote-Schleyer correlation was proposed as a means of resolving such problems. Indeed, it was applied to the norbornyl problem and it was concluded that *endo*-norbornyl was normal, so that *exo*-norbornyl must be fast, pointing to σ-participation.[14] A detailed examination of the Foote-Schleyer correlation should help us decide whether this conclusion can be accepted.

---

[16] For example, torsional effects have been suggested as a factor in the dominant *exo* stereochemistry. However, for the present reactions it can make only a minor contribution and so will not be considered in detail: P. von R. Schleyer, *J. Amer. Chem. Soc.*, **89,** 699 (1967).

4. Application of the Foote-Schleyer Correlation [17]

If it were possible to calculate the rate for *endo*-norbornyl tosylate, without allowing for steric hindrance to ionization, agreement of the calculated with the observed value would confirm the absence of such a steric-strain term. Then the faster rate for the *exo* derivative would support the proposal that this transition state is stabilized by major σ-participation. Alternatively, if the calculated rate for the *exo* agreed with the observed value, σ-participation could not be a factor, and we would be forced to consider a steric explanation for the slower *endo* rate.

The Foote-Schleyer correlation was proposed as a means of making such a calculation[14,18] (11.11).

$$\log k_{\text{rel}} = \frac{1715 - \nu_{\text{co}}}{8} + 1.32 \sum_i (1 + \cos 3\phi_i)$$

$$+ \text{ inductive term} + \frac{GS_{\text{strain}} - TS_{\text{strain}}}{1.36} \qquad 11.11$$

This correlation utilized the infrared carbonyl frequency of the ketone related to the secondary tosylate[19] under consideration as a method of estimating the contribution of angle strain to the solvolysis rate. The second term correlates for torsional effects. The third term corrects for the inductive effect of substituents, but is not required in the present discussion. The last term is important for the present discussion. It represents the effect of differences in steric strain in the ground (GS) and transition states (TS) upon the relative rate. In compounds where such nonbonded strains were not

[17] H. C. Brown, I. Rothberg, P. von R. Schleyer, M. M. Donaldson, and J. J. Harper, *Proc. Natl. Acad. Sci., U.S.*, **56**, 1653 (1966).

[18] C. S. Foote, *J. Amer. Chem. Soc.*, **86**, 1853 (1964).

[19] The correlation was originally restricted to the acetolysis of secondary tosylates upon the widely accepted position at that time that the solvolysis of such systems was close to limiting. Consequently, it does not contain a term for the contributions of solvent to the rate. This position is now undergoing reconsideration: J. L. Fry, C. J. Lancelot, L. K. M. Lam, J. M. Harris, R. C. Bingham, D. J. Raber, R. E. Hall, and P. von R. Schleyer, *J. Amer. Chem. Soc.*, **92**, 2538 (1970), fn. 21.

considered to be serious, good agreement was achieved between calculated and observed rates.[14] In this discussion we will consider cases where nonbonded interactions involving the substituent could be quite serious, if steric hindrance to ionization is a real phenomenon.

In principle the nonbonded strain term of the correlation should be capable of handling both steric assistance to ionization ($GS_{strain} >$ TS_{strain} for the leaving group) and steric hindrance to ionization ($GS_{strain} < TS_{strain}$). However, in practice it was recognized that the latter situation would be difficult to calculate, since neither the path of departure of a leaving group nor the proper nonbonded potential functions are known. Consequently, for all cases the simplest possible assumption was adopted, that $TS \approx 0$ for the leaving group.

This simplifying assumption was justified on the basis that the steric environment of the leaving group in the six model compounds used to define the correlation line, as well as in *endo*-norbornyl, was similar to the steric environment of a leaving group in an axial position of the cyclohexane molecule. In such flexible cyclohexane structures there is actually no evidence of any appreciable steric hindrance to ionization.[20] Accordingly, the assumption of $TS \approx 0$ was justified on the ground that "steric deceleration, in fact, is but rarely encountered, evidently because leaving groups are generally able to find a propitious avenue for departure."[14]

A major flaw in this argument may arise from the extrapolation from flexible alicyclic ring systems to rigid bicyclic structures. Steric effects are generally relatively small in flexible systems. The system can bend or rotate in such a manner as to minimize steric interactions. In rigid bicyclic structures, this avenue of escape is largely absent. Steric effects can be much larger.[21]

Accordingly, it was decided to test the generality of the assumption that $GS_{steric\ strain}$ involving the leaving group can be assumed to vanish in the transition state.[17] The 6,6-dimethylnorbornane and the *endo*-5,6-trimethylenenorbornane systems exaggerate the U-shaped structural feature of the *endo* side of the norborane molecule (Figure

[20] E. L. Eliel, N. L. Allinger, S. J. Angyal, and G. A. Morrison, *Conformation Analysis* (John Wiley, New York, 1965), pp. 84–85.

[21] H. C. Brown and J. Muzzio, *J. Amer. Chem. Soc.*, **88**, 2811 (1966).

XI-1). Consequently, the molecules, 6,6-dimethyl-*endo*-norbornyl tosylate (**11.6**), *endo*-5,6-trimethylene-*endo*-2-norbornyl tosylate (**11.7**), and *endo*-5,6-trimethylene-*endo*-8-norbornyl tosylate (**11.8**), were selected for study.

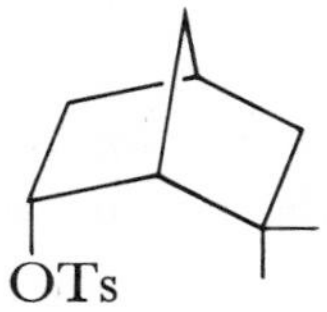

11.6 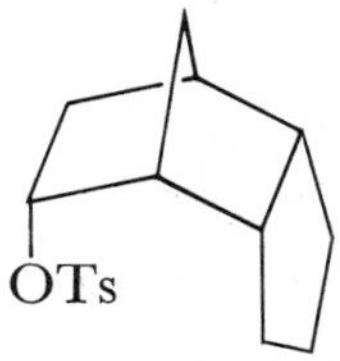   **11.7**

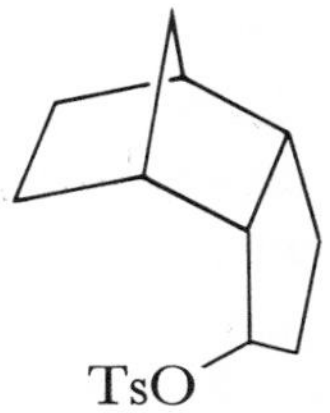

11.8 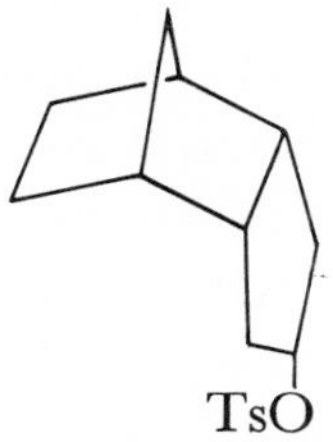   **11.9**

Obviously, if steric hindrance to ionization is a factor, it should be much more important in the *endo*-2 (**11.7**) and *endo*-8 (**11.8**) derivatives than in *endo*-norbornyl tosylate. If it is possible to estimate the rates of **11.6, 11.7**, and **11.8** and similar molecules with the assumption of $TS \approx 0$ (for strain involving the leaving group), the steric hindrance to ionization must be negligible in *endo*-norbornyl tosylate, and the Foote-Schleyer correlation could be used with a high degree of confidence to resolve the nonclassical ion problem, as already proposed.[14] On the other hand, if the observed rates are much smaller than those calculated in this manner, the assumption that $TS \approx 0$ cannot be universally correct. Such a result would not necessarily show that $TS > 0$ in the *endo*-norbornyl system, but it would lower the level of assurance that TS can be considered to be negligible in the solvolyses of *endo*-norbornyl tosylates and related rigid bicyclic derivatives.

The corresponding 9-derivative (**11.9**) was included to provide a test of the Foote-Schleyer correlation for a less rigid system. In this case the flexibility at the 9-position, greater than at the 8-, should serve to minimize steric hindrance to the departure of the leaving group.

The carbonyl frequency in the ketone corresponding to **11.7** is 1743 cm^{-1}, as compared to 1751 cm^{-1} for 2-norbornanone itself. According to the correlation (11.11), a shift of this magnitude should correspond to an increase in the rate of acetolysis of **11.7** over *endo*-norbornyl by a factor of 10.

Differences in torsional and inductive effects in the two systems are negligible. In the ground state, the strain in *endo*-norbornyl tosylate involving the leaving group is estimated as 1.3 kcal/mole, and that in *endo*-2 (**11.7**) is estimated roughly as 4.0 kcal/mole. If the usual assumption is made that steric strain is largely relieved in the transition state, this will result in another factor of 100 favoring *endo*-2 (**11.7**) over *endo*-norbornyl tosylate. Consequently, the correlation, with the usual assumption that $TS_{\text{strain}} \approx 0$ for the leaving group, suggests that *endo*-2 (**11.7**) will solvolyze at a rate 1000 times that of *endo*-norbornyl tosylate. The observed rate is 0.1. Thus there is a discrepancy of 10,000 (factors of 10 for the carbonyl frequency difference, 100 for the difference in GS_{strain}, and $1/0.1$ for the observed relative rates) between the rates observed and that calculated in this manner. Even if we assume that none of the ground-state nonbonded strain involving the leaving group in **11.7** is relieved in the transition state $(GS - TS = 0)$, a sizable discrepancy, a factor of 100 (due to rates and carbonyl frequencies), would remain. Using this approach, the observed rate of **11.7** can be calculated only if it is assumed that nonbonded strain involving the leaving group *increases* considerably in going from the ground to the transition state $(GS - TS = -2.8 \text{ kcal/mole})$.

The data are summarized in Table XI-1.

In 6,6-dimethyl-*endo*-norbornyl (**11.6**) the ground-state strain involving the leaving group is likewise estimated to be 4.0 kcal/mole.

Table XI–1. Predicted and Observed Rates of Acetolysis at 25°

Tosylate	ν_{co} for ketone (cm^{-1})	GS^a (kcal/ mole)	Rate constant ($k_1 \times 10^8$ sec^{-1})	Calcd.	Obsd.	Discrep.	TS^b (kcal/ mole)
endo-Norbornyl	1751	1.3	8.28		1.00		
endo-2-d (**11.7**)	1743	4.0	0.860	1000	0.10	10,000	6.8
6,6-Dimethyl-							
endo- (**11.6**)	1746	4.0	0.447	420	0.054	8,000	6.6
Cyclopentyl	1740	0	158		1.00		
endo-9-d (**11.9**)	1739	0	71.9	1.0	0.46	2	
exo-9-d	1739	0	44.8	1.0	0.28	3	
endo-8-d (**11.8**)	1732	3.0	1.56	1000	0.0099	100,000	7.2
				(200)^c	(0.18)^c	(1,000)^c	(4.1)^c
exo-8-d	1732	0	8.88	10	0.056	180	
					(1.0)^c		

^{a} Ground-state strain associated with the leaving group. This is the $(GS - TS_{strain})$ term of the Schleyer correlation, with the assumption that for the leaving group $TS_{strain} \approx 0$. ^{b} Value of leaving group TS required in the $(GS - TS)$ term to achieve agreement between the calculated and observed relative rates. ^{c} Values in parentheses are based on *exo*-8 as a model for *endo*-8. ^{d} *endo*-5,6-trimethylene-( )-norbornyl tosylate.

Again the assumption that this strain is relieved in the transition state leads to an estimated increase over *endo*-norbornyl by a factor of 100. The carbonyl frequency difference, $\Delta\nu_{co} = 5$ cm^{-1}, introduces another factor of 4.2. Thus the rate predicted on the basis of the Schleyer correlation, with the provisional assumption $TS \approx 0$, is 420. The observed rate is one-nineteenth that of *endo*-norbornyl,[12] resulting in a discrepancy between the predicted and observed rates of 8000. Clearly, such a comparison of the observed and calculated rates provides a better assessment of the magnitude of steric hindrance to ionization than does a mere comparison of the two actual rates, a procedure which ignores the potential driving force of the large ground-state energy present in 6,6-dimethyl-*endo*-norbornyl tosylate (**11.6**).[22]

For **11.8** and **11.9** it is appropriate to select cyclopentyl as the

[22] S. Winstein, *J. Amer. Chem. Soc.*, **87**, 381 (1965).

model system and to assume that differences in torsional and inductive effects between the model system and **11.8** and **11.9** are small and can be neglected. The carbonyl frequencies for the ketones corresponding to **11.8** and **11.9** are 1732 and 1739 cm^{-1}, respectively, compared to a value of 1740 cm^{-1} for cyclopentanone itself. On the basis of these carbonyl frequencies, we should anticipate a rate enhancement by a factor of 10 in the 8-derivative (**11.8**), with no significant effect in the 9-derivative (**11.9**). Only in the *endo*-8 compound should there also be a significant steric term in the ground state, roughly estimated as 3.0 kcal/mole.

In the case of the *endo*-8-, both the significant shift in the carbonyl frequency and the estimated strain result in a prediction by the correlation, with the usual assumption, that the rate will be enhanced by a factor of 1000. Since the observed rate is only one-hundredth that of cyclopentyl tosylate, *there is a discrepancy of 100,000, between the rate so calculated and that observed.*

On the other hand, in the *endo*-9 compound (**11.9**), where steric hindrance to ionization would not be expected to be a significant factor, the observed and calculated rates of acetolysis agree to within a factor of 2.

These results point to an important conclusion. Despite the large ground-state strain present in the U-shaped structures examined, relative to the model substances used for comparison, solvolysis involves not a decrease, but an actual increase in strain in proceeding to the transition state, producing not an increase but an actual decrease in rate. It is clear that it will be necessary in the future to give careful consideration to the precise model for the departure of the leaving group.

5. Steric Effects in U-Shaped Systems [23]

The results described in Section XI-4 encouraged my belief that steric hindrance to ionization in rigid bicyclic systems could be a major contributor to the observed *exo*:*endo* ratios in such systems. However, this represented a major revolution in chemical thought.

[23] H. C. Brown, W. J. Hammar, J. H. Kawakami, I. Rothberg, and D. L. Vander Jagt, *J. Amer. Chem. Soc.*, **89,** 6381 (1967).

For many years workers had been almost automatically attributing high *exo*:*endo* rate ratios to σ-participation in the *exo* isomer.

How could we proceed to test this idea further?

One promising approach might be to explore more widely the apparent *exo*:*endo* product ratio in the norbornyl system exhibited by many reactions, such as hydroboration-oxidation, epoxidation, oxymercuration-demercuration, and the *exo*:*endo* rate ratio in the norbornyl system. Accordingly, we decided to examine the stereo-chemistry of the present active reactions of a series of three bicyclic systems of increasing U-shaped character, *cis*-bicyclo[3.3.0] > norbornane > *endo*-5,6-trimethylenenorbornane. We utilized the olefins **11.10**, **11.11**, and **11.12**, as well as the ketones **11.13**,

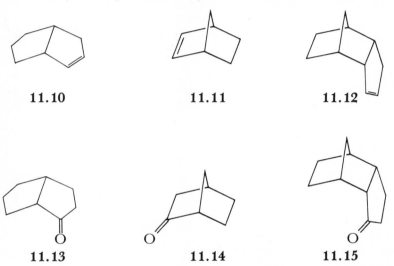

11.10	**11.11**	**11.12**
11.13	**11.14**	**11.15**

11.14, and **11.15**. These results are summarized in Table XI-2.

Although individual reactions evidently differ considerably in the stereoselectivities they exhibit, the results reveal a consistent pattern. In all cases, the *cis*-bicyclo[3.3.0]octane system (**11.10** and **11.13**) exhibits the least preference for *exo* attack, presumably because of its higher flexibility and more accessible *endo* face, and the *endo*-5,6-trimethylenenorbornane system (**11.12** and **11.15**) exhibits the highest stereoselectivity of the three systems examined. Indeed, an examination of a model reveals that in this rigid U-shaped structure

Table XI-2. Comparison of the Relative Stereoselectivities Exhibited by Three Representative U-Shaped Systems

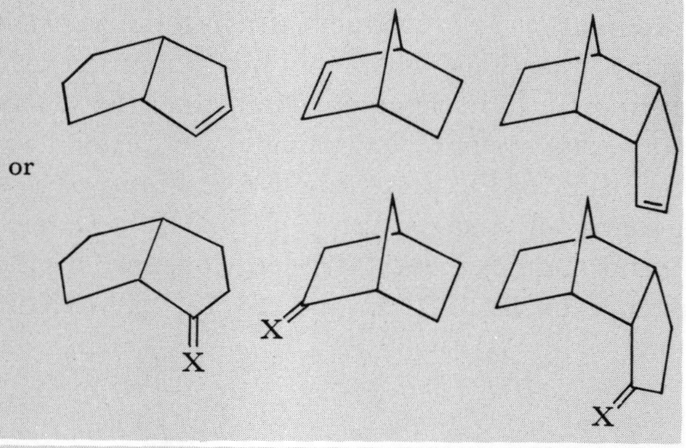

Reaction	*exo* :*endo* ratios		
Hydroboration-oxidation of olefin	24	200	>1000
Epoxidation of olefin	6.7	200	>1000
Oxymercuration-demercuration of olefin	8	>500	
Lithium aluminum hydride reduction of ketone (X = O)	3	8.1	>1000
Addition of CH_3MgX to ketone (X = O)	50	200	>1000
Oxymercuration-demercuration of methylene derivatives (X = CH_2)	8.1	200	>1000
Solvolysis of the tertiary methyl *p*-nitrobenzoates $\left(X = \begin{smallmatrix} CH_3 \\ \diagup \\ \diagdown \\ OPNB \end{smallmatrix} \right)$	17	885	4300

the *endo* face is highly hindered to the approach of reagents. In all cases, the stereospecificities indicated by the norbornane system are intermediate.

The corresponding tertiary methyl *p*-nitrobenzoates (**11.16, 11.17, 11.18**) were then synthesized and the rates of solvolysis in

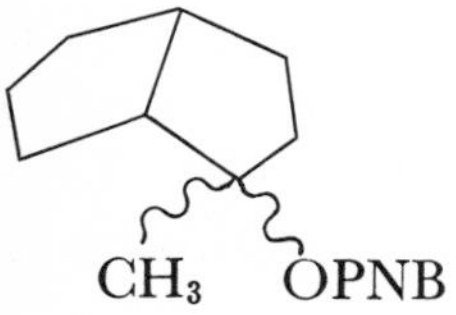

| **11.16** | **11.17** | **11.18** |

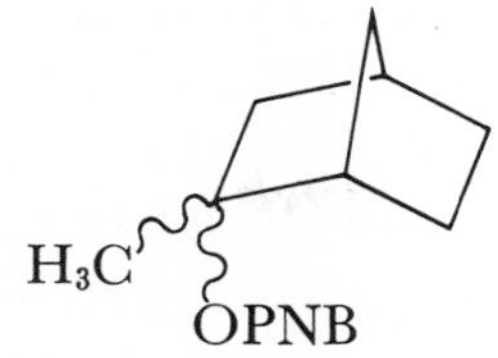

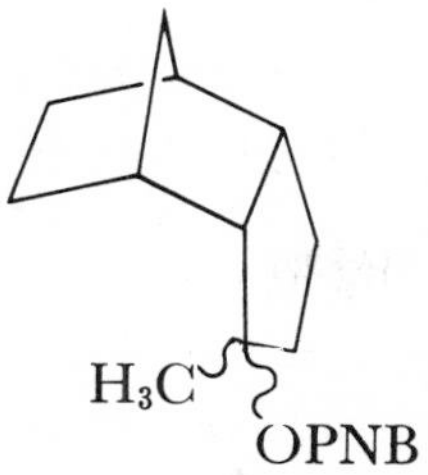

80 percent aqueous acetone determined. The tertiary esters derived from the bicyclo[3.3.0]octane system (**11.16**) revealed an *exo*:*endo* rate ratio of 17. The norbornyl system (**11.17**) revealed an *exo*:*endo* rate ratio of 885. Finally, the *endo*-5,6-trimethylenenorbornane system (**11.18**) revealed an *exo*:*endo* rate ratio of 4300. Consequently, there is an excellent correlation within these three systems as to the stereoselectivities they show towards various representative reagents and the stereoselectivities they reveal in the solvolyses of the tertiary esters.

This common pattern of reactivity for carbonium ion and non-carbonium ion reactions seems to support the position we have been led to: that the *exo*:*endo* rate ratio in norbornyl systems must be largely steric in origin.

6. Exo:Endo Rate Ratios as a Steric Phenomenon [24, 25]

The evidence from a wide variety of approaches favors the conclusion that tertiary norbornyl cations, such as 2-methylnorbornyl, are classical in nature. Yet this classical system exhibits an *exo*:*endo* rate ratio of 885, easily comparable to the titrimetric rate ratio of 350 (1600 if we correct for internal return) exhibited by 2-norbornyl brosylate.[1] Since apparently neither σ-participation nor torsional effects[16] make a significant contribution to this high *exo*:*endo* rate ratio in the tertiary derivative,[26] we appear to be left with steric hindrance to ionization as the major contributor.

[24] H. C. Brown and S. Ikegami, *J. Amer. Chem. Soc.*, **90,** 7122 (1968).

[25] S. Ikegami, D. L. Vander Jagt, and H. C. Brown, *J. Amer. Chem. Soc.*, **90,** 7124 (1968).

[26] See H. C. Brown and M.-H. Rei, *J. Amer. Chem. Soc.*, **90,** 6216 (1968) for pertinent references.

We have pointed out that steric effects in rigid bicyclic systems should be much more important than in flexible aliphatic or alicyclic systems. If the *exo*:*endo* rate ratio of tertiary norbornyl derivatives is indeed a steric phenomenon, we should be able to achieve large changes in the *exo*:*endo* rate ratio by placing methyl substituents in appropriate positions in the norbornane system.

We undertook to synthesize 2-methylnorbornyl *p*-nitrobenzoate containing *gem*-dimethyl groups individually in the 7-position and in the 6-position, and to determine their rates of solvolysis. The results are summarized in Table XI-3.

Table XI–3. Rates of Solvolysis of 2,7,7-Trimethyl-2-norbornyl and 2,6,6-Trimethyl-2-norbornyl *p*-Nitrobenzoates and Related Derivatives in 80 Percent Aqueous Acetone at 25°

p-Nitrobenzoate	$10^6 k$ sec^{-1}	Relative rate	Rate ratio *exo*:*endo*
1-Methylcyclopentyl	2.11×10^{-3}	1.00	
2-Methyl-*exo*-norbornyl	1.00×10^{-2}	4.74	885
2-Methyl-*endo*-norbornyl	1.13×10^{-5}	0.00536	
2,7,7-Trimethyl-*exo*-norbornyl	4.01×10^{-2}	19.0	6.1
2,7,7-Trimethyl-*endo*-norbornyl	6.54×10^{-3}	3.1	
2,6,6-Trimethyl-*exo*-norbornyl	7.26	3440	3,630,000
2,6,6-Trimethyl-*endo*-norbornyl	2.00×10^{-6}	0.000948	

Indeed, we observed that the presence of *gem*-dimethyls in the 7-position (**11.19**) decreases the *exo*:*endo* rate ratio from the 885 value observed in the parent compound (**11.20**) to a value of 6.1 (**11.19**). On the other hand, the presence of *gem*-dimethyls in the 6-position (**11.21**) increases the *exo*:*endo* rate ratio to 3,630,000.

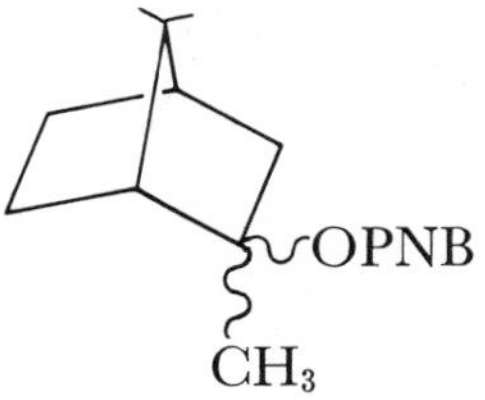

6.1

11.19

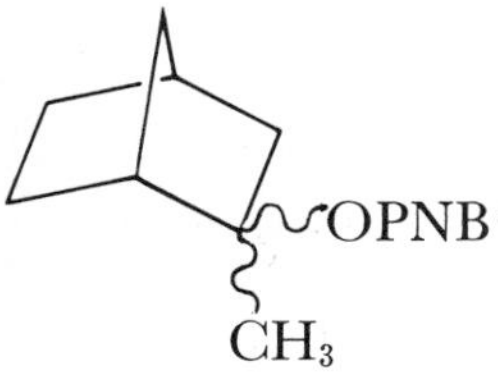

885

11.20

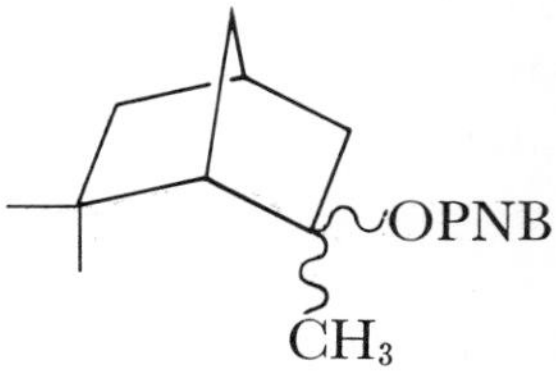

3,630,000

11.21

Thus the *exo*:*endo* rate ratio changes by a factor of 600,000 merely by a shift of the methyl substituents from the 7- to the 6-position—a truly remarkable effect.

The Goering-Schewene diagrams for these three systems are shown in Figures XI-7, XI-8, and XI-9. The wide variation in the ΔF values of each pair of transition states is noteworthy. Clearly it cannot be correlated with difference in σ-participation with car-

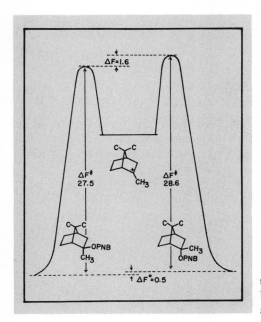

Figure XI-7. Free-energy diagram for the solvolysis of 2,7,7-trimethyl-2-norbornyl *p*-nitrobenzoates in 80 percent aqueous acetone at 25°.

bonium ions of such similar structures. Only major differences in steric effects provide a reasonable explanation.

The low *exo*:*endo* rate ratio in the 2,7,7-trimethylnorbornyl system arose from a small increase in the rate of the *exo* isomer ($\times$ 4) and a large increase in the rate of the *endo* isomer. Obviously, this marked increase in the rate of the *endo* isomer cannot be attributed to σ-participation. However, it is readily accounted for in terms of relief of steric strain accompanying the rotation of the 2-methyl substituent away from the *gem*-dimethyl group during the ionization process (**11.22, 11.23**).

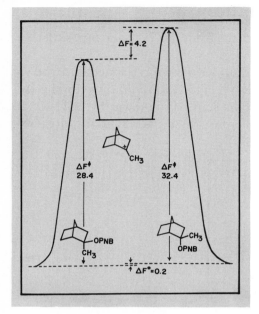

Figure XI-8. Free-energy diagram for the solvolysis of 2-methyl-2-norbornyl *p*-nitrobenzoates in 80 percent aqueous acetone at 25°.

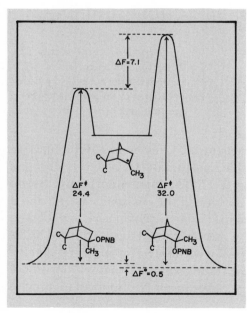

Figure XI-9. Free-energy diagram for the solvolysis of 2,6,6-trimethyl-2-norbornyl *p*-nitrobenzoates in 80 percent aqueous acetone at 25°.

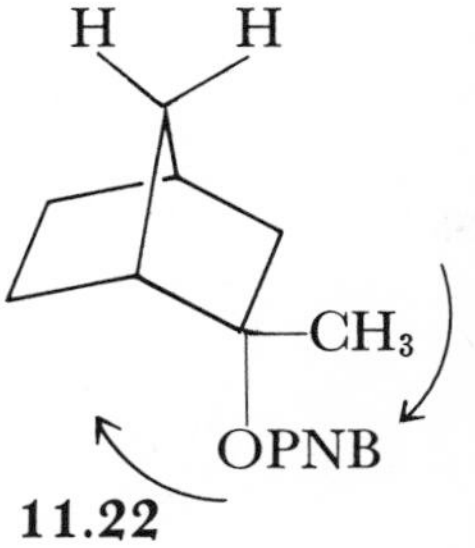

The very fast rate of 2,6,6-trimethyl-*exo*-norbornyl *p*-nitrobenzoate (726 times that of the parent compound) is again attributed to relief of steric strain (**11.24**). The steric requirements of a methyl and *p*-

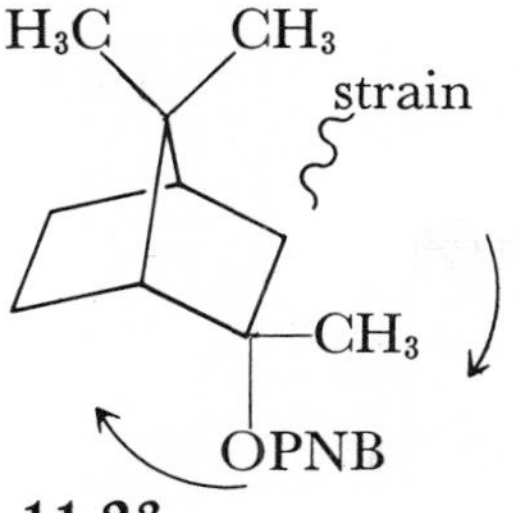

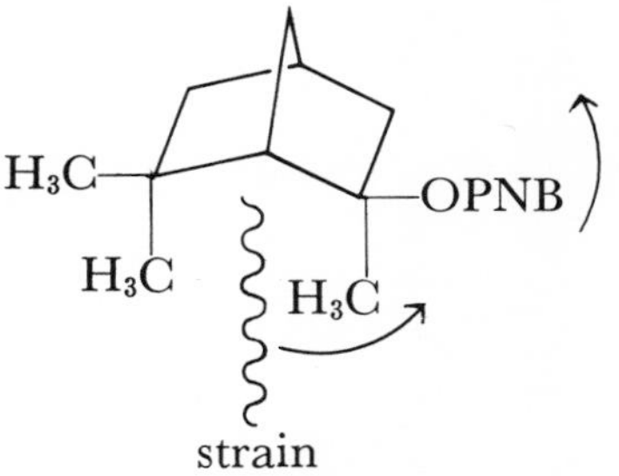

nitrobenzoate group are very similar.[26] Consequently, **11.25** should be just as strained as **11.24**. Yet **11.25** reacts not faster, but slower than the much less strained parent compound (**11.22**). This is readily accounted for by the fact that the natural reaction path for ionization of the *p*-nitrobenzoate does not allow for relief of steric strain.

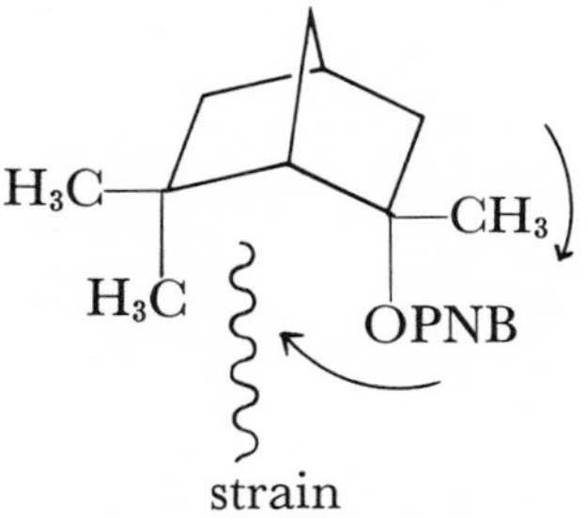

These results clearly support steric hindrance to ionization and a steric basis for the high *exo*:*endo* rate ratios in the norbornyl and related bicyclic systems.

PART FOUR

SELECTIVE REDUCTIONS

XII. Selective Reductions with Borohydrides and Aluminohydrides[1-3]

1. Background

I have already described the background for my interest in selective reductions utilizing hydrogen compounds of boron and aluminum (Chapter I). My Ph.D. thesis involved a study of the reduction of aldehydes and ketones by diborane (Chapter II). At this time diborane was a chemical rarity, so little interest was aroused among organic chemists by this new facile reduction procedure. In the course of developing a practical synthetic route to uranium borohydride we accomplished the first synthesis of sodium borohydride[4] (Chapter IV). We needed a solvent for this new chemical, and one that we tried was acetone. Thus we discovered the reducing properties of the alkali metal borohydrides.

We also initiated experiments to utilize the alkali metal hydrides for the synthesis of the corresponding hydrogen compounds of aluminum. It appeared that the reaction of aluminum hydride with diborane would provide an alternative route to aluminum borohydride required for the preparation of uranium (IV) borohydride. However, violent reactions were discouraging, and the successful synthesis of lithium aluminum hydride was not achieved until 1945.[5]

At Wayne University, in view of severe limitations, I decided to devote myself primarily to the program on steric strains. At Purdue in 1947, I continued the program on steric strains (Chapters V–

[1] H. C. Brown, *J. Chem. Educ.*, **38**, 173 (1961).

[2] H. C. Brown, "Moderni Sviluppi Della Sintesi Organica," *Conferenze X, Corso Estivo di Chimica, Academia Nationale dei Lincei* (Rome, 1968), pp. 31–50.

[3] E. Schenker, *Newer Methods of Preparative Organic Chemistry*, Vol. IV, W. Foerst, ed. (Verlag Chemie GmbH, Weinheim/Bergstr., 1968), pp. 196–335.

[4] H. I. Schlesinger, H. C. Brown, H. R. Hoekstra, and L. R. Rapp, *J. Amer. Chem. Soc.*, **75**, 199 (1953).

[5] A. E. Finholt, A. C. Bond, Jr., and H. I. Schlesinger, *J. Amer. Chem. Soc.*, **69**, 1199 (1947).

VIII) and initiated a program on a quantitative approach to aromatic substitution. It was not until 1951, after belatedly writing up our war research on borohydrides for publication,[6] that I decided to undertake studies on addition compounds of the alkali metal hydrides as an offshoot of the program on molecular addition compounds for the study of steric effects.

The results of the work with lithium aluminum hydride were published in 1947.[5] Organic chemists took to this new hydrogenating agent, and it rapidly came into wide use.[7,8] On the other hand, relatively little use was made of sodium borohydride. This annoyed me. As a co-discoverer of sodium borohydride, I had a deep personal interest in this fascinating material. The fact that I was the co-owner (with Professor H. I. Schlesinger) of a patent covering its manufacture may have augmented my interest. It seemed to me that the long delay in the publication of the sodium borohydride research was largely responsible for the neglect of this reagent. People were using lithium aluminum hydride even for mild selective reductions, where sodium borohydride should obviously be preferred. Accordingly I decided to undertake a research program in selective reductions that would attempt to define the characteristics of these two complex hydrides and the various derivatives to which they could be readily transformed.

Thus began the program that led to the discovery of hydroboration (Chapters XIV, XV) and from this to the discovery of the new universe of the organoboranes (Chapters XVI–XIX).

2. Objectives of the New Program

The discovery of sodium borohydride in 1942[4] and of lithium aluminum hydride in 1945[5] brought about a revolutionary change in procedures for the reduction of functional groups in organic chemistry.[7,8] Today, faced with the problem of reducing a carbonyl, ester, or nitrile group, the synthetic chemist will rarely undertake to use the Meerwein-Ponndorf-Verley reaction, the Bouveault-Blanc pro-

[6] H. I. Schlesinger, H. C. Brown, *et al.*, *J. Amer. Chem. Soc.*, **75,** 186, 191, 192, 195, 199, 205, 209, 213, 215, 219, 222 (1953).

[7] W. G. Brown, *Organic Reactions*, **6,** 469 (1951).

[8] N. G. Gaylord, *Reductions with Complex Hydrides* (Interscience Publishers, New York, 1956).

cedure, or catalytic hydrogenation. The complex hydrides provide a simple convenient route for the reduction of such functional groups, and they are invariably used in laboratory syntheses involving such reductions.

In spite of their great convenience, these two reagents suffer from certain deficiencies. Lithium aluminum hydride is an exceedingly powerful reducing agent, capable of reducing practically all of the functional groups used by the organic chemist. This great reactivity renders it relatively difficult to apply to selective reductions. On the other hand, sodium borohydride is a relatively mild reducing agent, one that reacts readily only with aldehydes, ketones, and acid chlorides. Consequently, it is useful only for selective reductions involving these relatively reactive groups.[7,8]

This situation made it desirable to develop means of controlling the reducing power of such reagents. If such control could be achieved, either by decreasing the reducing power of lithium aluminum hydride or by increasing that of sodium borohydride, or both, the organic chemist would then have a complete spectrum of reagents for selective reductions. With organic research undertaking the synthesis of structures of increasing complexity, there was an evident and growing need for reagents possessing a high degree of selectivity.

Five means of controlling the reducing power of the complex hydrides suggested themselves. First, it appeared that the reducing power might be influenced by the solvent. Second, the cation in the complex hydride might be expected to alter the reducing power. Third, substituents in the complex ion should exert marked steric and electronic influences upon the reactivity of the substituted complex ion. Fourth, since the complex hydride salts were basic in nature, the development of acidic reducing agents, such as BH_3 and AlH_3, might be expected to alter greatly the relative reactivities of different groups toward reduction (Chapter XIII). Finally, the introduction of substituent groups into these acidic reducing agents offered promise of additional variations in their reducing characteristics.

Accordingly, we undertook a program of research to investigate these possibilities. A lecture in 1961 presented our earlier results;[1] a lecture in 1968 reported additional developments.[2] The present two lectures (Chapters XII and XIII) will attempt to summarize all of our work in this area. Many other workers have made important contributions. Our present scope does not permit a complete review of all developments; it will focus on our investigations in "Selective Reductions."

For our examination of the reducing characteristics of each new reagent, we developed a group of 56 representative compounds, containing the more common functional groups of interest in reductions. It is impractical to consider here the results realized with the full range of substrates. I shall attempt, instead, to define the characteristics of each reagent in terms of its behavior toward twelve representative groups: aldehydes, ketones, acid chlorides, lactones, epoxides, esters, carboxylic acids, carboxylic acid salts, *tert*-amides, nitriles, aromatic nitro compounds, and olefins.

A word of caution is in order. The reactivity of any functional group can be modified greatly by the organic structure to which it is attached. For example, hydroboration of the carbon-carbon double bond appears to be a reaction of exceptional generality—yet it is possible so to bury a double bond within the steroid structure that the reaction proceeds exceedingly slowly, if at all. Similarly, aldehydes are generally highly reactive toward diborane, yet the three chlorine substituents in chloral make this otherwise reactive group relatively inert toward diborane. It is therefore essential to recognize that the relative reactivities established by the present study must be considered approximate estimates for simple representative groups and may be greatly altered or even inverted by major modifications in the molecular structure.

3. Lithium Aluminum Hydride as a Reducing Agent [9]

Lithium aluminum hydride is readily soluble in the usual hydroboration solvents—ethyl ether, tetrahydrofuran, and diglyme. In all of these solvents it is a powerful reducing agent, generally reducing such groups rapidly to the lowest reduced state.[7] This is indicated by the accompanying list of representative groups (12.1), arranged approximately in the order of their relative ease of reduction by the complex hydrides.[9]

[9] H. C. Brown, P. M. Weissman and N. M. Yoon, *J. Amer. Chem. Soc.*, **88**, 1458 (1966).

aldehyde	$\longrightarrow$ alcohol	
ketone	$\longrightarrow$ alcohol	
acid chloride	$\longrightarrow$ alcohol	
lactone	$\longrightarrow$ glycol	
epoxide	$\longrightarrow$ alcohol	
ester	$\longrightarrow$ alcohol	12.1
carboxylic acid	$\longrightarrow$ alcohol	
carboxylic acid salt	$\longrightarrow$ alcohol	
tert-amide	$\longrightarrow$ amine	
nitrile	$\longrightarrow$ amine	
nitro	$\longrightarrow$ azo, etc.	
olefin	$\longrightarrow$ no reaction	

The great power of lithium aluminum hydride as a reducing agent is indicated by the relatively rapid reduction of aromatic halides, such as bromobenzene, in refluxing tetrahydrofuran, as recently observed in our laboratories.[10]

4. Sodium Borohydride as a Reducing Agent [11]

Sodium borohydride is a very mild reducing agent, representing the opposite extreme. In hydroxylic solvents it reduces aldehyde and ketone groups rapidly at 25°, but it is essentially inert to the other groups except under more vigorous conditions or unusual structures (12.2).

aldehyde	$\longrightarrow$ alcohol	
ketone	$\longrightarrow$ alcohol	
acid chloride	$\longrightarrow$ reaction with solvent	
lactone	$\longrightarrow$ slow reaction	
epoxide	$\longrightarrow$ slow reaction	
ester	$\longrightarrow$ slow reaction	12.2
carboxylic acid	$\longrightarrow$ reacts, no reduction	
carboxylic acid salt	$\longrightarrow$ no reaction	
tert-amide	$\longrightarrow$ no reaction	
nitrile	$\longrightarrow$ no reaction	
nitro	$\longrightarrow$ no reaction	
olefin	$\longrightarrow$ no reaction	

[10] H. C. Brown and S. Krishnamurthy, *J. Org. Chem.*, **34,** 3918 (1969).
[11] S. W. Chaikin and W. G. Brown, *J. Amer. Chem. Soc.*, **71,** 122 (1949).

In aqueous solvents, sodium borohydride reacts with ionizable alkyl halides (and other derivatives) to give the corresponding hydrocarbon. The reaction mechanism appears to involve the capture of the carbonium ion formed in the solvolysis by the borohydride anion. The reaction has proven very useful for exploring the structure of carbonium ions produced in such solvolyses.[12]

5. Solvent Effects [13]

Could the reducing power of these complex hydrides be influenced by the solvent?

Sodium borohydride has a great advantage over lithium aluminum hydride in that it may be used in a much wider range of solvents. Thus it is highly soluble in water, reacting only slowly with the solvent to evolve hydrogen. Even this slow reaction is markedly decreased by the addition of alkali,[4] and stable solutions of sodium borohydride in concentrated sodium hydroxide are commercially available. Such aqueous solutions readily reduce aldehydes and ketones, even in cases where the solubilities of the compounds in the aqueous system are quite limited (12.3).

$$4R_2CO + NaBH_4 \longrightarrow Na[B(OCHR_2)_4]$$
$$\downarrow 4H_2O \qquad\qquad 12.3$$
$$NaB(OH)_4 + 4R_2CHOH$$

The reducing agent is readily soluble in methyl and ethyl alcohols.[4] It reacts rapidly with methyl alcohol, but only slowly with ethyl alcohol.[13] Thus 100 percent evolution of hydrogen is observed in methanol in 24 minutes at 60°, whereas under the same conditions in ethanol the evolution is less than 2 percent. It is unfortunate that early procedures emphasized the use of methanol as a solvent,[11] so that it is still commonly used, although ethanol possesses the obvious advantage of permitting reductions in homogeneous solution with relatively little loss of the reducing agent through this side reaction with the solvent.

[12] H. M. Bell and H. C. Brown, *J. Amer. Chem. Soc.*, **88**, 1473 (1966).

[13] H. C. Brown, E. J. Mead and B. C. Subba Rao, *J. Amer. Chem. Soc.*, **77**, 6209 (1955).

Sodium borohydride possesses only a modest solubility (0.1 M at 25°) in isopropyl alcohol. However, this solvent possesses the great advantage of being quite stable toward the reagent—standard solutions of sodium borohydride in this solvent have exhibited no measurable change in active hydride over periods of several weeks. Consequently, this solvent not only is convenient for reductions on a preparative scale, but it made possible kinetic studies of the reduction of aldehydes and ketones[14] used in our studies of I-strain (Section VIII-8) and of steric and electronic effects (e.g., Section IX-8).

The great difference in reactivity between benzaldehyde and acetophenone observed in these kinetic studies is noteworthy. Thus, the second-order rate constant for acetophenone at 0° is 2.05×10^{-4} mole^{-1} sec^{-1}, whereas the corresponding constant for benzaldehyde is 820×10^{-4} mole^{-1} sec^{-1}. With a factor of 400 in rate constants, it is evident that aldehyde groups should readily be reduced selectively in the presence of related ketone groupings.

Sodium borohydride is insoluble in ethyl ether, only slightly soluble in tetrahydrofuran, but readily soluble in diglyme (dimethyl ether of diethylene glycol) and triglyme (dimethyl ether of triethylene glycol). The solubility of sodium borohydride in diglyme reaches a maximum ($\sim$ 3 M) in the neighborhood of 40°, with greatly reduced solubilities at 0° (0.3 M) and 100° (0.1 M). At 0° a 1:1 solvate with diglyme separates from the solution; at 100° the unsolvated salt separates. Consequertly, this medium is a convenient one for the preparation of sodium borohydride of high purity. In triglyme the solubility is high ($\sim$ 3 M) and does not vary greatly with temperature.

These solutions in diglyme and triglyme have proven very useful in utilizing sodium borohydride for the generation of diborane[15] and for the hydroboration of olefins.[16,17] However, instead of enhancing

[14] H. C. Brown, O. H. Wheeler and K. Ichikawa, *Tetrahedron*, **1**, 214 (1957): H. C. Brown and K. Ichikawa, *J. Amer. Chem. Soc.*, **84**, 373 (1962).

[15] H. C. Brown and P. A. Tierney, *J. Amer. Chem. Soc.*, **80**, 1552 (1958).

[16] H. C. Brown and B. C. Subba Rao, *J. Amer. Chem. Soc.*, **78**, 2582 (1956); **81**, 6423 (1959).

[17] H. C. Brown and B. C. Subba Rao, *J. Org. Chem.*, **22**, 1136 (1957); *J. Amer. Chem. Soc.*, **81**, 6428 (1959).

the reducing power of sodium borohydride, these solvents appear to decrease it. No significant reduction of acetone was observed in diglyme solution under conditions where the ketone is quantitatively reduced in aqueous or alcohol solution in a matter of minutes.[13]

Aldehydes are reduced by solutions of sodium borohydride in diglyme,[18] so that such solutions may be effective for the selective reduction of aldehydes in the presence of ketones.

Although occasional references have appeared in the literature on the use of pyridine or acetonitrile as solvents for sodium borohydride reductions, no detailed study has been reported on the influence of these solvents on the reducing power of borohydride. However, in view of the greatly decreased reducing power of diglyme solutions as compared to solutions of borohydrides in hydroxylic solvents, it is to be anticipated that such aprotic solvents will not exhibit enhanced activity.

Lithium aluminum hydride is so reactive that possible solvents are essentially restricted to hydrocarbons, ethers, and tertiary amines. It is insoluble in hydrocarbons and readily soluble in ethers. Unfortunately, no detailed study of its reactions in amine solvents is available. It is generally utilized in ethyl ether, tetrahydrofuran, and diglyme, in which it is a powerful reducing agent, with no significant difference apparent in its reducing power.

6. Effect of Different Cations [12, 19]

Early observations indicated a marked difference in the reactivity of lithium and sodium borohydrides. Thus sodium borohydride reduces typical esters (such as ethyl acetate and ethyl benzoate) only very slowly, whereas lithium borohydride reduces such esters quite easily.[20]

In aqueous solution there is no measurable difference in the rate constants for the reactions of sodium and lithium borohydrides with

[18] H. C. Brown and G. Zweifel, unpublished observations.

[19] H. C. Brown and B. C. Subba Rao, *J. Amer. Chem. Soc.*, **78**, 2582 (1956).

[20] R. F. Nystrom, S. W. Chaikin and W. G. Brown, *J. Amer. Chem. Soc.*, **71**, 3245 (1949).

acetone. However, in isopropyl alcohol solution the rate constant for lithium borohydride is several times greater than that for the sodium salt.[21] This suggests that the enhancing effect of the lithium ion will be greatest in solvents of low dielectric constant. In such solvents the reaction presumably proceeds through the ion-pair ($Li^+BH_4^-$), rather than the dissociated ions.

Kollonitsch and his co-workers achieved the successful reduction of simple esters by sodium borohydride in tetrahydrofuran in the presence of lithium and magnesium iodides.[22] Later they synthesized calcium, strontium, and barium borohydrides, and demonstrated that these borohydrides exhibit an enhanced reducing power over sodium borohydride.[23]

In our own approach to this problem, we utilized homogeneous solutions of sodium borohydride in diglyme. Thus, the addition of an equivalent quantity of lithium chloride or lithium bromide to a 1.0 M solution of sodium borohydride in diglyme results in the formation of a precipitate of sodium halide and the formation *in situ* of lithium borohydride. The reagent can be utilized directly, without removing the precipitated salt. At 100°, we observed essentially quantitative uptake of "active hydride" in 1–3 hours by a number of representative esters: ethyl acetate, ethyl stearate, ethyl benzoate, ethyl cinnamate, and ethyl *p*-chlorobenzoate.[12] Under the same conditions, sodium borohydride alone brings about only slight reduction of such esters.

The results realized in reductions with lithium chloride-sodium borohydride in diglyme are summarized in the accompanying list[12,24] (12.4).

aldehyde	$\longrightarrow$ alcohol
ketone	$\longrightarrow$ alcohol
acid chloride	$\longrightarrow$ alcohol
lactone	$\longrightarrow$ glycol
epoxide	$\longrightarrow$ alcohol

[21] H. C. Brown and K. Ichikawa, *J. Amer. Chem. Soc.*, **83,** 4372 (1961).

[22] J. Kollonitsch, P. Fuchs, and V. Gabor, *Nature*, **173,** 125 (1954).

[23] J. Kollonitsch, P. Fuchs, and V. Gabor, *Nature*, **175,** 346 (1955).

[24] Research in progress with N. M. Yoon.

ester	$\longrightarrow$	alcohol	12.4
carboxylic acid	$\longrightarrow$	no reaction	
carboxylic acid salt	$\longrightarrow$	no reaction	
tert-amide	$\longrightarrow$	no reaction	
nitrile	$\longrightarrow$	no reaction	
nitro	$\longrightarrow$	reaction	
olefin	$\longrightarrow$	no reaction	

Ions of higher ionic potential would be expected to be even more effective. Accordingly, we explored the effectiveness of magnesium and aluminum halides in this connection.

Anhydrous magnesium chloride and bromide possess only a small solubility in diglyme (0.013 M for magnesium chloride and 0.078 M for the bromide, both at 100°). However, the addition of equivalent amounts of the solid salts to diglyme solutions of sodium borohydride brings about the reduction of esters. In this way ethyl stearate was reduced in four hours at 100° to 1-octadecanol in 74 percent yield, and ethyl *p*-nitrobenzoate was reduced in 3–4 hours at 40–50° to *p*-nitrobenzyl alcohol in 57 percent yield.[12]

In pursuing this line of investigation, it was observed that diglyme solutions of aluminum chloride and sodium borohydride, in a molar ratio of 1:3, mix to give a clear solution with no evidence for the formation of the insoluble sodium chloride. Nevertheless, such solutions exhibit markedly enhanced reducing powers.[19]

At room temperatures, such solutions rapidly reduce aldehydes, ketones, oxides, lactones, acid chlorides, acid anhydrides, esters, acids, nitriles, *t*-amides, azo compounds, and disulfides. No reduction was observed for sodium salts of carboxylic acids, primary amides, nitro derivatives, sulfones, and aromatic nuclei. Consequently, the presence of aluminum chloride has transformed the sodium borohydride from a reagent which in diglyme fails to reduce even ketones at any significant rate to one that exhibits greatly increased reducing power, approaching that exhibited by lithium aluminum hydride itself.

The results at 25° with this interesting reagent, $AlCl_3 + 3NaBH_4$ in diglyme,[19] may be summarized in the accompanying list (12.5). It should be mentioned that in the process of exploring the reagent we encountered an anomaly in the quantitative results we obtained for ethyl oleate; the exploration of this anomaly led to the discovery of hydroboration (Chapter XV).

aldehyde	$\longrightarrow$ alcohol	
ketone	$\longrightarrow$ alcohol	
acid chloride	$\longrightarrow$ alcohol	
lactone	$\longrightarrow$ glycol	
epoxide	$\longrightarrow$ alcohol	
ester	$\longrightarrow$ alcohol	12.5
carboxylic acid	$\longrightarrow$ alcohol	
carboxylic acid salt	$\longrightarrow$ no reaction	
tert-amide	$\longrightarrow$ amine	
nitrile	$\longrightarrow$ amine	
nitro	$\longrightarrow$ no reaction	
olefin	$\longrightarrow$ organoborane	

The failure to obtain a precipitate of sodium chloride upon mixing sodium borohydride and aluminum chloride argues against the complete formation of aluminum borohydride in solution (12.6).

$$AlCl_3 + 3NaBH_4 \longrightarrow Al(BH_4)_3 + 3NaCl \qquad 12.6$$

However, it is possible that the reaction proceeds to produce small equilibrium amounts of aluminum borohydride, with the equilibrium being shifted to completion as the aluminum borohydride reacts with the organic compound (12.7).

$$AlCl_3 + 3NaBH_4 \rightleftharpoons Al(BH_4)_3 + 3NaCl \qquad 12.7$$

Other polyvalent metal halides, such as gallium trichloride and titanium tetrachloride, likewise enhance the reducing power of sodium borohydride. Studies of their full scope are not yet available.

A number of aluminohydrides containing cations other than lithium have been synthesized.[8,25] These all appear to be very powerful reducing agents. The available data do not allow us to draw any conclusions as to the effect of the cation on the reducing power of the aluminohydride anion.

[25] E. Wiberg, *Angew. Chem.*, **65,** 16 (1953).

7. Effect of Substituents in Sodium Borohydride [26]

The reaction of a typical ketone, such as acetone, with sodium borohydride must involve four successive stages (12.8).

$$R_2CO + BH_4^- \xrightarrow{k_2} [H_3BOCHR_2]^-$$
$$R_2CO + [H_3BOCHR_2]^- \xrightarrow{k_2'} [H_2B(OCHR_2)_2]^-$$
$$R_2CO + [H_2B(OCHR_2)_2]^- \xrightarrow{k_2''} [HB(OCHR_2)_3]^- \qquad 12.8$$
$$R_2CO + [HB(OCHR_2)_3]^- \xrightarrow{k_2'''} [B(OCHR_2)_4]^-$$

The reaction exhibits simple second-order kinetics, first-order in each component.[13,27] The observed kinetics are consistent with a slow rate-determining reaction for the first stage, with successive stages being considerably faster.

This postulate was confirmed by the preparation of the intermediate, sodium triisopropoxyborohydride, $NaBH[OCH(CH_3)_2]_3$, and the demonstration that its reaction is indeed much faster than the reaction of the ketone with sodium borohydride itself. Thus, the reaction of acetone with sodium borohydride in diglyme is very slow —with no measurable reaction being observed in several hours at room temperature. However, in the same solvent the reaction of acetone with sodium triisopropoxyborohydride is too fast to follow, being complete in a matter of seconds at 0°.[26]

These results clearly demonstrate that the effect of the alkoxy substituents is to increase the reducing power of the borohydride. Reduction of esters by the reagent was achieved. Unfortunately, the simple alkoxyborohydrides (sodium trimethoxyborohydride and triethoxyborohydride), readily synthesized from sodium hydride and the borate ester in the absence of a solvent (12.9), undergo rapid disproportionation in solvents to sodium borohydride and sodium tetraalkoxyborohydride[28] (12.10).

[26] H. C. Brown, E. J. Mead, and C. J. Shoaf, *J. Amer. Chem. Soc.*, **78**, 3616 (1956).

[27] E. R. Garrett and D. A. Lyttle, *J. Amer. Chem. Soc.*, **75**, 6051 (1953).

[28] H. C. Brown, E. J. Mead, and P. A. Tierney, *J. Amer. Chem. Soc.*, **79**, 5400 (1957).

$$(RO)_3B + NaH \longrightarrow NaBH(OR)_3 \qquad\qquad 12.9$$

$$4NaBH(OR)_3 \xrightarrow{\text{THF}} NaBH_4 \downarrow + 3NaB(OR)_4 \qquad 12.10$$

This disproportionation renders difficult a quantitative estimation of the effectiveness of these reagents in reducing less reactive functional groups and markedly lessens their utility for such relatively slow reductions.

Alkyl substituted borohydrides, such as lithium trimethylboro-hydride and sodium triethylborohydride, are readily synthesized[29] and prove to be quite stable in solution.[30] They have interesting possibilities as selective reducing agents[30,31] and are especially interesting for achieving steric control of the reduction of ketones, as will be discussed in Chapter XIII.

8. Effect of Substituents in Lithium Aluminum Hydride [32, 33, 34]

In view of the marked influence of alkoxy substituents upon the reducing power of sodium borohydride, it was of interest to examine their effect on the reducing power of lithium aluminum hydride. Whereas alkyl borates react readily with the alkali metal hydrides to produce the corresponding trialkylborohydrides,[26,29] the related reaction with aluminum alkoxides proved to be quite sluggish. However, this reaction has been applied as a synthetic route to sodium triethoxyaluminohydride by treating the "monomeric α-form" of aluminum ethoxide with sodium hydride in tetrahydrofuran in an autoclave at 70–90°.[35,36] In our studies, the treatment of lithium aluminum hydride with alcohols appeared to offer a more convenient route to the trialkoxyaluminohydrides, and so we used this procedure.[32,35]

[29] H. C. Brown, H. I. Schlesinger, I. Sheft, and D. M. Ritter, *J. Amer. Chem. Soc.*, **75,** 192 (1953).

[30] H. C. Brown and A. Khuri, unpublished observations.

[31] Research in progress with S. Krishnamurthy.

[32] H. C. Brown and R. F. McFarlin, *J. Amer. Chem. Soc.*, **80,** 5372 (1958).

[33] H. C. Brown and P. M. Weissman, *Israel J. Chem.*, **1,** 430 (1963).

[34] H. C. Brown and P. M. Weissman, *J. Amer. Chem. Soc.*, **87,** 5614 (1965).

[35] H. C. Brown and C. J. Shoaf, *J. Amer. Chem. Soc.*, **86,** 1079 (1964).

[36] P. M. Weissman and H. C. Brown, *J. Org. Chem.*, **31,** 283 (1966).

Treatment of lithium aluminum hydride in ethyl ether solution with four moles of methyl, ethyl, or isopropyl alcohol at 25° results in the evolution of four moles of hydrogen and the precipitation of the corresponding lithium tetraalkoxyaluminohydride (12.11). However, the addition of four moles of t-butyl alcohol results in the formation of only three moles of hydrogen (12.12). The fourth mole of hydrogen is evolved only on extended treatment at elevated temperatures.

$$\text{LiAlH}_4 + 4\text{CH}_3\text{OH} \xrightarrow[25°]{\text{EE}} \text{LiAl(OCH}_3)_4 \downarrow + 4\text{H}_2 \qquad 12.11$$

$$\text{LiAlH}_4 + 3(\text{CH}_3)_3\text{COH} \xrightarrow[25°]{\text{EE}} \text{LiAlH[OC(CH}_3)_3]_3 + 3\text{H}_2 \qquad 12.12$$

The reaction product, lithium tri-t-butoxyaluminohydride, is only slightly soluble in ethyl ether but readily soluble in tetrahydrofuran and diglyme. Accordingly, the reaction with t-butyl alcohol was examined in these solvents. Here also the reaction ceases with the evolution of three moles of hydrogen (12.13), the fourth mole being produced only at elevated temperatures (12.14).

$$\text{LiAlH}_4 + 3(\text{CH}_3)_3\text{COH} \xrightarrow[25°]{\text{THF,}}$$
$$\text{LiAlH[OC(CH}_3)_3]_3 + 3\text{H}_2 \quad 12.13$$

$$\text{LiAlH[OC(CH}_3)_3]_3 + (\text{CH}_3)_3\text{COH} \xrightarrow[\text{slow}]{\text{THF, reflux}}$$
$$\text{LiAl[OC(CH}_3)_3]_4 + \text{H}_2 \quad 12.14$$

Consequently, the tendency for the reaction to halt with the formation of the tri-t-butoxy derivative cannot be attributed to its low solubility in ether.

Lithium tri-t-butoxyaluminohydride proved to be surprisingly stable. A sample in diglyme solution, heated to 165° for 5 hours, retained 92 percent of its active hydrogen. It failed to melt below 400° and exhibited signs of decomposition above that temperature. It could be sublimed at 280° at 2 mm pressure.

The failure of lithium tri-t-butoxyaluminohydride to react with excess t-butyl alcohol at 25° suggested that this reagent should have reducing properties markedly different from that of the parent compound. Accordingly, a systematic examination was made of its reducing capabilities in tetrahydrofuran at 0° (12.15).

aldehyde	$\longrightarrow$ alcohol	
ketone	$\longrightarrow$ alcohol	
acid chloride	$\longrightarrow$ alcohol	
lactone	$\longrightarrow$ glycol (slow)	
epoxide	$\longrightarrow$ alcohol (slow)	
ester	$\longrightarrow$ slow reaction	12.15
carboxylic acid	$\longrightarrow$ no reduction	
carboxylic acid salt	$\longrightarrow$ no reaction	
tert-amide	$\longrightarrow$ no reaction	
nitrile	$\longrightarrow$ no reaction	
nitro	$\longrightarrow$ no reaction	
olefin	$\longrightarrow$ no reaction	

Thus, the three *t*-butoxy groups have greatly diminished the reducing potential of the parent reagent. Indeed, the reducing characteristics are so mild that the reagent resembles sodium borohydride much more closely than it does lithium aluminum hydride.

Although the reagent is rather sluggish in its reaction with alkyl esters, such as ethyl caproate and ethyl benzoate, it is much more reactive towards aryl esters, such as phenyl caproate, and can be used to reduce such esters to aldehydes.[36] The reagent is also useful in permitting the reduction of acid chlorides to aldehydes.[32]

For this purpose the acid chloride in diglyme or tetrahydrofuran at low temperatures (preferably −78°) is treated with one equivalent of the reagent in the same solvent. The temperature is permitted to rise to room temperature, the reaction mixture is poured onto crushed ice, and the aldehyde is isolated from the aqueous mixture.

A detailed study of the scope of this study was made.[37] Aromatic acid chlorides containing substituents in the *meta* and *para* positions form aldehydes in yields of 60–90 percent. Many types of substituents can be tolerated, including those that might be considered sensitive to reduction, such as nitro, cyano, and carbethoxy (12.16).

[37] H. C. Brown and B. C. Subba Rao, *J. Amer. Chem. Soc.*, **80,** 5377 (1958).

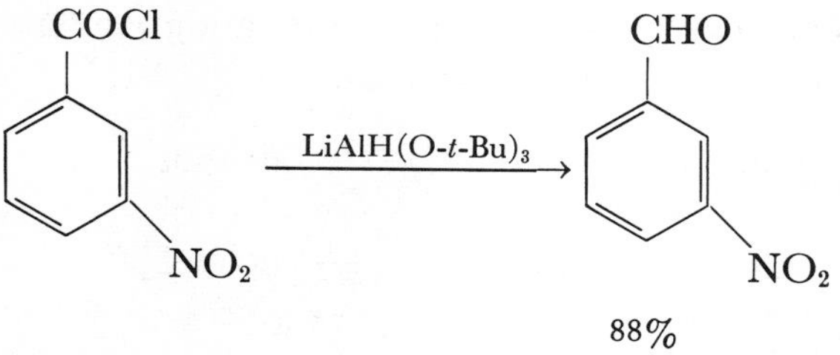

$$12.16$$

88%

In some cases, substituents in the *ortho* position tend to reduce the yield: *o*-nitro, 77 percent, *o*-chloro, 41 percent.

Polycyclic (α- and β-naphythyl), heterocyclic (nicotinyl), unsaturated (cinnamoyl), and polyfunctional (terephthalyl) acid chlorides may be used in this synthesis (12.17).

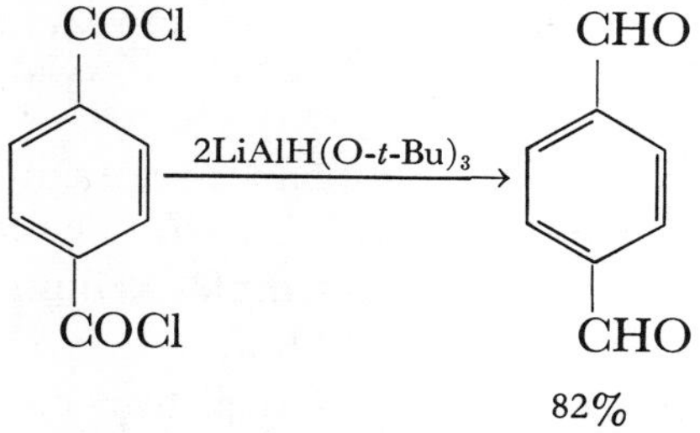

$$12.17$$

82%

In the case of aliphatic and alicyclic acid chlorides (isobutyryl, adipyl, cyclopropanecarboxyl, fumaryl), the yields fall in the range of 40–60 percent.

In synthetic work it is frequently desirable to proceed from some carboxylic acid derivative, other than the acid chloride, to the aldehyde. Consequently, we explored the possibility of converting the dimethylamides and nitriles to aldehydes. In both cases direct treatment with lithium aluminum hydride results in poor yields. Moreover, in neither case does lithium tri-*t*-butoxyaluminohydride react with these derivatives. Fortunately, lithium trimethoxyaluminohydride reacted readily, with moderate yields of aldehyde. Better results were realized with lithium diethoxyaluminohydride.[28] Finally, we have realized optimum results by utilizing lithium triethoxyaluminohydride in ether solutions.[38,39]

[38] H. C. Brown and A. Tsukamoto, *J. Amer. Chem. Soc.*, **86,** 1089 (1964).
[39] H. C. Brown and C. P. Garg, *J. Amer. Chem. Soc.*, **86,** 1085 (1964).

The procedure is very simple. A 1.0 M solution of lithium aluminum hydride in ethyl ether is treated with three moles of ethanol or 1.5 moles of ethyl acetate to form the reagent *in situ* at 0°. To the solution is added one equivalent of the N,N-dimethylamide or the nitrile. After one hour, the reaction mixture is hydrolyzed and the aldehyde is isolated. From the *t*-amide, the aldehydes are generally obtained in yields of 80–90 percent for both aliphatic and aromatic derivatives. In the case of the nitrile reduction, the yields vary from 70 percent for aliphatic derivatives to 90 percent for aromatic derivatives.

It may also be mentioned here that reduction of the 1-acylaziridines with lithium aluminum hydride also provides a convenient synthetic route to the aldehyde[40] (12.18).

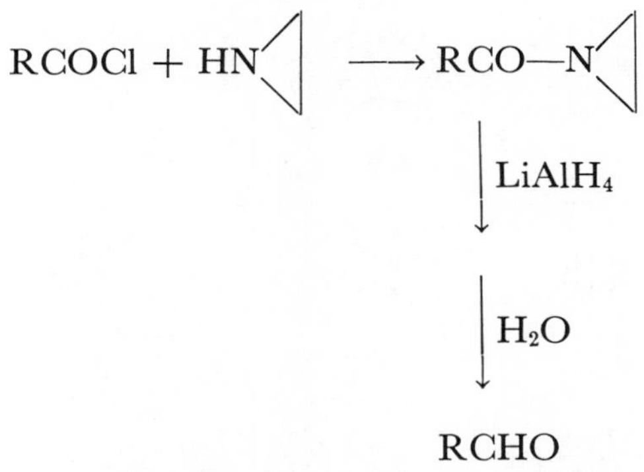

$$\text{RCOCl} + \text{HN}\!\!\triangleleft \longrightarrow \text{RCO}\!-\!\text{N}\!\!\triangleleft$$

$$\Big\downarrow \text{LiAlH}_4$$

$$\Big\downarrow \text{H}_2\text{O}$$

$$\text{RCHO}$$

12.18

The simplicity and wide applicability of these new synthetic procedures should provide valuable new routes from the carboxylic acids to the corresponding aldehydes.

It was of interest to explore systematically the reducing characteristics of lithium trimethoxyaluminohydride.[34] In contrast to the behavior of lithium tri-*t*-butoxyaluminohydride, its reducing characteristics are quite powerful, resembling closely those of lithium aluminum hydride itself (12.19).

aldehyde	$\longrightarrow$ alcohol
ketone	$\longrightarrow$ alcohol
acid chloride	$\longrightarrow$ alcohol

[40] H. C. Brown and A. Tsukamoto, *J. Amer. Chem. Soc.*, **83**, 4549 (1961).

lactone	$\longrightarrow$ glycol	
epoxide	$\longrightarrow$ alcohol (slow)	
ester	$\longrightarrow$ alcohol	12.19
carboxylic acid	$\longrightarrow$ alcohol	
carboxylic acid salt	$\longrightarrow$ alcohol	
tert-amide	$\longrightarrow$ amine	
nitrile	$\longrightarrow$ amine	
nitro	$\longrightarrow$ reaction	
olefin	$\longrightarrow$ no reaction	

In one case, the reduction of bicyclic ketones, lithium trimethoxya-luminohydride exhibited a major advantage over lithium aluminum hydride in having a much higher stereoselectivity[41] (12.20).

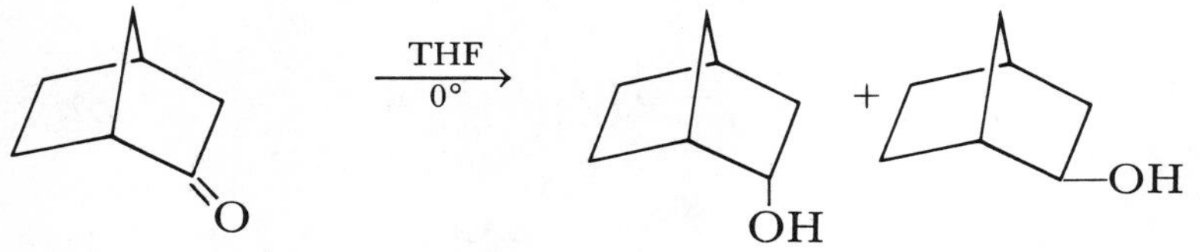

LiAlH$_4$	89%	11%	
LiAlH(OMe)$_3$	98%	2%	12.20

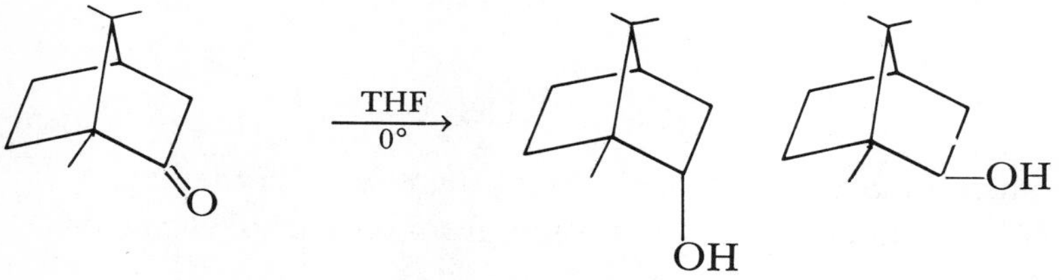

LiAlH$_4$	8%	92%
LiAlH(OMe)$_3$	1%	99%

Dialkylboranes and certain trialkylborohydrides have proven to possess highly valuable properties for the steric control of such reductions.[42,43] This we shall consider in detail in Chapter XIII.

[41] H. C. Brown and H. R. Deck, *J. Amer. Chem. Soc.*, **87,** 5620 (1965).
[42] H. C. Brown and V. K. Varma, *J. Amer. Chem. Soc.*, **88,** 1271 (1966).
[43] H. C. Brown and W. C. Dickason, *J. Amer. Chem. Soc.*, **92,** 709 (1970).

XIII. Selective Reductions with Boranes and Alanes[1-3]

1. Electrophilic vs. Nucleophilic Reducing Agents

The reaction of lithium aluminum hydride with cyclic epoxides corresponds to a typical S_N2 displacement by hydride with inversion at the carbon atom undergoing reaction[4] (13.1). Furthermore, a systematic study of the reaction of lithium aluminum hydride with a

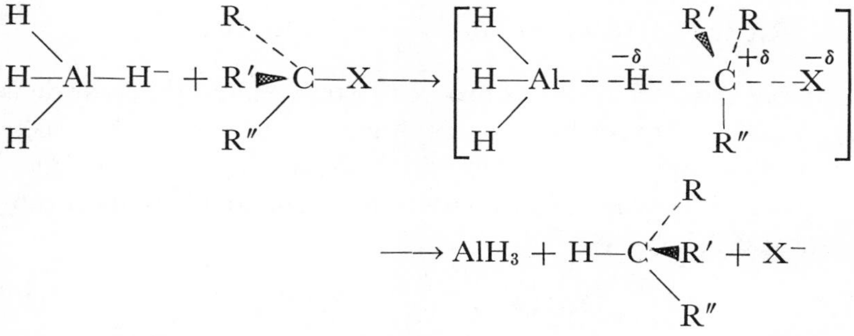

$$\underset{H}{\overset{H}{\diagdown}}Al-H^- + R'\!\!-\!\!\!C-X \longrightarrow \left[\underset{H}{\overset{H}{\diagdown}}Al\text{-}\,\text{-}\,\text{-}\overset{-\delta}{H}\text{-}\,\text{-}\,\text{-}\overset{R'\ R}{\underset{R''}{\overset{+\delta}{C}}}\text{-}\,\text{-}\,\overset{-\delta}{X} \right]$$

$$\longrightarrow AlH_3 + H-\underset{R''}{\overset{R}{C}}\!\!-\!\!R' + X^- \qquad 13.1$$

wide range of alkyl halides reveals a structural dependence that parallels that present in typical S_N2 substitutions[5] (13.2).

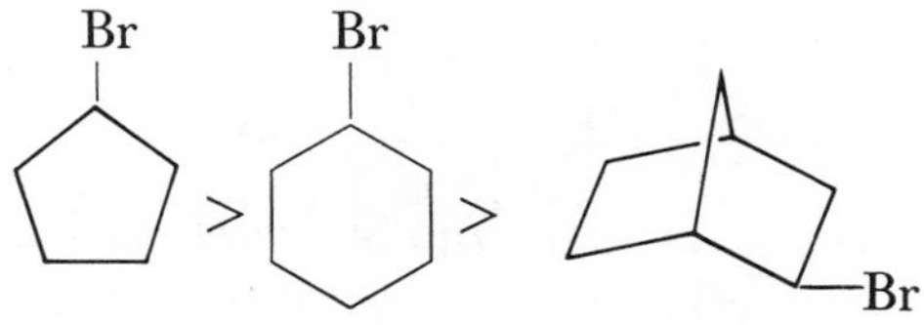

$$\underset{}{\overset{Br}{\bigcirc}} > \underset{}{\overset{Br}{\bigcirc}} > \underset{Br}{\diagup\!\!\!\!\bigtriangleup} \qquad 13.2$$

[1] H. C. Brown, *J. Chem. Educ.*, **38,** 173 (1961).

[2] H. C. Brown, "Moderni Sviluppi Della Sintesi Organica," *Conferenze X, Corso Estivo di Chimica, Academia Nazionale dei Lincei* (Rome, 1968), pp. 31–50.

[3] E. Schenker, *Newer Methods of Preparative Organic Chemistry*, Vol. IV, W. Foerst, ed. (Verlag Chemie GmbH, Weinheim/Bergstr., 1968), pp. 196–335.

[4] L. W. Trevoy and W. G. Brown, *J. Amer. Chem. Soc.*, **71,** 1675 (1949).

[5] Research with S. Krishnamurthy.

The reaction of sodium borohydride with ketones in isopropyl alcohol[6] is first-order in ketone and first-order in borohydride, although four moles of ketone react per mole of the complex hydride. It was concluded that the rate-determining step involved a transfer of hydride from the borohydride to the carbon atom of the carbonyl group. This is followed by three successive transfers at faster rates (13.3).

$$R_2CO + BH_4^- \xrightarrow{k_2} [H_3BOCHR_2]^-$$
$$R_2CO + [H_3BOCHR_2]^- \xrightarrow{k_2'} [H_2B(OCHR_2)_2]^- \qquad 13.3$$
$$R_2CO + [H_2B(OCHR_2)_2]^- \xrightarrow{k_2''} [HB(OCHR_2)_3]^-$$
$$R_2CO + [HB(OCHR_2)_3]^- \xrightarrow{k_2'''} [B(OCHR_2)_4]^-$$

(Presumably the later transfers proceed faster because the hydride is transferring from weaker Lewis acids. Alkyl borates are much weaker acids than borane.) In agreement with this picture the introduction of electron-withdrawing substituents into the ketone results in enhanced rates[7] (13.4).

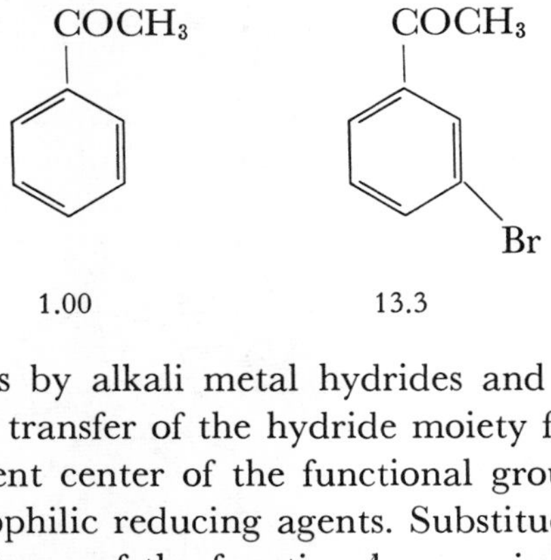

| 1.00 | 13.3 |

13.4

Thus reductions by alkali metal hydrides and aluminohydrides appear to involve transfer of the hydride moiety from the anion to an electron-deficient center of the functional group. That is, they function as nucleophilic reducing agents. Substituents that increase the electron deficiency of the functional group increase the ease of reaction. In this way it is easy to account for the fact that molecules

[6] H. C. Brown, O. H. Wheeler, and K. Ichikawa, *Tetrahedron*, **1**, 214 (1957).
[7] K. Bowden and M. Hardy, *Tetrahedron*, **22**, 1169 (1966).

such as chloral and acetyl chloride are much more rapidly reduced in diglyme than are simple aldehydes and ketones (13.5).

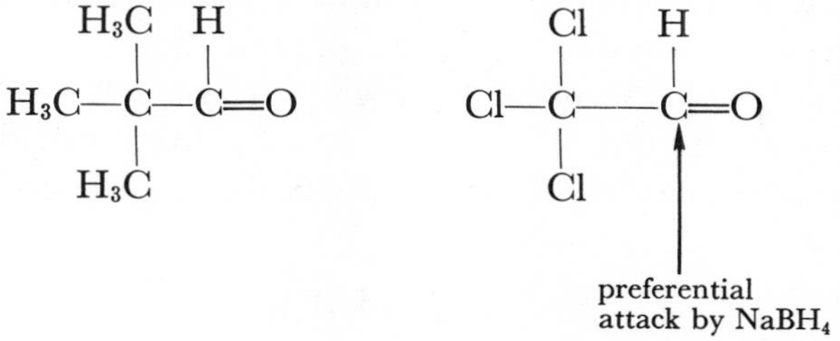

$$
\begin{array}{cc}
\underset{\displaystyle H_3C-\underset{\displaystyle H_3C}{\overset{\displaystyle H_3C}{|}}{C}-\overset{\displaystyle H}{\overset{|}{C}}{=}O}{} &
\underset{\displaystyle Cl-\underset{\displaystyle Cl}{\overset{\displaystyle Cl}{|}}{C}----\overset{\displaystyle H}{\overset{|}{C}}{=}O}{}
\end{array}
\qquad 13.5
$$

preferential
attack by $NaBH_4$

On the other hand, diborane is a strong Lewis acid. It forms stable addition compounds with tertiary amines. It would therefore be expected to involve a preferred electrophilic attack on the centers of highest electron densities. Indeed, diborane reduces trimethyl-acetaldehyde readily, but fails to react with chloral under the usual mild conditions[8] (0 to 25°) (13.6).

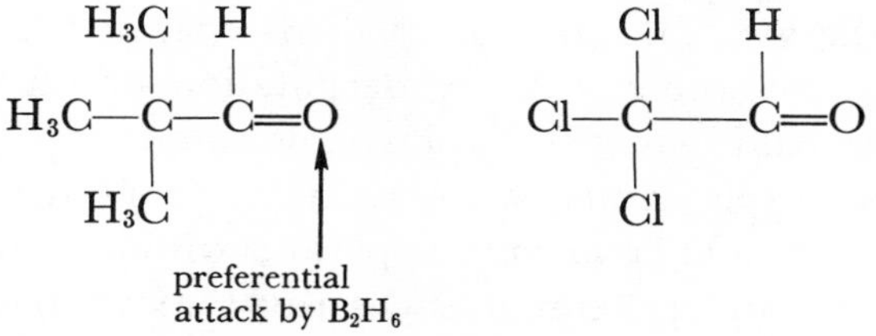

$$
\begin{array}{cc}
\underset{\displaystyle H_3C-\underset{\displaystyle H_3C}{\overset{\displaystyle H_3C}{|}}{C}-\overset{\displaystyle H}{\overset{|}{C}}{=}O}{} &
\underset{\displaystyle Cl-\underset{\displaystyle Cl}{\overset{\displaystyle Cl}{|}}{C}----\overset{\displaystyle H}{\overset{|}{C}}{=}O}{}
\end{array}
\qquad 13.6
$$

preferential
attack by B_2H_6

Similarly aluminum hydride (alane) is a Lewis acid, forming stable addition compounds with tertiary amines.[9]

In contrast to the nucleophilic character of sodium borohydride and lithium aluminum hydride, it is evident that borane and alane should function as electrophilic hydrogenating agents. Reagents with such different reducing characteristics might be exceedingly useful in selective reductions, and so an extensive study of their character-istics was undertaken.

[8] H. C. Brown, H. I. Schlesinger, and A. B. Burg, *J. Amer. Chem. Soc.*, **61**, 673 (1939).

[9] E. Wiberg, *Angew. Chem.*, **65**, 16 (1953).

2. Diborane as a Reducing Agent [10]

Diborane is a gas (bp $-92.5°$), highly reactive to air and moisture, and difficult to handle in simple organic operations. It is only slightly soluble in the usual hydrocarbon and ether solvents, with one major exception: it dissolves readily in tetrahydrofuran, in which it evidently exists as the tetrahydrofuran addition compound[11]

13.1

(**13.1**). Such solutions are readily prepared by treating sodium borohydride in diglyme with boron trifluoride etherate and passing the gas, as generated, into tetrahydrofuran.[12] Alternatively, the solution may be prepared by treating a suspension of sodium borohydride in tetrahydrofuran with the calculated quantity of boron trifluoride etherate. The solution is then separated from the precipitated sodium borofluoride by decantation or filtration under nitrogen. Such solutions are now available commercially.

The original experiments on the reducing properties of diborane utilized the gas in a high-vacuum apparatus without solvents.[8] Then we utilized diborane, generated as described above, passed into the compound dissolved in a suitable solvent.[13,14] However, the use of standardized solutions of borane in tetrahydrofuran has proved so convenient, both for reductions[1] and for hydroborations (Chapter XIV), that this has become the procedure of choice.

Both aliphatic and aromatic ketones are rapidly reduced at room temperature. Such reductions generally involve the rapid reaction of two moles of carbonyl compound per mole of borane to form the

[10] H. C. Brown, P. Heim, and N. M. Yoon, *J. Amer. Chem. Soc.*, **92**, 1637 (1970).

[11] J. R. Elliott, W. L. Roth, G. F. Roedel, and I. M. Boldebuck, *J. Amer. Chem. Soc.*, **74**, 5211 (1952).

[12] G. Zweifel and H. C. Brown, *Org. Reactions*, **13**, 1 (1963).

[13] H. C. Brown and B. C. Subba Rao, *J. Amer. Chem. Soc.*, **82**, 681 (1960).

[14] H. C. Brown and W. Korytnyk, *J. Amer. Chem. Soc.*, **82**, 3866 (1960).

dialkoxyborane. The third hydride reacts with difficulty. Hydrolysis yields the alcohol (13.7).

$$2RCHO + BH_3 \longrightarrow (RCH_2O)_2BH$$
$$(RCH_2O)_2BH + 3H_2O \longrightarrow 2RCH_2OH + B(OH)_3 + H_2 \qquad 13.7$$

An unexpected feature is the high stereoselectivity revealed in the reduction of norcamphor: 2 percent *exo*- and 98 percent *endo*-norbornanol. This is comparable to the stereospecificity realized with lithium trimethoxyaluminohydride (12.20), and is to be contrasted to the 11:89 distribution provided by lithium aluminum hydride. On the other hand, borane provides a 52:48 *exo*:*endo* distribution for the two alcohols from camphor, as compared to a 92:8 distribution with lithium aluminum hydride (12.20).

The reduction of acid chlorides under these conditions is quite slow. As pointed out in Chapter II, the electron-withdrawing effect of the chlorine substituent greatly reduces the donor properties of the carbonyl oxygen atom. Consequently, borane does not coordinate with it, and reduction takes place only relatively slowly.[15]

Esters and lactones are reduced relatively slowly. On the other hand, carboxylic acids are reduced rapidly. This was unexpected. Normally carboxylic acids are much more resistant to hydrogenation or reduction than the corresponding esters. However, investigation revealed a simple explanation,[13] to be discussed in the next section. Epoxides are also reduced, but at a relatively slow rate. In this case the reduction can often be directed into different paths by the use of catalytic quantities of boron trifluoride or sodium borohydride.

t-Amides are rapidly reduced. The reaction of the reagent is somewhat slower, but the reduction to the amine is readily achieved. The nitrile group is highly inert to nucleophilic attack by sodium borohydride. However, diborane evidently reacts through an electrophilic attack on the nitrile nitrogen (13.8).

$$R—C{\equiv}N + BH_3 \longrightarrow R—C{\equiv}N{:}BH_3 \qquad 13.8$$

The reaction then proceeds to the formation of the corresponding N,N',N''-trialkylborazoles[16] (**13.2**).

---

[15] However, see S. L. Ioffe, V. A. Jartakovskii, and S. S. Novikov, *Izv. Akad. Nauk SSSR, Ser. Khim.*, 622 (1964), and discussion in ref. 10.

[16] H. J. Emeleus and K. Wade, *J. Chem. Soc.*, 2615 (1960).

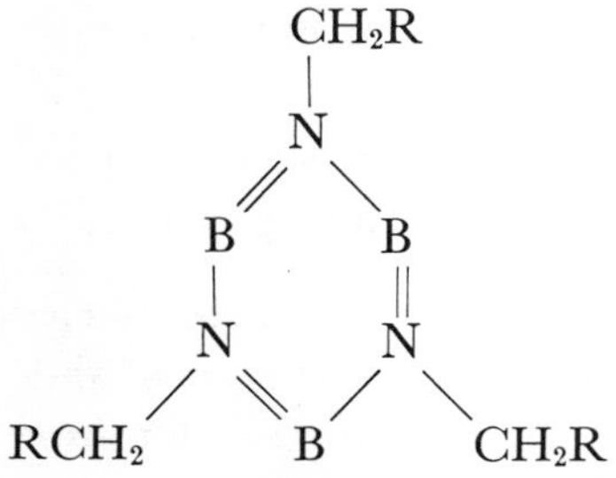

$$CH_2R$$

$$N$$

$$B \qquad B$$

$$N \qquad N$$

$$RCH_2 \qquad B \qquad CH_2R$$

13.2

This intermediate is readily hydrolyzed by acid to the corresponding amine.[13]

Nitro compounds, as might be expected because of their very weakly basic properties, show no sign of reaction with borane-tetrahydrofuran under the mild test conditions over many hours. Sulfones, sulfonic acids, disulfides, and organic halides also are stable to the reagent. Unsaturated carbon-carbon bonds, such as olefins and acetylenes, react remarkably rapidly with it (Chapter XIV).

The results with the standard substrates may be summarized as follows (13.9).

aldehyde	$\longrightarrow$ alcohol (fast)	
ketone	$\longrightarrow$ alcohol	
acid chloride	$\longrightarrow$ very slow reaction	
lactone	$\longrightarrow$ glycol	
epoxide	$\longrightarrow$ alcohol (slow)	
ester	$\longrightarrow$ alcohol (slow)	13.9
carboxylic acid	$\longrightarrow$ alcohol (fast)	
carboxylic acid salt	$\longrightarrow$ no reduction	
t-amide	$\longrightarrow$ amine	
nitrile	$\longrightarrow$ amine	
nitro	$\longrightarrow$ no reduction	
olefin	$\longrightarrow$ organoborane (fast)	

A study of the relative reactivity of a number of representative groups toward diborane in tetrahydrofuran indicated the following order of reactivity:[14] carboxylic acids > olefins > ketones > nitriles > epoxides > esters > acid chlorides. On the other hand,

toward alkali metal borohydride the order is: acid chlorides >
ketones > epoxides > esters > nitriles > carboxylic acids.

With such markedly different reactivities, the judicious use of
either diborane or alkali metal borohydrides permits the reduction
of one group in the presence of a second, or the reverse. For example,
in the half ester of a dicarboxylic acid it should be possible to use
borane in tetrahydrofuran to reduce the free carboxylic acid group
preferentially. Alternatively, by converting the free acid to the
carboxylate salt, it should then be possible to reduce the ester group
preferentially with lithium borohydride.

3. Applications of Diborane

The rapid reaction of borane in tetrahydrofuran with carboxylic
acids is noteworthy. Normally, carboxylic acids are much more
resistant to hydrogenation or reduction than the corresponding esters.
This facile reduction suggests that borane in tetrahydrofuran should
be the reagent of choice for such reductions in systems not containing
other highly reactive features, such as carbon-carbon multiple bonds
or aldehyde groups.

The precise explanation for the high reactivity of carboxylic acid
was explored.[13] The first step in the reaction is the formation of the
triacyloxyborane (13.10).

$$3RCO_2H + BH_3 \longrightarrow 3(RCO_2)_3B + 3H_2 \qquad 13.10$$

This product contains a highly reactive carbonyl group, as shown by
its ready reduction to the alcohol stage by sodium borohydride as
well as by diborane. It was proposed that the electron deficiency of
the boron atom in the triacyloxyborane exerts a powerful demand
on the electron-pairs of the acyloxy oxygen. Consequently, resonance
will involve interaction of this oxygen atom with the boron atom,
rather than the usual resonance with the carbonyl group (13.11).

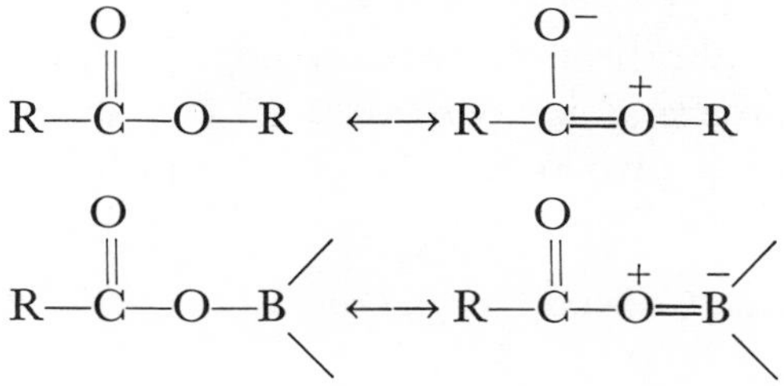

13.11

According to this interpretation, the carbonyl group in the tri-acyloxyboranes should resemble that in a simple aldehyde or ketone far more than the resonance-stabilized, less active carbinyl group in an ester. Indeed, the carbonyl groups in the triacyloxyborane react readily with sodium borohydride[13] as well as with diborane. Typical aldehydes and ketones are likewise highly reactive towards both sodium borohydride and diborane.

t-Amides are readily reduced to *t*-amines by diborane, and considerable application has been made of this reaction[17] (13.12, 13.13, 13.14).

$$(CH_3)_3CCON(CH_3)_2 \longrightarrow (CH_3)_3CCH_2N(CH_3)_2 \qquad 13.12$$
$$92\%$$

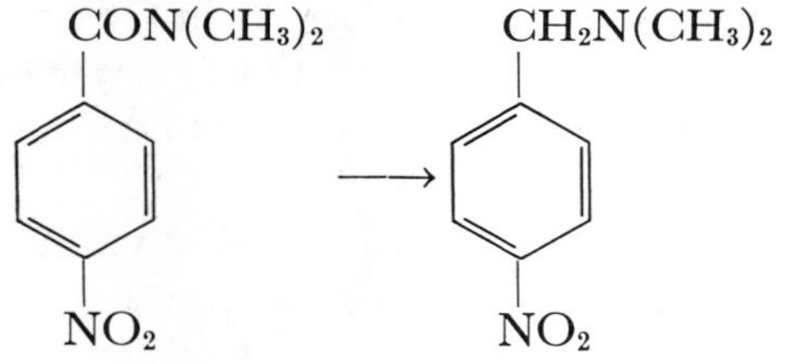

13.13

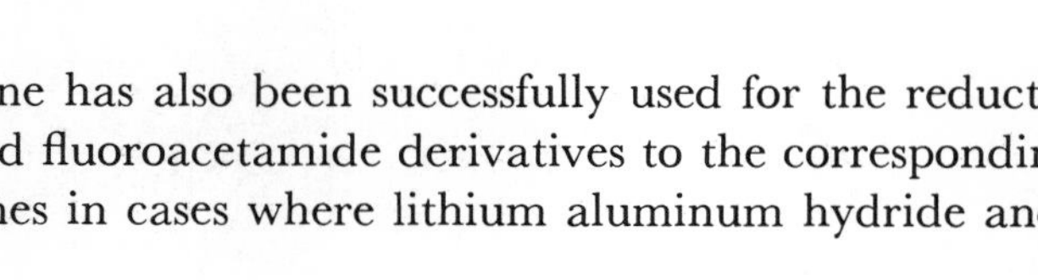

13.14

Diborane has also been successfully used for the reduction of N-substituted fluoroacetamide derivatives to the corresponding fluoro-ethylamines in cases where lithium aluminum hydride and lithium

---

[17] H. C. Brown and P. Heim, *J. Amer. Chem. Soc.*, **86,** 3566 (1964).

aluminum hydride-aluminum chloride cause hydrogenolysis of the flourine-carbon bond.[18] It has proven useful also for reducing the salts of nitroalkanes, and oximes, to the corresponding N-substituted hydroxylamines or to the corresponding amines.[19]

The reaction of diborane with epoxides reveals some fascinating features. First, the reaction is quite slow. Thus, at 0° 1,2-butylene oxide required three days to complete the uptake of one hydride per mole of epoxide. Yet at this time there was present only 48 percent of 2- and 1-butanols, in the ratio of 96:4.[10] Apparently under these conditions the solvent participates in part in the opening of the epoxide ring.[20] The reactions are also complex for styrene oxide and 1-methyl-1,2-cyclohexene oxide, although in different manners. Thus, in the case of styrene oxide, the uptake goes beyond the stoichiometric point. Apparently addition to the aromatic nucleus is occurring in some manner. In the case of 1-methyl-1,2-cyclohexene oxide nearly two hydride equivalents are used, with a mole of hydrogen evolved. Hydrolysis followed by oxidation with alkaline hydrogen peroxide gives a glycol as the main product, 1-hydroxy-methyl-2-cyclohexanol.

These complexities are avoided by introducing small catalytic quantities of sodium or lithium borohydride.[21] As shown in Figure XIII-1, the effect of a small amount of sodium borohydride on the rate of reduction is fantastic. Moreover, the reaction yields in one hour nearly a quantitative yield of 2-butanol.

In the case of trisubstituted epoxides, such as 1-methylcyclohexene oxide, the rate of the reaction is also increased. Moreover, the course of the reaction is modified tremendously. One mole of hydride per mole of compound is taken up, no hydrogen is evolved, and there takes place a predominant anti-Markovnikov opening of the epoxide ring (13.15, 13.16).

---

[18] Z. B. Papanastassion and R. J. Bruni, *J. Org. Chem.*, **29,** 2870 (1964).

[19] H. Feuer, R. S. Bartlett, B. F. Vincent, Jr., and R. S. Anderson, *J. Org. Chem.*, **30,** 2880 (1965). H. Feuer, B. F. Vincent, Jr., and R. S. Bartlett, *J. Org. Chem.*, **30,** 2877 (1965). H. Feuer and D. M. Braustein, *J. Org. Chem.*, **34,** 1817 (1969).

[20] D. J. Pasto, C. C. Cumbo, and J. Hickman, *J. Amer. Chem. Soc.*, **88,** 2201 (1966).

[21] H. C. Brown, N. M. Yoon, *J. Amer. Chem. Soc.*, **90,** 2686 (1968).

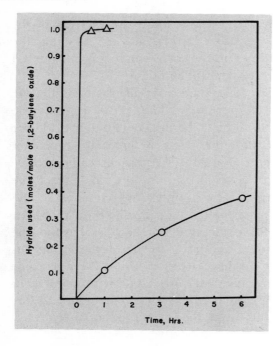

Figure XIII-1. The reaction of borane (0.33 M) with 1,2-butylene oxide (0.25 M) in the presence (△) and absence (○) of sodium borohydride (0.036 M).

$$\underset{O}{\overset{CH_3}{\underset{\diagdown\diagup}{CH_3C\text{---}CHCH_3}}} \xrightarrow[\text{LiBH}_4]{\text{BH}_3} \underset{OH}{\overset{CH_3}{CH_3C\text{---}CH_2CH_3}}$$

25%

$$+ \underset{OH}{\overset{CH_3}{CH_3CH\text{---}CHCH_3}} \quad 13.15$$

75%

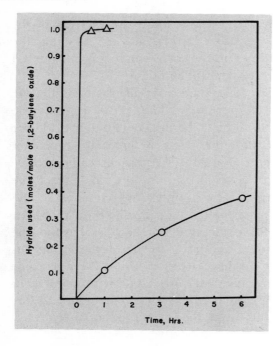 $\xrightarrow[\text{NaBH}_4]{\text{BH}_3}$ $+$ 13.16

26% 74%

The reaction presumably involves preferentially the attack by borohydride at the polarized tertiary position, with inversion, to form the *cis*-2-methylcycloalkanol, whereas electrophilic attack by diborane occurs at the methyl group (13.17).

13.17

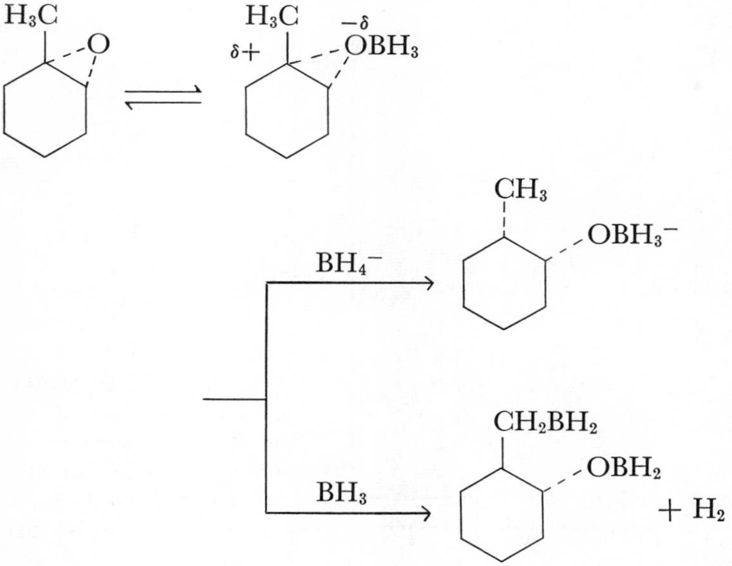

It was pointed out originally that diborane reacts with carbon monoxide at elevated pressures and temperatures to form borane-carbonyl, $H_3B{:}CO$, and that this product is not further reduced by diborane (Chapter II). Carbon monoxide placed over a solution of borane in tetrahydrofuran is not reduced by the solution. However, here also a small quantity of sodium borohydride brings about a rapid reaction[22] (Figure XIII-2). The product is trimethylboroxine (13.18).

$$3BH_3 + 3CO \xrightarrow{NaBH_4}$$

13.18

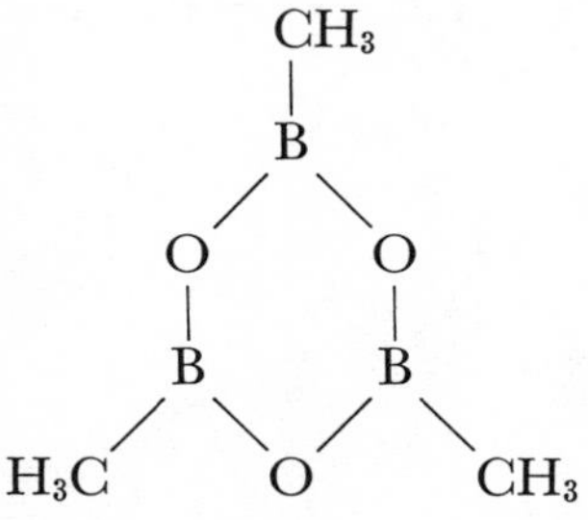

[22] M. W. Rathke and H. C. Brown, *J. Amer. Chem. Soc.*, **88**, 2606 (1966).

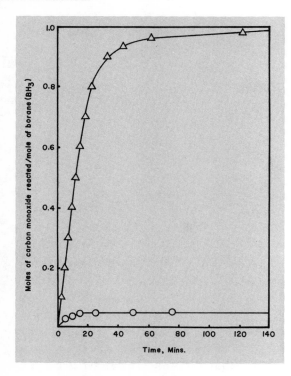

Figure XIII-2. Reaction of carbon monoxide at atmospheric pressure with borane in tetrahydrofuran (0.5 M) at 25° in the presence (△) and absence (○) of sodium borohydride (2 mole percent).

Finally, the introduction of small quantities of boron trifluoride greatly facilitates the reduction of styrene oxide and related epoxides in an anti-Markovnikov direction[23] (13.19, 13.20).

$$\xrightarrow[\text{BF}_3]{\text{BH}_3}$$

CH₂CH₂OH

98%

13.19

$$\xrightarrow[\text{BF}_3]{\text{BH}_3}$$

)₂CHCH₂OH   13.20

95%

[23] H. C. Brown and N. M. Yoon, *Chem. Commun.*, 1549 (1968).

Consequently, it is now evident that the electrophilic reducing properties of diborane, interesting as they are, can be made more versatile by the introduction of catalytic amounts of boron trifluoride or sodium borohydride.

4. Disiamylborane as a Reducing Agent [24]

Hydroboration of certain hindered olefins or structurally suited dienes can be controlled to produce dialkylboranes (Chapters XIV, XV). Typical examples are disiamylborane (**13.3**), diisopinocampheylborane (**13.4**), and 9-borabicyclo[3.3.1]nonane or 9-BBN (**13.5**).

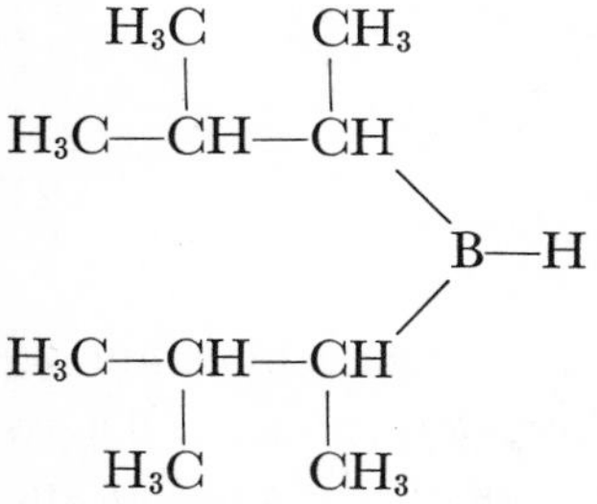

13.3 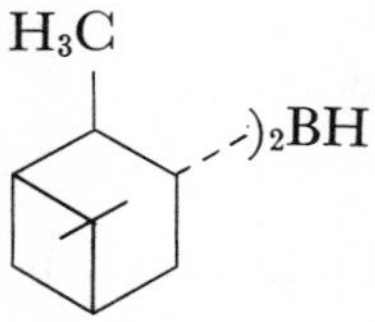   **13.4** 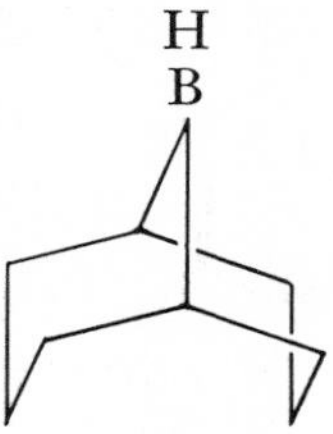   **13.5**

Up to the present time, the greatest attention has been paid to disiamylborane,[24] which therefore will be emphasized in this discussion. However, diisopinocampheylborane is promising as an optically active dialkylborane (prepared as it can be from optically active α-pinene). Finally, 9-BBN is sufficiently stable in air so that it can be weighed and handled rapidly, without the usual precautions required by other boranes, so it will be especially valuable for laboratory operations when it becomes commercially available. Unfortunately, these reagents have not yet been adequately explored.

It should be pointed out that for application in selective reductions there are major advantages to having only one active hydride in the reagent. A reagent, such as lithium aluminum hydride or diborane, goes through a number of stages in a reduction, providing successively a number of species of different activities. Consequently,

[24] H. C. Brown, D. B. Bigley, S. K. Arora, and N. M. Yoon, *J. Amer. Chem. Soc.*, **92**, 7161 (1970).

it becomes difficult to control such reactions to achieve optimum results in a preferential reduction of one group in the presence of another group. Cyclic reactions of such intermediate species also constitute a problem with suitably situated functional groups. These difficulties are largely avoided by the use of monofunctional reagents, such as lithium trimethoxyaluminohydride or disiamylborane.

Aldehydes and ketones are rapidly reduced by disiamylborane. The reaction of disiamylborane with many internal carbon-carbon double bonds is relatively slow, so the selective reduction of α,β-unsaturated aldehydes and ketones, such as cinnamaldehyde, becomes feasible (13.21).

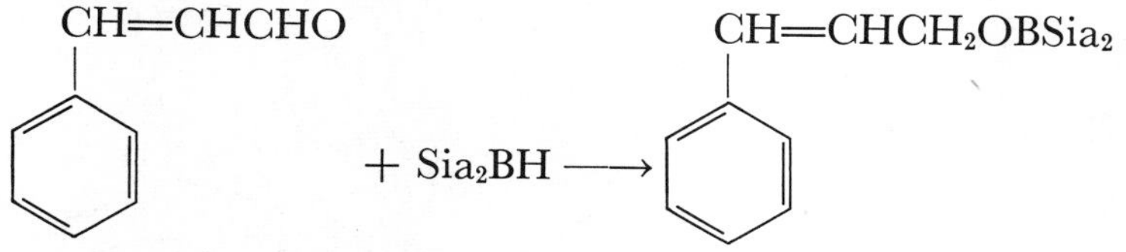

$$CH{=}CHCHO \qquad\qquad CH{=}CHCH_2OBSia_2$$

$$+ Sia_2BH \longrightarrow \qquad\qquad 13.21$$

The reduction of cyclic ketones, such as 2-methylcyclohexanone, proved particularly promising. Whereas diborane produces 74 percent of the *trans* and 26 percent of the *cis* alcohol, disiamylborane reversed the ratio: 21 percent *trans*, and 79 percent *cis*. This observation stimulated a study of the phenomenon, and even better results were realized with diisopinocampheylborane (**13.4**). The steric control of such reductions is discussed in Section XII-6.

Acid chlorides are stable to disiamylborane, as are simple esters. Lactones, however, react readily with one mole of the reagent, providing a valuable route to the corresponding hydroxyaldehydes (13.22).

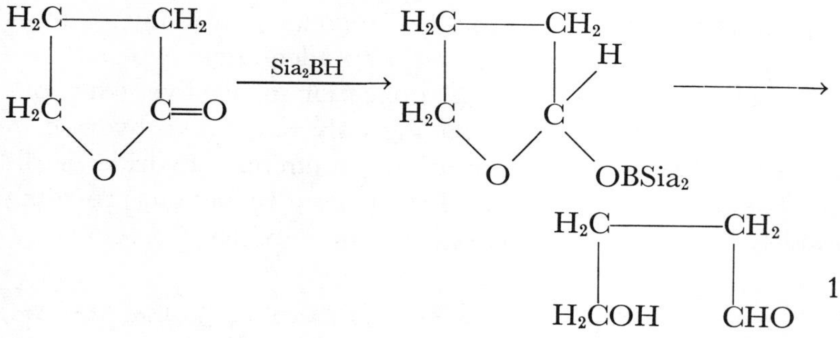

$$H_2C{-}CH_2 \qquad\qquad H_2C{-}CH_2$$
$$\xrightarrow{Sia_2BH}$$
$$H_2C \quad C{=}O \qquad\qquad H_2C \quad C$$
$$O \qquad\qquad O \quad OBSia_2$$

$$\longrightarrow$$

$$H_2C{-}CH_2$$
$$\qquad\qquad 13.22$$
$$H_2COH \quad CHO$$

This development will be discussed in more detail in the following section.

Carboxylic acids react rapidly to liberate one mole of hydrogen, but reduction does not occur. This is in contrast to the fast reduction by diborane. It would appear that the steric requirements of di-siamylborane are so great that it fails to react with the first intermediate (13.23, 13.24), even though it must contain an activated carbonyl group (13.11).

$$R\text{—}\overset{\overset{\displaystyle O}{\|}}{C}\text{—OH} + Sia_2BH \xrightarrow{0°} R\text{—}\overset{\overset{\displaystyle O}{\|}}{C}\text{—O—BSia}_2 + H_2 \quad 13.23$$

$$R\text{—}\overset{\overset{\displaystyle O}{\|}}{C}\text{—O—BSia}_2 + Sia_2BH \xrightarrow{0°} \text{no reaction} \quad 13.24$$

The reactions with epoxides are quite slow, with styrene oxide and 1-methyl-1,2-cyclohexene oxide showing the same peculiarities noted in their reactions with diborane.[10] Perhaps the most useful feature of these reactions is their making possible many hydroborations in the presence of epoxide groups.

t-Amides react readily with one mole of reagent, providing a promising new route to aldehydes (13.25).

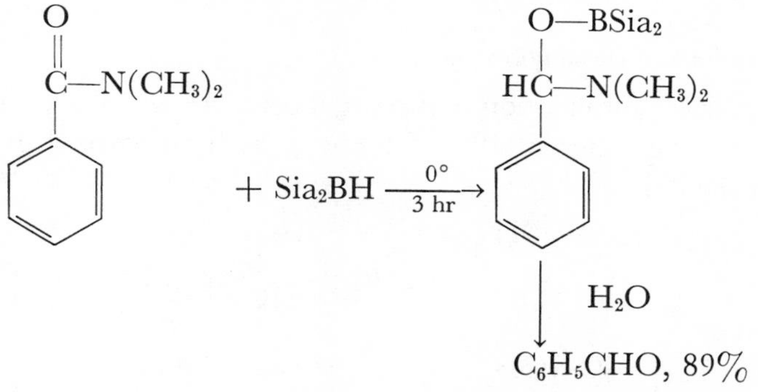

$$+ Sia_2BH \xrightarrow[3\ hr]{0°}$$

13.25

$$\downarrow H_2O$$

$$C_6H_5CHO, 89\%$$

The reaction of nitriles with the reagent is very slow. Presumably, the large steric requirements of the two siamyl groups greatly reduce the electrophilic properties of the reagent compared to those of borane itself.

Cyclohexanone oxime reacts at the active hydrogen, but is not reduced. Nitroalkanes are not reduced, but nitrobenzene undergoes a slow reaction. Sulfones, sulfonic acids, sulfides, and disulfides are not reduced.

These results may be summarized in the usual way (13.26).

aldehyde	$\longrightarrow$ alcohol	
ketone	$\longrightarrow$ alcohol	
acid chloride	$\longrightarrow$ no reaction	
lactone	$\longrightarrow$ hydroxyaldehyde	
epoxide	$\longrightarrow$ very slow reaction	
ester	$\longrightarrow$ no reaction	13.26
carboxylic acid	$\longrightarrow$ no reduction	
carboxylic acid salt	$\longrightarrow$ no reduction	
t-amide	$\longrightarrow$ aldehyde	
nitrile	$\longrightarrow$ very slow reaction	
nitro	$\longrightarrow$ very slow reaction	
olefin	$\longrightarrow$ organoborane	

It was pointed out that diisopinocampheylborane is an optically active dialkylborane and is of interest for that reason. It has achieved the reduction of ketones to optically active alcohols with optical purities in the 11 to 30 percent range.[25, 26] However, it has proven far more valuable in the achievement of asymmetric hydroboration (Chapter XV).

5. Applications of Disiamylborane

One possible application of disiamylborane is as a blocking group. Thus it reacts preferentially with the primary hydroxyl groups of glycerol (13.27).

$$\begin{array}{ccc} H_2C-OH & & H_2C-OBSia_2 \\ | & & | \\ HC-OH & \xrightarrow{2Sia_2BH} & HC-OH \\ | & & | \\ H_2C-OH & & H_2C-OBSia_2 \end{array} \qquad 13.27$$

The secondary hydroxyl group can now be acylated with an acid chloride and a tertiary amine. The disiamylboryl groups are readily

[25] H. C. Brown and D. B. Bigley, *J. Amer. Chem. Soc.*, **83,** 3166 (1961).

[26] E. Caspi and K. R. Varma, *J. Org. Chem.*, **33,** 2181 (1968): K. R. Varma and E. Caspi, *Tetrahedron*, **24,** 6365 (1968).

removed under mild conditions with hydrogen peroxide at slightly alkaline conditions ($pH \backsim 8$).[27]

The reduction of lactones to hydroxyaldehydes is another area of utility,[28] and many applications have already been reported.[29]

However, the most valuable application of the reagent would appear to be its use as a hydroborating agent in the presence of substituents highly reactive toward diborane. Thus, 11-undecenoic acid is readily converted by disiamylborane into the 11-hydroxyundecanoic acid without reduction of the carboxylic acid group[28] (13.28).

$$H_2C{=}CH(CH_2)_8CO_2H \xrightarrow{\text{2Sia}_2\text{BH}} \xrightarrow{\text{[O]}} \underset{\overset{|}{OH}}{H_2C(CH_2)_9CO_2H} \quad 13.28$$

Such applications will be discussed further in Chapter XV.

6. Steric Control of Reductions [30]

Since their introduction, the complex hydrides, such as lithium aluminum hydride and sodium borohydride, have been exceedingly valuable for the convenient conversion of ketones to alcohols. One aspect of this application, however, has led to frustrating ambiguities. The reduction of monocyclic ketones, such as 2-methylcyclopentanone and 2-methylcyclohexanone, appears to involve preferential attack of the reagent from the side of the methyl group, presumably the more hindered direction, to yield predominantly the more stable of the two possible alcohols (*trans*). On the other hand, the reduction of bicyclic systems involves preferential approach of the reagent from the less hindered side of the carbonyl group to yield the less stable of the two possible alcohols.

Considerable attention has been paid to possible explanations to these apparently conflicting reaction preferences.[31, 32] It now appears

[27] Unpublished research with D. B. Bigley.

[28] H. C. Brown and D. B. Bigley, *J. Amer. Chem. Soc.*, **83,** 486 (1961).

[29] See fn. 9 of ref. 24 for a number of references to such applications.

[30] H. C. Brown and V. Varma, *J. Amer. Chem. Soc.*, **88,** 2871 (1966).

[31] W. G. Dauben, G. J. Fonken, and D. S. Noyce, *J. Amer. Chem. Soc.*, **78,** 2579 (1956).

[32] J. C. Richer, *J. Org. Chem.*, **30,** 324 (1965).

Table XIII–1. Reduction of Representative Ketones by Various Reducing Agents at 0°

	Less stable alcohol epimer	Percent alcohol involving reduction to form the less stable epimer						
		LiAlH$_4$ in THF	LiAlH(OMe)$_3$ in THF	BH$_3$ in THF	Sia$_2$BH in THF	(IPC)$_2$BH[a] in DG	LiBH(n-Bu)$_3$[b] in THF	LiPBPH[c] in THF
2-Methylcyclobutanone	cis	25		41	74	83		
2-Methylcyclopentanone	cis	21	44	25	80	94	67	94
2-Methylcyclohexanone	cis	25	69	26	79	94	85	97
2-Methylcycloheptanone	cis	73		74	64	98		
2-Methylcyclooctanone	cis	73		82	d			
3-Methylcyclohexanone	trans	16				35		59
3-t-Butylcyclohexanone	trans	15				28		72
4-Methylcyclohexanone	cis	17				33		52
4-t-Butylcyclohexanone	cis	8				37		54
3,3,5-Trimethylcyclohexanone	trans	82						99
Norcamphor	endo	89	98	98	92	94	98	99
Camphor	exo	92	99	52	65e	100e	98	99

[a] Diisopinocampheylborane in diglyme. [b] Lithium tri-n-butylborohydride in tetrahydrofuran. [c] Lithium perhydro-9b-boraphenalylhydride. [d] Very slow reaction. [e] Slow reaction.

that this behavior can be accounted for in terms of a preference for axial approach of the reagent in simple ring systems involving small steric effects. With increasing steric effects the course of approach of the reagent will be altered to minimize the steric interaction.[33,34] However, at present it is not possible to predict when a given system will shift from one preferred to another. Consequently, it is not possible to assign the structure of an alcohol produced in such a reduction from arguments based on the steric situation around the carbonyl group.

For example, we have pointed out that in such reductions 2-methylcyclopentanone and 2-methylcyclohexanone form the *trans* alcohols preferentially. Yet 2-methylcycloheptanone and 2-methylcyclooctanone yield the *cis* alcohol preferentially (Table XIII-1). There would be obvious advantages in a reagent that provided consistent steric control. The dialkylboranes appeared promising for this purpose. Several of them were tested, and diisopinocampheylborane appeared to be the most favorable (Table XIII-1).

All of the 2-methylcyclanones examined underwent reduction to give the *cis* alcohol predominantly. On the other hand, where the alkyl substituent is relatively remote from the reaction center, as in

[33] J. Klein, E. Dunkelblum, E. L. Eliel, and Y. Senda, *Tetrahedron Letters*, 6127 (1968).

[34] E. L. Eliel and Y. Senda, *Tetrahedron*, **26**, 2411 (1970).

3- and 4-methyl- and *t*-butylcyclohexanones, the dialkylboranes exert only a minor influence on the direction taken by the reduction—in all cases the product is predominantly the more stable of the two possible epimers. Thus it is quite clear that the 2-methyl substituent in the 2-methylcyclanones must be exerting a dominant steric effect on the direction taken in the reductions involving the highly hindered dialkylboranes.

One difficulty was observed. The reactions of diisopinocampheyl-borane with the more hindered ketones became very slow—evidently a reflection of the large steric interactions.

Fortunately, it has been possible to solve this problem. Lithium hydride reacts readily with *cis,cis,trans*-perhydro-9b-boraphenalene[35, 36] (13.29).

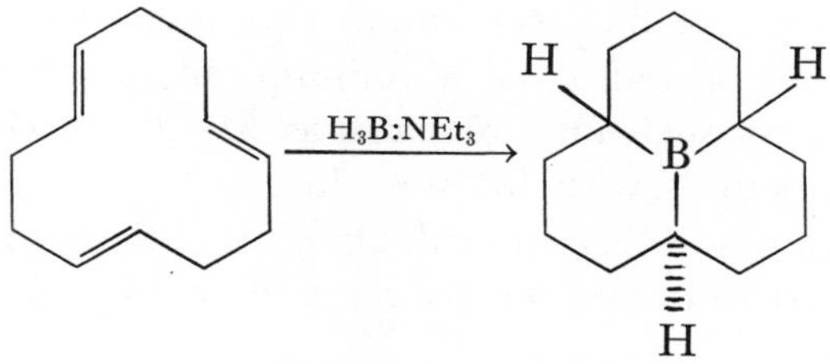

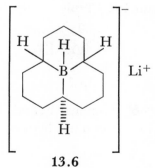

13.29

This organoborane readily reacts with lithium hydride in tetrahydrofuran to form the new reagent, lithium perhydro-9b-boraphenalylhydride[37] (**13.6**).

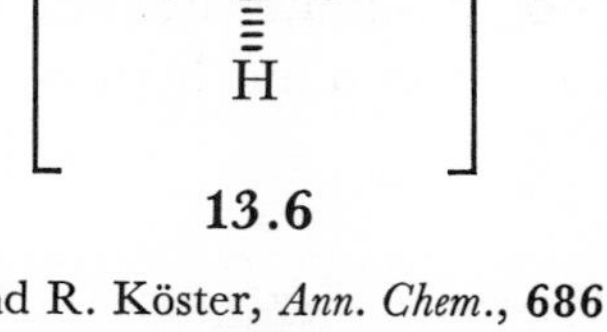

13.6

[35] G. W. Rotermund and R. Köster, *Ann. Chem.*, **686**, 153 (1965).

[36] Köster's assignment of configuration (*cis, cis, cis*) has been revised: H. C. Brown and W. C. Dickason, *J. Amer. Chem. Soc.*, **91**, 1226 (1969).

[37] H. C. Brown and W. C. Dickason, *J. Amer. Chem. Soc.*, **92**, 709 (1970).

This reagent undergoes a rapid reaction with all of the ketones examined. Even the most hindered ketone, camphor, is completely converted within one-half hour at 0°. Yet the reagent achieves steric control of the reaction at least as complete as that realized with di-isopinocampheylboranes (Table XIII-1).

7. Alane as a Reducing Agent [38]

Aluminum hydride may be prepared by treating lithium aluminum hydride in ether solution with aluminum chloride. The lithium chloride precipitates, and an ether solution of the reagent is obtained. Unfortunately, such solutions are metastable; on standing, the aluminum hydride associates and precipitates from solution. This behavior has discouraged any significant study of its reducing characteristics. Instead, the practice has been adopted to utilize solutions stabilized by the presence of additional aluminum chloride.[39] Such solutions doubtless contain such intermediates as AlH_2Cl, $AlHCl_2$, and their addition compounds with lithium chloride. Consequently, the results, although interesting and valuable, are difficult to interpret, and cannot be used to define the reducing characteristics of aluminum hydride itself.

In tetrahydrofuran aluminum hydride is stable and shows no sign of association and precipitation, such as occurs in ethyl ether. Presumably, the aluminum hydride exists as the 1:1 addition compound[40] (**13.7**). Unfortunately, lithium chloride is soluble in tetra-

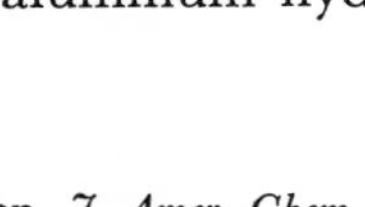

$$\text{O}:AlH_3$$

13.7

hydrofuran, so solutions of aluminum hydride in this solvent cannot be obtained using lithium aluminum hydride and aluminum chloride.

[38] H. C. Brown and N. M. Yoon, *J. Amer. Chem. Soc.*, **88**, 1464 (1966).

[39] E. E. Eliel, *Rec. Chem. Progr.*, **22**, 129 (1961).

[40] E. Wiberg and W. Gösel, *Z. Naturforsch.*, **716**, 485 (1956).

The addition of the calculated quantity of 100 percent sulfuric acid to solutions of lithium aluminum hydride in tetrahydrofuran provides a convenient synthesis of aluminum hydride. Lithium sulfate precipitates, and a clear, relatively stable solution of aluminum hydride is obtained[38] (13.30).

$$2LiAlH_4 + H_2SO_4 \xrightarrow[THF]{O°} Li_2SO_4 \downarrow + 2AlH_3 + 2H_2 \qquad 13.30$$

Aldehydes and ketones are reduced rapidly, with no complications introduced by the presence of a conjugated double bond as in cinnamaldehyde (Section 13.8). Norcamphor is reduced with good stereoselectivity, yielding 7 percent *exo-* and 93 percent *endo-*norbornanol.

Acids, acid anhydrides, acid chlorides, esters, and lactones are rapidly reduced to the alcohol stage. Epoxides are also reduced rapidly, with 1,2-butylene oxide and 1-methyl-1,2-cyclohexene oxide yielding 2-butanol and 1-methylcyclohexanol exclusively. However, styrene oxide exhibits more opening at the secondary center than is observed with lithium aluminum hydride, yielding 76 percent secondary and 24 percent primary alcohol.

Tertiary amides are reduced rapidly, primary amides more slowly. Nitriles, oximes, phenyl isocyanate, and pyridine N-oxide are all reduced rapidly. On the other hand, both aliphatic and aromatic nitro compounds, azobenzene, and azoxybenzene are relatively stable to the reagent, although pyridine undergoes slow reaction. Diphenyl disulfide and dimethyl sulfoxide are rapidly reduced, whereas sulfones, sulfonic acids, and cyclohexyl tosylate are stable under the mild reaction conditions used (0°, THF).

The results with the standard list of compounds are summarized as follows (13.31):

aldehyde	$\longrightarrow$ alcohol
ketone	$\longrightarrow$ alcohol
acid chloride	$\longrightarrow$ alcohol
lactone	$\longrightarrow$ glycol
epoxide	$\longrightarrow$ alcohol

ester	$\longrightarrow$ alcohol	13.31
carboxylic acid	$\longrightarrow$ alcohol	
carboxylic acid salt	$\longrightarrow$ alcohol	
t-amide	$\longrightarrow$ amine	
nitrile	$\longrightarrow$ amine	
nitro	$\longrightarrow$ very slow reaction	
olefin	$\longrightarrow$ no reaction	

The characteristics of aluminum hydride offer promise for certain applications; these are considered in the next section.

8. Applications of Alane [41]

The powerful reducing characteristics of aluminum hydride together with its decreased nucleophilic characteristics (relative to lithium aluminum hydride) give it promising possibilities for many selective reductions. For example, the reduction of halogen substituted carboxylic acids and esters with lithium aluminum hydride often results in the serious loss of halogen. Thus, the yield of 2-chloroethanol from such a reduction of chloroacetic acid and ester is in the range of 5 to 37 percent. Aluminum hydride gives 2-chloroethanol in a yield of 83 percent. Similarly, ethyl 3-bromobutyrate is readily reduced to 3-bromobutanol in 93 percent yield (13.32).

$$CH_3CHCH_2CO_2C_2H_5 \xrightarrow[\text{THF, } 0°]{AlH_3} CH_3CHCH_2CH_2OH \quad 13.32$$
$$\hspace{1.0cm} | \hspace{4.5cm} |$$
$$\hspace{1.0cm} Br \hspace{4.3cm} Br$$

Similarly, the reagent readily reduces a wide variety of group in the presence of nitro substituents, without attack on the latter. For example, methyl 4-nitropentanoate is converted to 4-nitropentanol in 80 percent yield (13.33).

$$CH_3CHCH_2CH_2CO_2C_2H_5 \longrightarrow CH_3CHCH_2CH_2CH_2OH \quad 13.33$$
$$\hspace{0.5cm} | \hspace{5.0cm} |$$
$$\hspace{0.5cm} NO_2 \hspace{4.7cm} NO_2$$

Reduction of oximes by lithium aluminum hydride is relatively slow, with yields from moderate to low. Reduction with aluminum

[41] N. M. Yoon and H. C. Brown, *J. Amer. Chem. Soc.*, **90**, 2927 (1968).

hydride was complete in one-half hour in refluxing tetrahydrofuran, with yields in the 90 percent range (13.34).

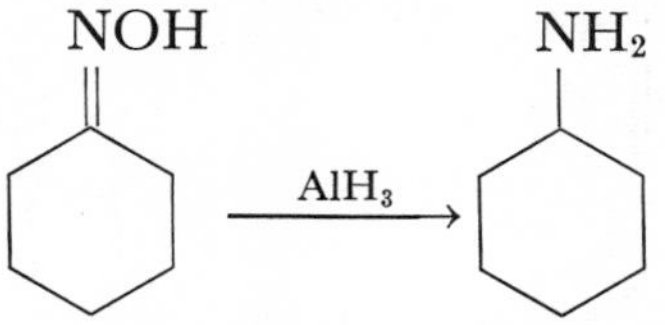

13.24

The reduction of aliphatic nitriles by lithium aluminum hydride is often complicated by concurrent attack of the base on the activated α-hydrogen atoms. This is particularly serious for molecules such as allyl cyanide and benzyl cyanide, where the α-position is doubly activated. For example, treatment of allyl cyanide with lithium aluminum hydride gives none of the desired amine. Aluminum hydride gives the amine in a yield of 83 percent (13.35). Diphenylacetonitrile gives the corresponding amine in a yield of 98 percent (13.36).

$$CH_2{=}CHCH_2CN \xrightarrow[\text{O°, THF}]{\text{AlH}_3} CH_2{=}CHCH_2CH_2NH_2 \quad 13.35$$

83%

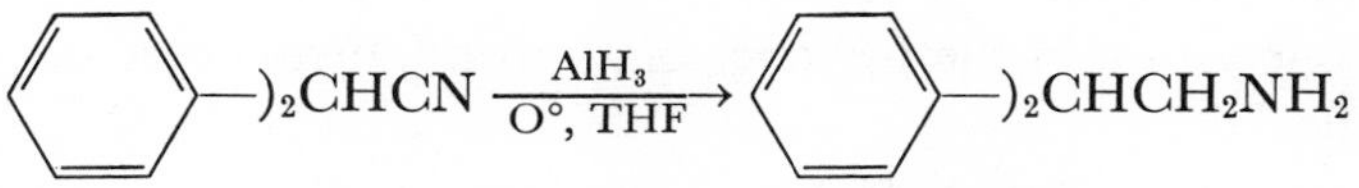

98%

There is also a major advantage in utilizing aluminum hydride for the reduction of carbonyl groups conjugated with activated double bonds.[42] For example, the reduction of α-cyclopenten-1-one (13.37) and of 5,6-dihydro-*endo*-dicyclopentadien-1-one (13.38) was examined with a wide variety of reducing agents. Aluminum hydride provided the most consistently high yield of the desired allylic alcohol.

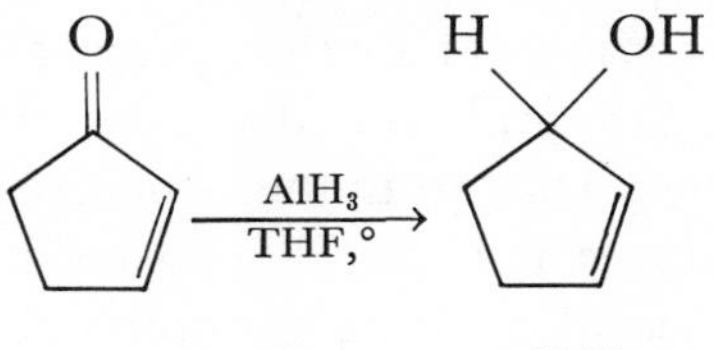

13.37

90%

[42] H. C. Brown and H. M. Hess, *J. Org. Chem.*, **34,** 2206 (1969).

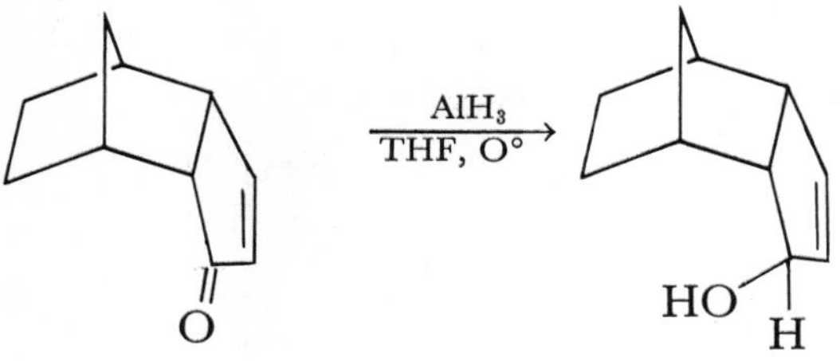

13.38

9. Dialkylalanes as Reducing Agents

A number of highly interesting individual applications of diiso-butylalane as a selective reducing agent have appeared. For example, the reduction of carboxylic acid esters to aldehydes by this reagent is an interesting development of considerable importance.[43] Unfortunately, a systematic examination of the reducing characteristics under standardized conditions does not appear to be available as yet.

10. Selective Reductions

These new reagents permit many selective reductions. Moreover, by a judicious choice of reducing agent, it is frequently possible to reduce one group in the presence of a second, or to carry out the reverse operation.

The amount of data that has accumulated in this area is appalling To assist the chemist in becoming acquainted with the major possibilities, Table XIII-2 summarizes the observed reactivities of the main functional groups toward these reagents. The symbol $(+)$ indicates a rapid reaction at 0–25°; correspondingly, the symbol $(-)$ indicates slow or insignificant reaction under these conditions.

It is evident that sodium borohydride will permit the reduction of an acid chloride group in the presence of the acid grouping, whereas diborane will reduce the ester in the presence of the acid chloride grouping. Similarly, lithium borohydride will reduce the ester group in the presence of the nitrile grouping, whereas diborane will do the opposite. With diborane it is possible to reduce a carboxylic acid group in the presence of an ester grouping. Lithium borohydride

[43] L. I. Zakharkin and I. M. Khorlina, *Tetrahedron Letters*, 619 (1962).

Table XIII–2. Summary of Behavior of Various Functional Groups toward the Hydride Reagents

	NaBH₄ in ethanol	NaBH₄ + LiCl in diglyme	NaBH₄ + AlCl₃ in diglyme	NaBH₄ + BF₃ in diglyme	BH₃ in THF	Disiamyl-borane in THF	AlH₃ in THF	LiAlH-(O-t-Bu)₃ in THF	LiAlH-(OMe)₃ in THF	LiAlH₄ in THF or ether
Aldehyde	+	+	+	+	+	+	+	+	+	+
Ketone	+	+	+	+	+	+	+	+	+	+
Acid chloride	a	+	+	+	−	−	+	+	+	+
Lactone	−	+	+	+	+	+	+	±ᶜ	+	+
Epoxide	−	+	+	+	+	±ᶜ	+	±ᶜ	+	+
Ester	−	+	+	+	±ᵇ	−	+	±ᶜ	+	+
Carboxylic acid	−	−	+	+	+	−	+	−	+	+
Carboxylic acid salt	−	−	−	−	−	−	+	−	+	+
tert-Amide	−	−	+	+	+	+	+	−	+	+
Nitrile	−	−	+	+	+	−	+	−	+	+
Nitro	−	−	+	+	−	−	−	−	+	+
Olefin	−	−	+	+	+	+	−	−	−	+

ᵃ Reacts with solvent. Reduced in nonhydroxylic solvents. ᵇ Slow reaction at variable rate, depending on structure. ᶜ Some derivatives react only very slowly, others at a moderate rate.

will do the opposite. The possibilities for numerous selective reductions are indicated by the summary. However, the reactivities of the various functional groups can be greatly altered by the structures containing them, so these generalizations must be applied with caution in predicting the behavior of greatly modified systems.

We have seen rapid progress in the development of new selective reducing agents and the exploration of their potentialities. These developments are both simplifying the synthetic problems of the organic chemist and complicating his task of selecting the preferred reagent and conditions for a given transformation. It is clearly possible that continued research in this area will make available specific functional groups in the presence of any other functional group and vice versa. As we understand better the influence of structure on reactivity in this area, hopefully we shall be able to design reducing agents to perform desired reductions, much as Nature has developed enzymes to accomplish highly specific reactions.

PART FIVE

HYDROBORATION

XIV. Hydroboration with Diborane[1]

1. What Is Hydroboration?

In 1956 and in 1957 Dr. B. C. Subba Rao and I reported that olefins can be readily and quantitatively converted into organoboranes under exceedingly mild experimental conditions, providing a major new route to these promising derivatives[2,3] (14.1).

$$9RCH{=}CH_2 + 3NaBH_4 + AlCl_3 \longrightarrow$$
$$3(RCH_2CH_2)_3B + AlH_3 + 3NaCl$$
$$12RCH{=}CH_2 + 3NaBH_4 + 4BF_3 \longrightarrow$$
$$4(RCH_2CH_2)_3B + 3NaBF_4 \qquad 14.1$$
$$6RCH{=}CH_2 + B_2H_6 \longrightarrow$$
$$2(RCH_2CH_2)_3B$$

Not only does the boron-hydrogen bond add rapidly and quantitatively to carbon-carbon double and triple bonds, but as we have seen (Chapter XIII) it adds also with remarkable ease to carbon-oxygen double bonds and to carbon-nitrogen double and triple bonds (14.2).

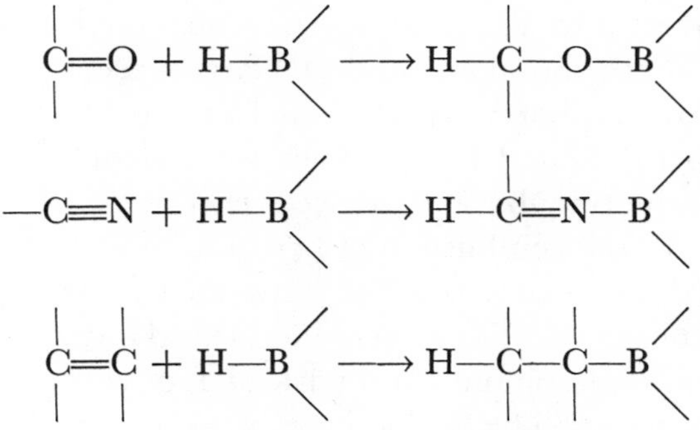

$$\overset{|}{\underset{|}{C}}{=}O + H{-}B\diagup_{\diagdown} \longrightarrow H{-}\overset{|}{\underset{|}{C}}{-}O{-}B\diagup_{\diagdown}$$

$$-C{\equiv}N + H{-}B\diagup_{\diagdown} \longrightarrow H{-}\overset{|}{C}{=}N{-}B\diagup_{\diagdown} \qquad 14.2$$

$$\overset{|}{\underset{|}{C}}{=}\overset{|}{\underset{|}{C}} + H{-}B\diagup_{\diagdown} \longrightarrow H{-}\overset{|}{\underset{|}{C}}{-}\overset{|}{\underset{|}{C}}{-}B\diagup_{\diagdown}$$

[1] H. C. Brown, *Hydroboration*, (W. A. Benjamin, New York, 1962).
[2] H. C. Brown and B. C. Subba Rao, *J. Amer. Chem. Soc.*, **78**, 5694 (1956).
[3] H. C. Brown and B. C. Subba Rao, *J. Org. Chem.*, **22**, 1136 (1957).

Hydroboration

Consequently, the addition of the boron-hydrogen linkage to multiple bonds between carbon and carbon, nitrogen, or oxygen appears to be a reaction of very wide generality—as general as the addition of hydrogen, but proceeding under much milder conditions. The term hydrogenation is used for this general addition of hydrogen to multiple bonds. The related term hydroboration, with the symbol HB, has been proposed for this equally general reaction. In practice, however, there has been a tendency to restrict application of the term only to the last of the three processes, the formation of organoboranes from structures containing double or triple bonds (14.3).

$$\overset{|}{\underset{|}{C}}{=}\overset{|}{\underset{|}{C}} + H{-}B\diagup_{\diagdown} \longrightarrow H{-}\overset{|}{\underset{|}{C}}{-}\overset{|}{\underset{|}{C}}{-}B\diagup_{\diagdown} \qquad 14.3$$

(Sometimes one's parents seem prescient in choosing his initials.—H. C. B.)

2. The Discovery

Early observations on the reaction of diborane with olefins indicated that the reaction was quite slow, requiring elevated temperatures and long reaction periods.[4] For example, in a typical experiment ethylene containing 2 percent by weight of diborane was heated under pressure in heavy-walled Pyrex tubes for 4 days at 100°. There was isolated a small amount of a clear liquid that flamed spontaneously on exposure to air. Mass spectroscopic analysis indicated the presence of triethylborane. Similarly large excesses of isobutylene were heated with diborane at 100° for 24 hours. Two fractions were obtained, bp 181.5° and 188.5°. They were identified as tri-*t*-butylborane and triisobutylborane, respectively.

In a study of the combustion of hydrocarbons induced by aluminum borohydride, a slow reaction between 1-butene and the borohydride was observed.[5] This observation led to a study of the kinetics of the reaction of aluminum borohydride vapor with ethylene, propylene, and 1-butene.[5,6] The reaction was first-order in aluminum boro-

[4] D. T. Hurd, *J. Amer. Chem. Soc.*, **70**, 2053 (1948).

[5] R. S. Brokaw, E. J. Badin, and R. N. Pease, *J. Amer. Chem. Soc.*, **72**, 1793 (1950).

[6] R. S. Brokaw and R. N. Pease, *J. Amer. Chem. Soc.*, **72**, 5263 (1950).

hydride and independent of the olefin concentration. At 140° the formation of triethylborane and an ethylaluminum hydride was noted (14.4).

$$Al(BH_4)_3 + 11C_2H_4 \xrightarrow{140°} 3(C_2H_5)_3B + \tfrac{1}{2}(C_2H_5)_4Al_2H_2 \qquad 14.4$$

Finally, a kinetic study was made of the reaction of diborane with ethylene in the temperature range 120 to 175°.[7]

These results certainly were not promising for achieving a fast quantitative addition of diborane to olefins in ether solvents, at 0°, complete in a matter of seconds. How was the discovery made?

In line with one of the special objectives of this book (Section I-1), I shall attempt to describe the actual occurrences that led to this development.

Dr. B. C. Subba Rao received his Ph.D. degree from Purdue University in 1955, working with me, and decided to remain with me for two years of postdoctoral work before returning to India. He was in the midst of a study of the reducing powers of sodium borohydride, enhanced by aluminum chloride (Section XII-6). In these studies we undertook to define the properties of each new reagent by running standard test experiments with representative organic compounds. He would add a standard solution of the reagent (containing about 10 mmoles of $NaBH_4$ and 3.3 mmoles of $AlCl_3$ in diglyme) to the compound (about 10 mmoles) under test. After a standard interval, usually 1 hour at 25°, aqueous hydrochloric acid was added to transform the residual "hydride" into hydrogen, which was collected and measured. At the same time a blank was run (the standard solution of the reagent without the compound under examination). The decrease in the mmoles of hydrogen obtained from that realized in the blank determination gave the moles of hydride used for reduction of the compound.

In this manner he established that representative aldehydes and ketones utilized one "hydride" per mole. Obviously, reduction was occurring to the alcohol stage. Nitriles utilized 2 moles of "hydride," corresponding to reduction to the amine. Esters, such as ethyl acetate

---

[7] A. T. Whatley and R. N. Pease, *J. Amer. Chem. Soc.*, **76**, 835 (1954).

and ethyl stearate, revealed the uptake of 2 moles of hydride. Evidently, reduction to the alcohol was occurring here also. One of the compounds Dr. Subba Rao examined was ethyl oleate. It showed a "hydride" uptake of 2.4 moles per mole of compound.

Dr. Subba Rao was an exceedingly productive postdoctoral researcher. I recall his bringing to my office his summary of results—tables and tables of data. All of the results appeared simple and easy to interpret, but the value for ethyl oleate stuck out like a sore thumb.

I asked how we might account for this nonstoichiometric value. Dr. Subba Rao ventured that impurity of the ethyl oleate he had used might have caused the peculiar result. The ethyl oleate he had obtained from the stockroom had been somewhat discolored, but he had used it anyhow for this first exploratory experiment. He suggested that the material might contain sufficient impurities, possibly peroxides, to account for the high results.

We discussed what should be done. Dr. Subba Rao was of the opinion that ethyl oleate was not really pertinent to our study—we already had a wide range of esters that established an uptake of 2.0 hydrides per mole. He recommended that we drop ethyl oleate. Its addition to the list had really been an afterthought.

Fortunately, the research director is in an enviable position to insist on high standards—he does not have to do the actual experimental work! I persuaded Dr. Subba Rao to return to the bench and repeat the experiment with purified ethyl oleate.

The repetition yielded the same result! The original sample of ethyl oleate had been of satisfactory purity. It required only a little more effort to establish that we were achieving the simultaneous reduction and hydroboration of ethyl oleate.

But we were still not home safe. It was a tortuous path to the final recognition that diborane adds to olefins at $0°$ with remarkable speed in ether solvents.

We argued that the successful hydroboration of olefins must involve aluminum borohydride. We prepared aluminum borohydride, placed it in diglyme, and demonstrated that it converted olefins into organoboranes (14.5).

$$9RCH{=}CH_2 + Al(BH_4)_3 \xrightarrow{DG} 3(RCH_2CH_2)_3B + AlH_3 \qquad 14.5$$

Note that aluminum hydride is a product of this reaction. It had been previously established that diborane adds to aluminum hydride to form aluminum borohydride.[8] Clearly we had a two-stage process for adding diborane to olefins under mild conditions (14.6).

$$9RCH{=}CH_2 + Al(BH_4)_3 \xrightarrow{\text{DG}} 3(RCH_2CH_2)_3B + AlH_3$$

$$AlH_3 + \tfrac{3}{2} B_2H_6 \xrightarrow{\text{DG}} Al(BH_4)_3 \qquad\qquad 14.6$$

It then occurred to us that we might develop this into a catalytic process. All we had to do was place the olefin in the ether solvent together with a small catalytic quantity of aluminum hydride. Then, diborane could be introduced. They would form aluminum borohydride, and the latter would react with the olefin present to form the desired organoborane, regenerating the aluminum hydride (14.6). This catalytic process was tested. It worked perfectly.

We were on the point of submitting a publication reporting this new catalytic process, when we decided to run a blank experiment. We knew that diborane could not possibly react with olefins under these mild conditions. The previous work made that quite clear.[4-7] However, we thought it dramatic to include an experiment showing the remarkable change in the rate of reaction of diborane with olefins achieved by small catalytic quantities of aluminum hydride. We tried the experiment. The addition of the diborane to the olefin (14.7) proceeded just as rapidly in the absence of the aluminum hydride!

$$6RCH{=}CH_2 + B_2H_6 \longrightarrow 2(RCH_2CH_2)_3B \qquad 14.7$$

Investigation soon revealed the reaction to be rapid, quantitative, and of wide generality. A number of attempts were made to measure the rate, but without success. The reaction appeared to be complete as fast as the two reagents could be brought into contact. In its speed, generality, and quantitative nature, the reaction resembles the addition of bromine to simple olefins.

These observations were exceedingly puzzling in view of the earlier reports. Accordingly, we undertook a reexamination of the reaction of olefins with diborane in the high-vacuum apparatus. It was ob-

[8] E. Wiberg, *Angew. Chem.*, **65,** 16 (1953).

served that the reaction of diborane with pure olefins is indeed quite slow, extending over periods of many hours. However, the addition is markedly catalyzed by ethers and similar weak bases. Addition of mere traces of ethers changed the initially slow reaction to a fast one —too fast to measure by the usual manometric techniques.[9] Consequently, the inadvertent use of the ether solvents for the reactions involving sodium borohydride appears to have been largely responsible for the fast, quantitative hydroboration procedures.

How is it possible that a reaction of such generality was not uncovered and utilized long ago?

As I look back from my present knowledge, it appears to me that I should have recognized the reaction much earlier. In 1936 I began my doctoral work in diborane chemistry. I was cautioned not to use stopcocks in any part of the vacuum line exposed to diborane gas. Stopcocks so exposed became very gummy and subsequently froze. At that time the stopcock greases had a rubber base and contained carbon-carbon double bonds. Later, when Apiezon greases, prepared from a petroleum base, became available, the difficulty disappeared. I recall thinking about the phenomenon, but did not put two and two together and recognize that diborane must be cross-linking the unsaturated centers in the rubber-base grease.

Again, some time later, we utilized trimethylborane as a means of obtaining thermodynamic data for the dissociation of addition compounds[10] (Section V-4). We encountered no difficulty in purifying and storing trimethylborane. However, later when I attempted to utilize triethylborane for similar studies, I encountered major difficulties. A sample of triethylborane would be purified thoroughly until it exhibited constant vapor pressure. It would be placed in storage while other reactants were purified. But when the time came to use it, the material was impure. It appeared to evolve small amounts of ethylene. Today, I should interpret the reaction as involving a minute amount of a reversible dissociation of the triethylborane into diethylborane and ethylene (14.8).

[9] H. C. Brown and L. Case, unpublished observations.

[10] H. C. Brown, M. D. Taylor, and M. Gerstein, *J. Amer. Chem. Soc.*, **66,** 431 (1944).

$$(C_2H_5)_3B \rightleftharpoons (C_2H_5)_2BH + C_2H_4 \qquad 14.8$$

Finally, triethylborane was used only in qualitative estimates of the base strengths of amines.[11]

When one is young, he is impatient with these little discrepancies and tends to hurry past them to the main objective. Gaining experience, he begins to appreciate their importance, but increasing professional obligations tend to reduce his direct contact with experimental observations, where there is an opportunity to recognize these nuggets.

3. Hydroboration Procedures [12]

The most convenient procedure today for achieving hydroboration is to add boron trifluoride etherate to a suspension of sodium borohydride in a solution of the unsaturated compound in tetrahydrofuran (14.9)

$$12RCH{=}CH_2 + 3NaBH_4 + 4BF_3{:}O(C_2H_5)_2 \xrightarrow[0° \text{ or } 25°]{THF}$$
$$4(RCH_2CH_2)_3B + 3NaBF_4 \downarrow + 4(C_2H_5)_2O \qquad 14.9$$

If the commercially available solution of borane in tetrahydrofuran is available, this provides a particularly simple procedure. In this case the hydroboration is achieved simply by adding the equivalent quantity of the borane solution to the unsaturated compound that is undergoing hydroboration (14.10).

$$3RCH{=}CH_2 + BH_3 \xrightarrow{0°} (RCH_2CH_2)_3B \qquad 14.10$$

Or the reverse procedure can be followed.

With the background acquired through our study of routes to diborane and borohydrides (Chapter IV), it was possible to develop alternate procedures that can circumvent the use of any particular solvent, or any specific hydride reagent. Thus, in place of sodium borohydride or lithium borohydride, one can utilize the simple hydrides (14.11).

[11] H. C. Brown, *J. Amer. Chem. Soc.*, **67**, 374 (1945).

[12] H. C. Brown, K. J. Murray, L. J. Murray, J. A. Snover, and G. Zweifel, *J. Amer. Chem. Soc.*, **82**, 4233 (1960).

Hydroboration

$$3RCH{=}CH_2 + 3LiH + BF_3 \longrightarrow$$
$$(RCH_2CH_2)_3B + 3LiF \qquad 14.11$$
$$3RCH{=}CH_2 + 3NaH + 4BF_3 \longrightarrow$$
$$(RCH_2CH_2)_3B + 3NaBF_4$$

Lithium aluminum hydride can be used[13] (14.12).

$$12RCH{=}CH_2 + 3LiAlH_4 + 4BF_3 \longrightarrow$$
$$4(RCH_2CH_2)_3B + 3LiAlF_4 \quad 14.12$$

Finally, at elevated temperatures the amine boranes can be used[14] (14.13).

$$3RCH{=}CH_2 + R_3'N{:}BH_3 \xrightarrow{\ 100°\ } (RCH_2CH_2)_3B + R_3'N \quad 14.13$$

4. Scope of the Hydroboration Reaction [15]

The hydroboration reaction has been shown to be widely applicable. Alkenes containing two, three, or four alkyl substituents on the double bond readily undergo hydroboration. Cyclic and bicyclic olefins, such as 1,2-dimethylcyclopentene and α-pinene, readily react. Aryl groups on the double bond are accommodated, as in 1-phenylcyclohexene, 1,1-diphenylethylene, *trans*-stilbene, and triphenylethylene. Finally, a number of steroids with highly hindered double bonds have been demonstrated to react.[16,17] Consequently, the hydroboration reaction appears to be fully as applicable as the addition of bromine to the carbon-carbon double bond.

The great majority of olefins undergo complete reaction to form the corresponding trialkylborane (14.14). However, in the case of olefins with a large degree of steric hindrance, the reaction appears to proceed rapidly only to the dialkylborane (14.15) or monoalkylborane stage[15] (14.16).

---

[13] S. Wolfe, M. Nussim, Y. Mazur, and F. Sondheimer, *J. Org. Chem.*, **24,** 1034 (1958).

[14] R. Köster, *Angew. Chem.*, **68,** 684 (1957).

[15] H. C. Brown and B. C. Subba Rao, *J. Amer. Chem. Soc.*, **81,** 6428 (1959).

[16] W. J. Wechter, *Chem. and Ind.* (London), 294 (1959).

[17] M. Nussim, Y. Mazur, and F. Sondheimer, *J. Org. Chem.*, **29,** 1120, 1131 (1964).

$$3 \; \underset{\underset{H}{|}}{\overset{\overset{H_3C}{|}}{C}} = \underset{\underset{H}{|}}{\overset{\overset{CH_3}{|}}{C}} \; + BH_3 \longrightarrow H - \underset{\underset{H}{|}}{\overset{\overset{H_3C}{|}}{C}} - \underset{\underset{H}{|}}{\overset{\overset{CH_3}{|}}{C}} -)_3 B \qquad 14.14$$

$$2 \; \underset{\underset{H_3C}{|}}{\overset{\overset{H_3C}{|}}{C}} = \underset{\underset{H}{|}}{\overset{\overset{CH_3}{|}}{C}} \; + BH_3 \longrightarrow H - \underset{\underset{H_3C}{|}}{\overset{\overset{H_3C}{|}}{C}} - \underset{\underset{H}{|}}{\overset{\overset{CH_3}{|}}{C}} -)_2 BH \qquad 14.15$$

$$\underset{\underset{H_3C}{|}}{\overset{\overset{H_3C}{|}}{C}} = \underset{\underset{CH_3}{|}}{\overset{\overset{CH_3}{|}}{C}} \; + BH_3 \longrightarrow H - \underset{\underset{H_3C}{|}}{\overset{\overset{H_3C}{|}}{C}} - \underset{\underset{CH_3}{|}}{\overset{\overset{CH_3}{|}}{C}} - BH_2 \qquad 14.16$$

This partial alkylation of diborane makes available a series of monoalkylboranes and dialkylboranes of interest as selective hydroborating agents, as discussed later (Chapter XV).

5. Directive Effects [18, 19]

The hydroboration of simple terminal olefins, such as 1-hexene, proceeds to place the boron predominantly on the terminal position. Consequently, the hydroboration of such olefins, followed by the oxidation *in situ* of the resulting organoborane, provides a very convenient procedure for the anti-Markovnikov hydration of double bonds[18] (14.17).

$$RCH=CH_2 \xrightarrow{HB} RCH_2CH_2B \overset{/}{\underset{\backslash}{}} \xrightarrow{[O]} RCH_2CH_2OH \qquad 14.17$$

A more detailed study of directive effects in the hydroboration of representative olefins[19] reveals that the simple 1-alkenes, such as

[18] H. C. Brown and B. C. Subba Rao, *J. Amer. Chem. Soc.*, **81**, 6423 (1959).
[19] H. C. Brown and G. Zweifel, *J. Amer. Chem. Soc.*, **82**, 4708 (1960).

1-pentene and 1-hexene, react to place 94 percent of the boron on the terminal position, 6 percent becoming attached at the 2-position (**14.1**). This distribution is not influenced significantly by branching of the alkyl group (**14.2**).

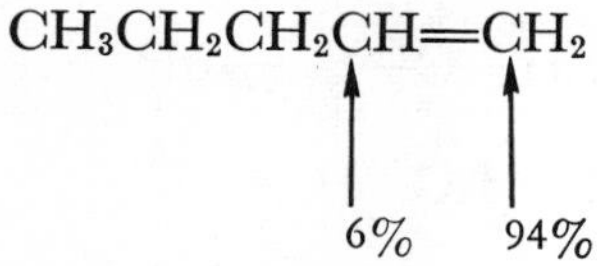

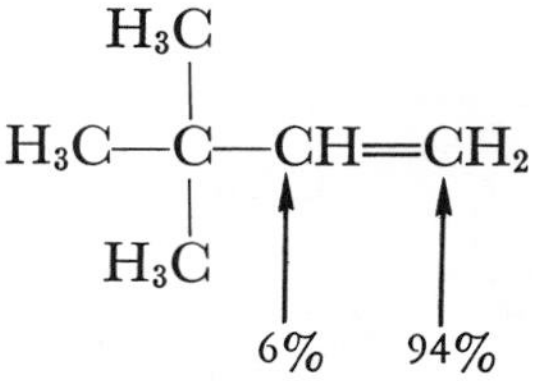

An aryl substituent, as in styrene, causes increased substitution in the nonterminal position (**14.3**).

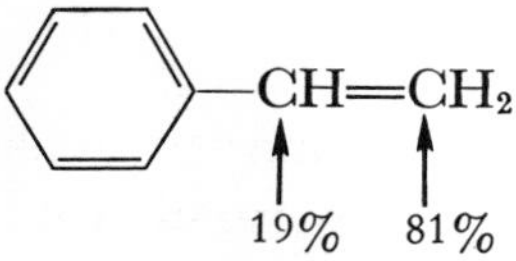

It is interesting that the distribution can be altered considerably by substituents in the *para* position. For example, the presence of a *p*-methoxy group decreases this to 7 percent (**14.4**), whereas a *p*-chloro substituent raises it to 27 percent[20] (**14.5**).

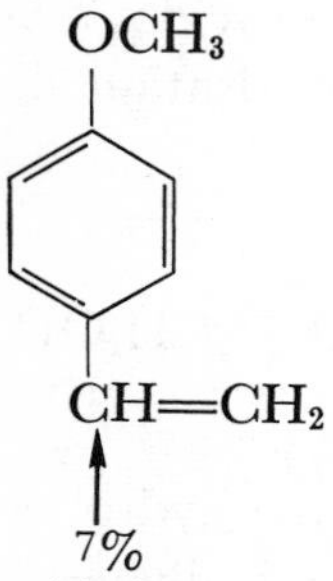

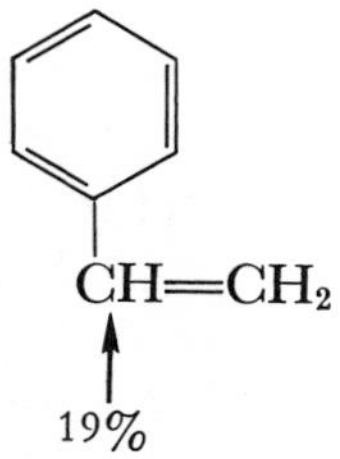

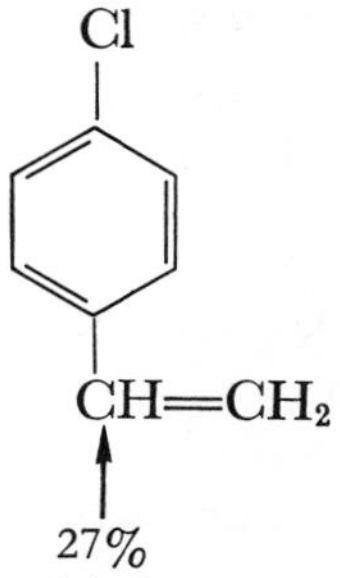

[20] H. C. Brown and R. L. Sharp, *J. Amer. Chem. Soc.*, **88**, 5851 (1966).

An alkyl substituent in the 2-position favors attachment of the boron to the terminal carbon atom[19] (**14.6**).

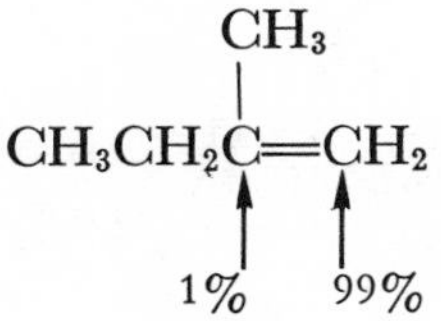

$$CH_3CH_2\underset{\underset{1\%}{\uparrow}}{\overset{\overset{CH_3}{|}}{C}}=\underset{\underset{99\%}{\uparrow}}{CH_2}$$

14.6

A similar preference for the less substituted position is demonstrated in internal olefins (**14.7, 14.8**).

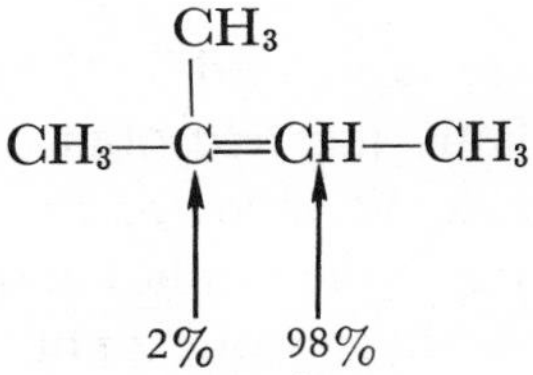

$$CH_3-\underset{\underset{2\%}{\uparrow}}{\overset{\overset{CH_3}{|}}{C}}=\underset{\underset{98\%}{\uparrow}}{CH}-CH_3$$

14.7

$$H_3C-\underset{\underset{2\%}{\uparrow}}{\overset{\overset{CH_3}{|}}{C}}=\underset{\underset{98\%}{\uparrow}}{CH}-\overset{\overset{CH_3}{|}}{\underset{\underset{CH_3}{|}}{C}}-CH_3$$

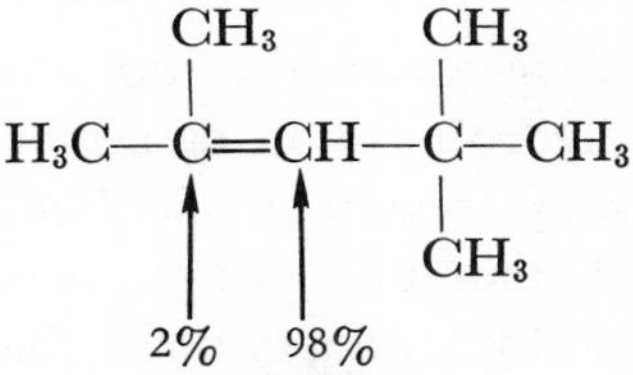

14.8

There is no significant discrimination between the two positions of an internal olefin containing groups of markedly different steric requirements (**14.9**). (However, as discussed later, it has been possible

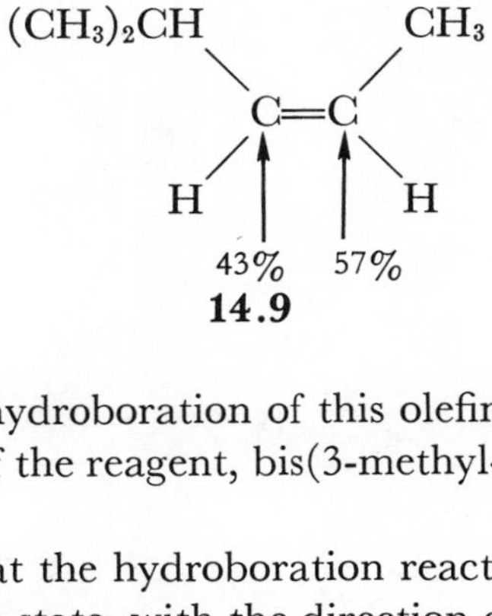

$$\underset{H}{\overset{(CH_3)_2CH}{\diagdown}}\underset{\underset{43\%}{\uparrow}}{C}=\underset{\underset{57\%}{\uparrow}}{C}\underset{H}{\overset{CH_3}{\diagup}}$$

14.9

to achieve selective hydroboration of this olefin at the less hindered position by the use of the reagent, bis(3-methyl-2-butyl)borane, disiamylborane.)

It is concluded that the hydroboration reaction involves a simple four-center transition state, with the direction of addition controlled primarily by the polarization of the boron-hydrogen bond, $\overset{\delta+}{\underset{}{B}}\!\!-\!\!\overset{\delta-}{H}$ (14.18).

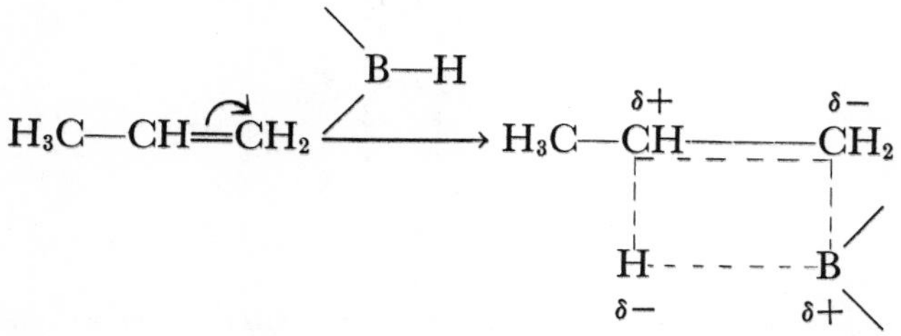

$$H_3C-CH\!\!=\!\!CH_2 \xrightarrow{\quad\overset{\diagdown}{B-H}\diagup\quad} \begin{array}{ccc} & \overset{\delta+}{H_3C-CH} & \cdots\overset{\delta-}{\cdots}CH_2 \\ & | & | \\ & \underset{\delta-}{H}\cdots\cdots\underset{\delta+}{B} \end{array} \qquad 14.18$$

6. Stereochemistry of Hydroboration [21]

This four-center transition state is likewise supported by the results of the hydroboration of cyclic olefins. Thus the hydroboration of 1-methylcyclopentene and 1-methylcyclohexene, followed by oxidation with alkaline hydrogen peroxide, results in the formation of pure *trans*-2-methylcyclopentanol and *trans*-2-methylcyclohexanol. Since the hydrogen peroxide oxidation evidently proceeds with retention of configuration, the hydroboration reaction must involve a *cis* addition of the hydrogen-boron bond to the double bond of the cyclic olefin (14.19, 14.20).

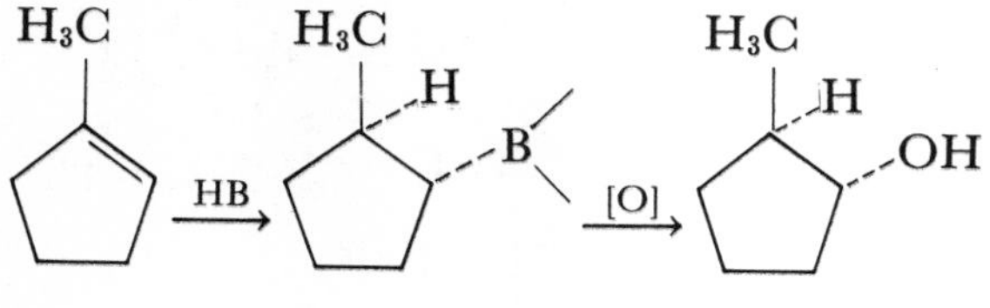

14.19

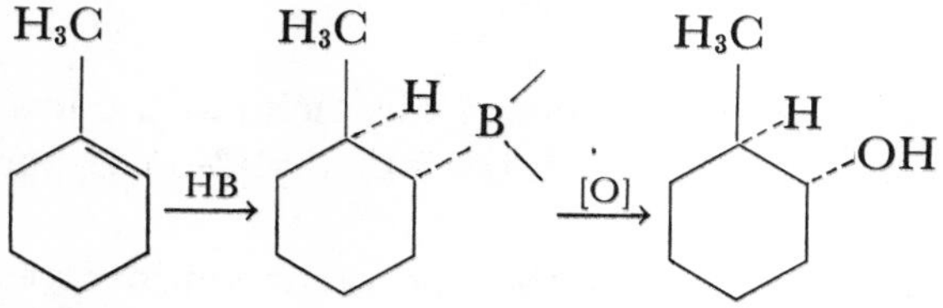

14.20

[21] H. C. Brown and G. Zweifel, *J. Amer. Chem. Soc.*, **81**, 247 (1959).

The *cis* hydration has been utilized to achieve a convenient synthesis of diastereomeric alcohols[22] (14.21, 14.22).

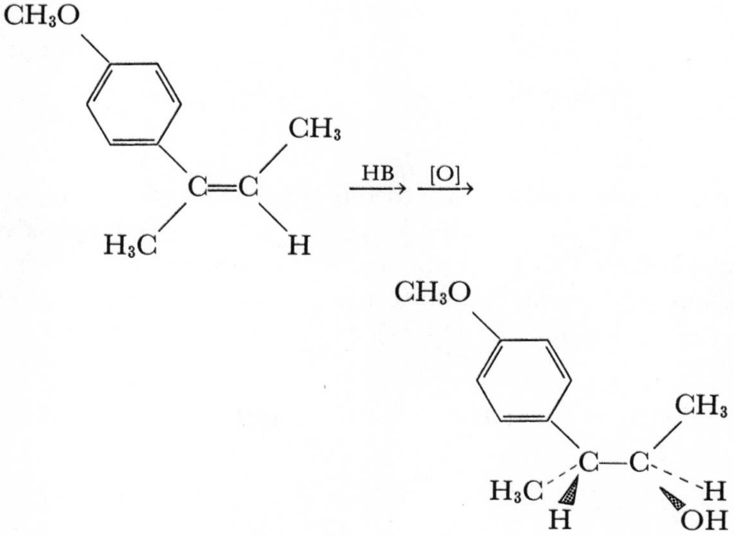

14.21

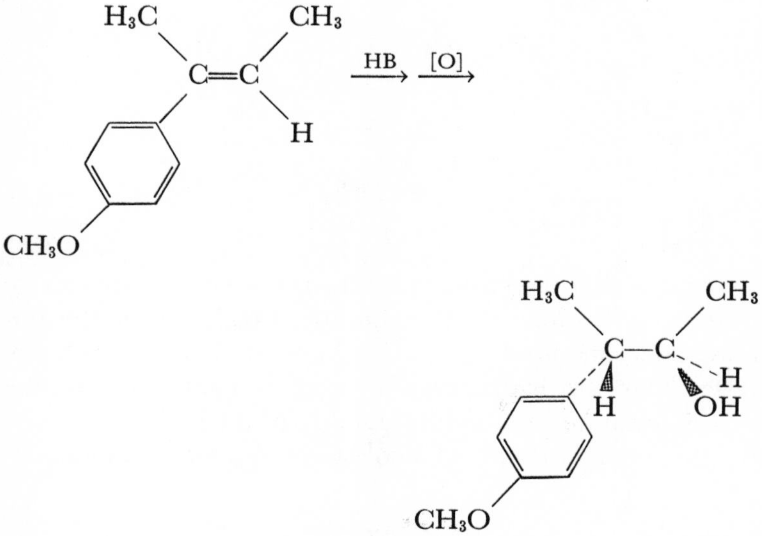

14.22

[22] E. L. Allred, J. Sonnenberg, and S. Winstein, *J. Org. Chem.*, **25,** 26 (1960).

The hydroboration of norbornene proceeds to give *exo*-2-norbornanol almost exclusively[21,23] (14.23).

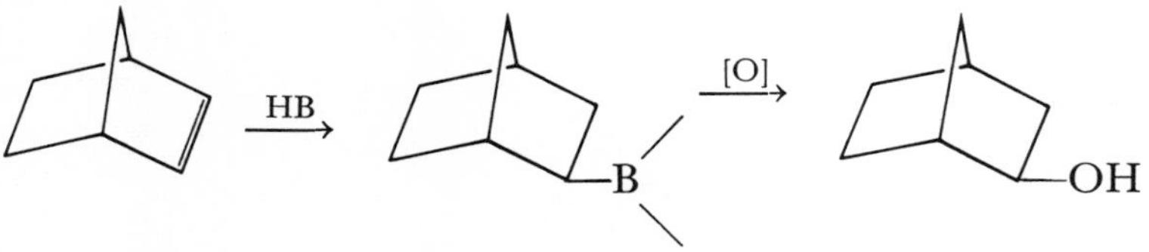

14.23

The generalization has been proposed that hydroboration proceeds by way of a *cis* addition, preferentially from the less hindered side of the double bond.[21] This generalization is now supported by a large number of independent observations. For example, α-pinene is converted into isopinocampheol[21] (14.24) and β-pinene into *cis*-myrtanol[24] (14.25).

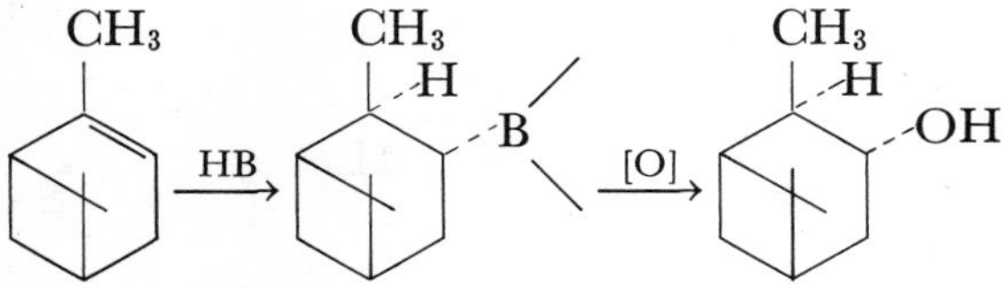

14.24

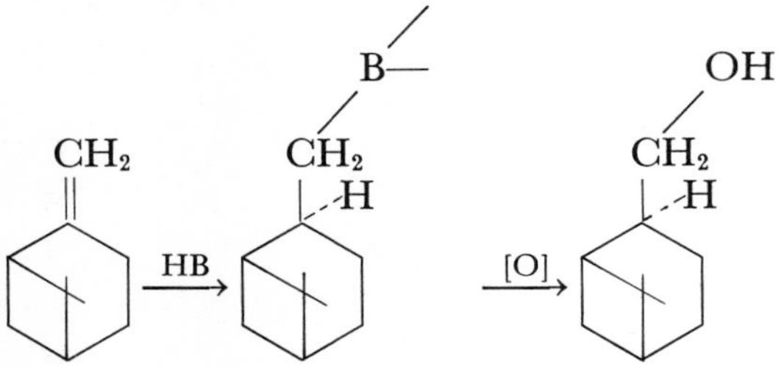

14.25

It is interesting that heat isomerizes the borane from β-pinene to the less hindered tris-*trans*-myrtanylborane. Oxidation of the isomerized organoborane produces *trans*-myrtanol.[25] This result clearly demonstrates that the hydroboration stage is controlled by the reaction mechanism and not by the stability of the product.

Finally, the hydroboration of cholesterol produces cholestane-

[23] H. C. Brown and J. H. Kawakami, *J. Amer. Chem. Soc.*, **92,** 1990 (1970).

[24] H. C. Brown and G. Zweifel, *J. Amer. Chem. Soc.*, **83,** 2544 (1961).

[25] J. C. Braun and G. S. Fisher, *Tetrahedron Letters*, 9 (1960).

$3\beta,6\alpha$-diol preferentially, clearly the result of a preferential *cis* addition from the less hindered underside of the molecule[16] (14.26).

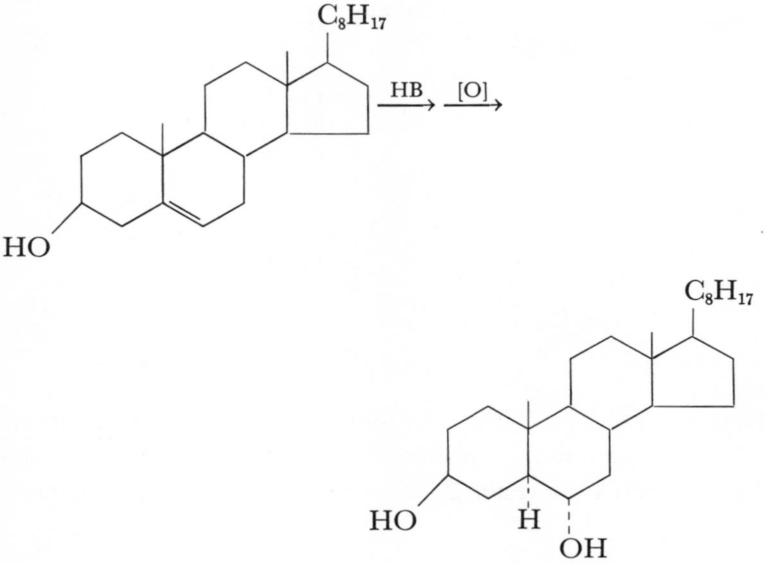

14.26

The clean stereochemical results achieved with the hydroboration reaction have made this synthetic route of major value for the chemist interested in the synthesis of stereochemically defined products.

7. Hydroboration of Hindered Olefins [26]

It was pointed out earlier that the reaction of hindered olefins with diborane appears to proceed rapidly only to the dialkylborane stage in some cases, and to the monoalkylborane stage in others. A detailed study of the hydroboration of hindered olefins has demonstrated that the reaction can be utilized for the synthesis of a number of mono- and dialkylboranes in high purity and essentially quantitative yield.

Thus the reaction of 2-methyl-2-butene at 0° readily produces bis(3-methyl-2-butyl)borane (disiamylborane) (14.27), and the reaction of the more hindered olefin, 2,4,4-trimethyl-2-pentene, is readily controlled to form 2,2,4-trimethyl-3-pentylborane (14.28).

[26] H. C. Brown and A. W. Moerikofer, *J. Amer. Chem. Soc.*, **84**, 1478 (1962).

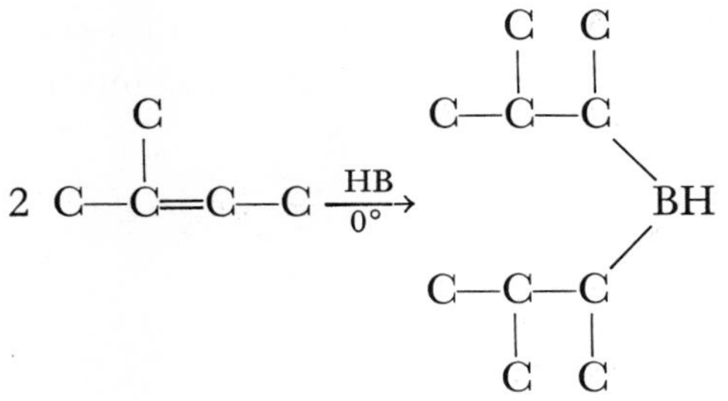

$$2 \underset{\underset{\text{C}}{|}}{\text{C}-\text{C}}=\text{C}-\text{C} \xrightarrow[0°]{\text{HB}}$$

14.27

$$\underset{\underset{\text{C}}{|}}{\text{C}}-\underset{\underset{\text{C}}{|}}{\text{C}}=\text{C}-\underset{\underset{\text{C}}{|}}{\text{C}}-\text{C} \xrightarrow[0°]{\text{HB}}$$

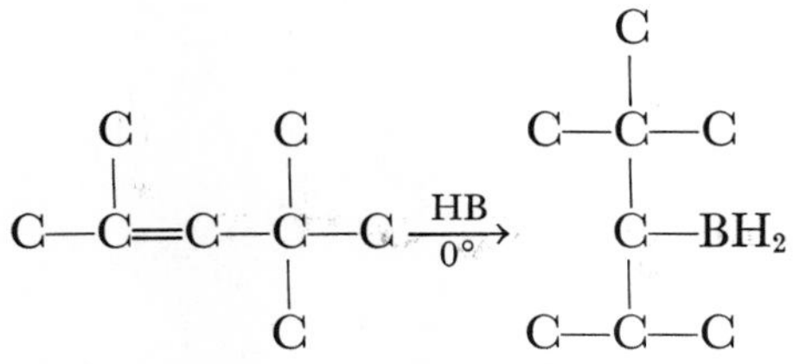

14.28

Similarly, the reaction of the tetrasubstituted olefin, 2,3-dimethyl-2-butene, can be directed quantitatively to the formation of the corresponding monoalkylborane, 2,3-dimethyl-2-butylborane (thexylborane) (14.29).

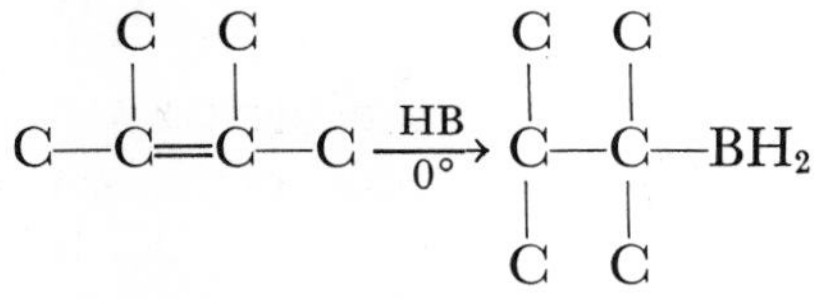

$$\underset{\underset{\text{C}}{|}}{\text{C}}-\underset{\underset{\text{C}}{|}}{\text{C}}=\underset{\text{C}}{\text{C}}-\text{C} \xrightarrow[0°]{\text{HB}} \underset{\underset{\text{C}}{|}}{\text{C}}-\underset{\underset{\text{C}}{|}}{\text{C}}-\text{BH}_2$$

14.29

In the case of the simple cyclic olefins, the reaction of cyclopentene proceeds rapidly to the tricyclopentylborane stage. However, the lower reactivity of cyclohexene, as well as the insolubility of the intermediate, dicyclohexylborane, permits the ready synthesis of this derivative (14.30).

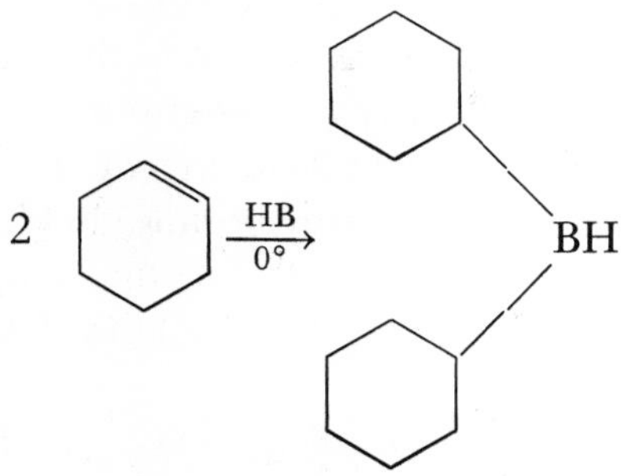

14.30

Both 1-methylcyclohexene (14.31) and α-pinene (14.32) are easily converted into the corresponding dialkylboranes.

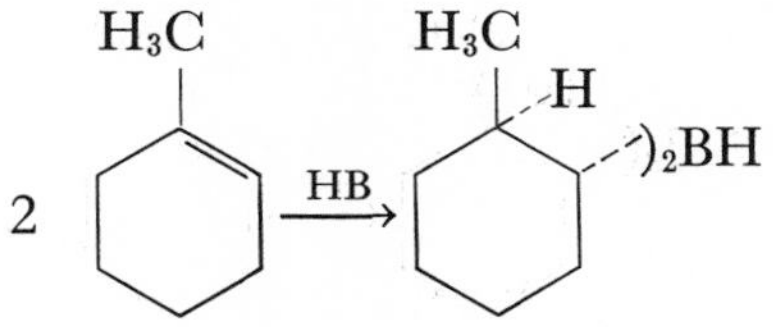

14.31

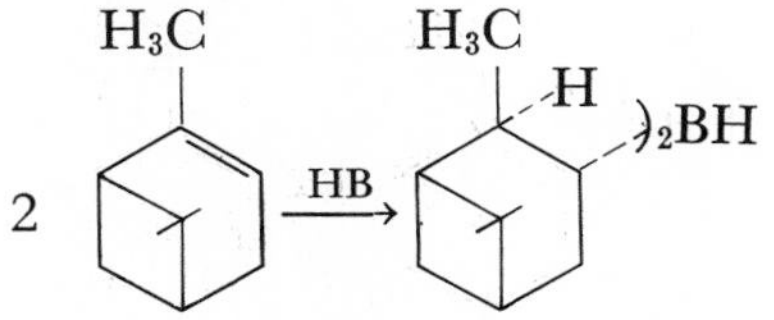

14.32

The reaction with α-pinene makes available an asymmetric hydroborating agent (Chapter XV).

Finally, the reaction of 2-methyl-1-propene, 2-methyl-1-butene, 2,4,4-trimethyl-1-pentene, and β-pinene with borane in tetrahydrofuran in the 2:1 molar ratio at 0° gives 60 to 70 percent of the dialkylboranes, 10 percent of free borane, and 20 percent of the corresponding trialkylborane (14.33).

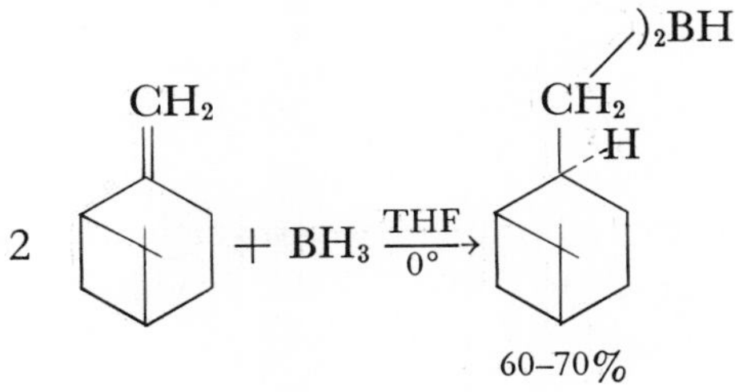

14.33

60–70%

8. Alkylboranes [27]

The ready availability of these mono- and dialkylboranes via the hydroboration reaction prompted a detailed study of their physical and chemical properties. Both molecular-weight determinations (in ethyl ether and tetrahydrofuran) and infrared examination reveal that both disiamylborane (**14.10**) and monothexylborane (**14.11**)

[27] H. C. Brown and G. J. Klender, *Inorg. Chem.*, **1**, 204 (1962).

exist as the dimers. Consequently, they should be considered derivatives of diborane.

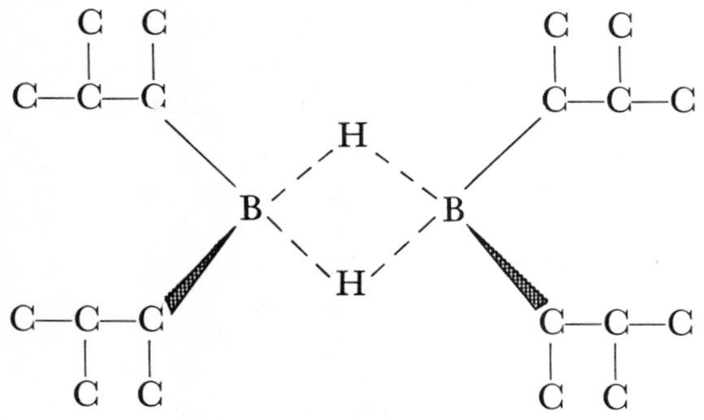

14.10

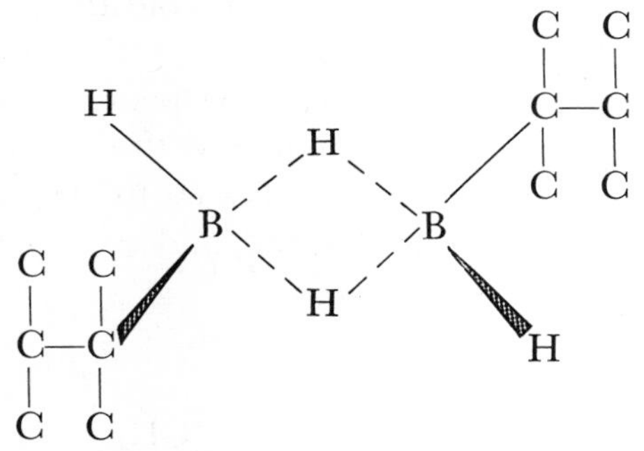

14.11

Both *sym*-tetrasiamyldiborane and *sym*-dithexyldiborane add diborane reversibly at 0° to form 1,1-disiamyldiborane (14.34) and monothexyldiborane (14.35), respectively.

$$(Sia_2BH)_2 + (BH_3)_2 \rightleftharpoons 2Sia_2BH_2BH_2 \qquad 14.34$$
$$(t\text{-}HexBH_2)_2 + (BH_3)_2 \rightleftharpoons 2t\text{-}HexHBH_2BH_2 \qquad 14.35$$

The reaction proceeds much further toward completion in the case of *sym*-tetrasiamyldiborane, presumably because the large steric strains in this derivative are largely relieved in the formation of the product.

sym-Tetrasiamyldiborane reacts with trimethylamine to form an addition compound that is stable at low temperatures but is largely dissociated into its components at room temperature (14.36). *sym*-

Dithexyldiborane forms an addition compound that is stable at room temperature (14.37).

$$(Sia_2BH)_2 + 2(CH_3)_3N \rightleftharpoons 2Sia_2BH:N(CH_3)_3 \qquad 14.36$$
$$(t\text{-}HexBH_2)_2 + 2(CH_3)_3N \longrightarrow 2t\text{-}HexBH_2:N(CH_3)_3 \qquad 14.37$$

The treatment of borane in tetrahydrofuran at 0° with one equivalent of cyclohexene ($C_6H_{10}/BH_3 = 1$) results in the predominant formation of 1,1-dicyclohexyldiborane. This observation suggests the following reaction path for hydroborations in tetrahydrofuran.

In this solvent diborane exists as the monomeric species, tetrahydrofuran-borane. The first step in the hydroboration reaction must involve the addition of this component to the olefin to give the corresponding monoalkylborane (14.38).

$$\underset{|\quad|}{\overset{|\quad|}{C}}{=}C + BH_3:THF \longrightarrow H\underset{|\quad|}{\overset{|\quad|}{-}C}{-}C{-}BH_2 + THF \qquad 14.38$$

In the case of highly hindered olefins, such as 2,3-dimethyl-2-butene and 2,4,4-trimethyl-2-pentene, a second reaction to place an additional alkyl group becomes relatively slow, and the monoalkylborane moieties dimerize to form the *sym*-dialkyldiboranes, which may be identified in the reaction solution and isolated therefrom.

However, with less hindered olefins, such as cyclohexene, the reaction with the monoalkylborane is faster than the reaction with tetrahydrofuran-borane (14.39).

$$\underset{|\quad|}{\overset{|\quad|}{C}}{=}C + RBH_2 \longrightarrow R_2BH \qquad 14.39$$

Consequently, in such cases the monoalkylborane is converted into the dialkylborane, in spite of the presence of free tetrahydrofuran-borane.

$$R_2BH + H_3B:THF \rightleftharpoons R_2BH_2BH_2 + THF \qquad 14.40$$

The results show that the dialkylborane is capable of removing the borane group from tetrahydrofuran (14.40).

Continued reaction with olefin produces the tetraalkyldiborane and, finally, the trialkylborane, in the case of the less hindered derivatives.

9. Hydroboration of Dienes [28] and Acetylenes [29]

The hydroboration of dienes with the trifunctional molecule, borane, obviously opens up numerous possibilities for the formation of cyclic and polymeric organoboranes as intermediates.[30] In the initial phases of our exploration[28] we avoided this question. Oxidation with alkaline hydrogen peroxide provided a convenient route to the glycols, and this satisfied our initial objectives (14.41, 14.42).

$$
\underset{\substack{| \quad | \\ H_2C=C-C=CH_2}}{\overset{\substack{H_3C \quad CH_3 \\ | \quad |}}{}} \xrightarrow{HB} \xrightarrow{[O]} \underset{\substack{| \quad | \\ HOH_2CCHCHCH_2OH}}{\overset{\substack{H_3C \quad CH_3}}{}} \quad 14.41
$$

$$
\underset{\substack{| \\ H_2CCH=CH_2}}{\overset{H_2CCH=CH_2}{}} \xrightarrow{HB} \xrightarrow{[O]} \underset{\substack{| \\ H_2CCH_2CH_2OH}}{\overset{H_2CCH_2CH_2OH}{}} \quad 14.42
$$

However, certain anomalies were observed, indicating that cyclic species were being formed as intermediates. For example, the product from 1,5-hexadiene (14.42) contained only 69 percent of the 1,6-diol; 22 percent of the 1,5- and 9 percent of the 2,5- were also present. These amounts are far larger than would have been anticipated for the independent hydroboration of two terminal bonds (Section XIV-5). Indeed, 1,4-pentadiene yielded the 1,4-pentanediol as the major product, 62 percent, with only 38 percent of the 1,5-product. Clearly a five-membered boron heterocycle must have been involved. Similarly, the hydroboration of cyclooctadiene with diborane yields only cis-1,4- and cis-1,5-cyclooctanediols.[31] Clearly, bicyclic derivatives, such as 9-borabicyclo[3.3.1]nonane (**14.12**) and its [4.2.1] isomer, must be involved.

[28] G. Zweifel, K. Nagase, and H. C. Brown, *J. Amer. Chem. Soc.*, **84**, 183 (1962).

[29] H. C. Brown and G. Zweifel, *J. Amer. Chem. Soc.*, **83**, 3834 (1961).

[30] R. Köster, *Advan. Organometal. Chem.*, **2**, 257 (1964).

[31] H. C. Brown and C. D. Pfaffenberger, *J. Amer. Chem. Soc.*, **89**, 5475 (1967).

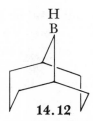

14.12

It is possible to circumvent many of these problems by the use of monofunctional hydroboration agents, as discussed in the next chapter.

More recently we have returned to a study of the specific structures formed in the reaction with diborane. For example, it was established that the reaction of diborane with butadiene in the stoichiometric ratio proceeds to place 70 percent of the boron in the 1,4-positions and 30 percent in the 1,3-positions (14.43).

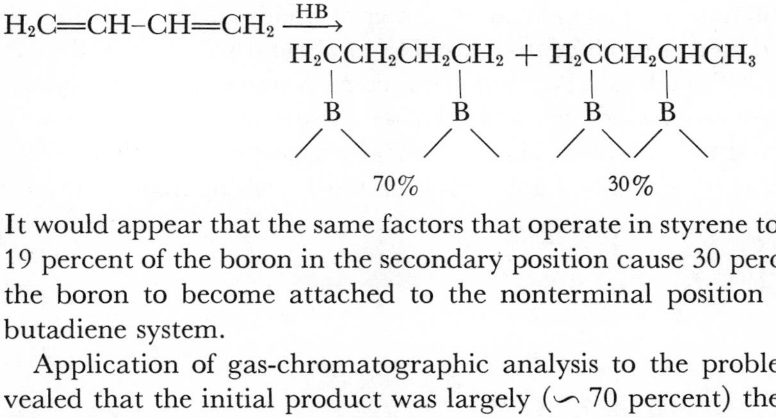

$$H_2C=CH-CH=CH_2 \xrightarrow{\text{HB}}$$

$$\underset{\underset{\text{70\%}}{\underset{| \quad\quad |}{\overset{| \quad\quad |}{B \quad\quad B}}}}{H_2CCH_2CH_2CH_2} + \underset{\underset{\text{30\%}}{\underset{| \quad\quad |}{\overset{| \quad\quad |}{B \quad\quad B}}}}{H_2CCH_2CHCH_3} \quad 14.43$$

It would appear that the same factors that operate in styrene to place 19 percent of the boron in the secondary position cause 30 percent of the boron to become attached to the nonterminal position of the butadiene system.

Application of gas-chromatographic analysis to the problem revealed that the initial product was largely (∼ 70 percent) the compound **14.13**. The minor product (∼ 30 percent) was **14.14**. Oxida-

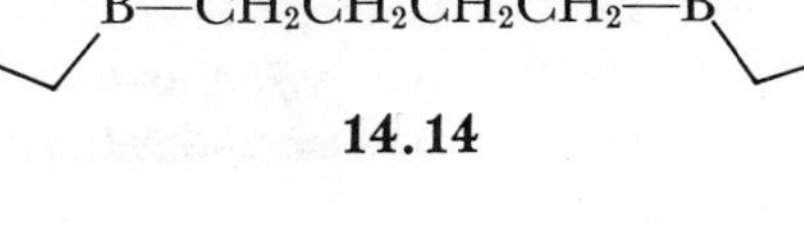

$$\overset{\displaystyle CH_3}{\underset{}{\bigg|}}$$
$$\bigcirc B-CH_2CH_2CH-B \bigcirc$$

14.13

$$\bigcirc B-CH_2CH_2CH_2CH_2-B \bigcirc$$

14.14

tion would produce the 1,3- and 1,4-diols in approximately the ratio realized.[32]

If one treats 1,3-butadiene with borane in tetrahydrofuran in a molar ratio of $1:1$, the product is a polymer, with an unusual type of repeating unit involving a pseudocyclic 1,2-tetramethylene diborane structure[33] (**14.15**). (In this structure the C_4H_8 unit contains both 1,4- and 1,3-butane moieties.)

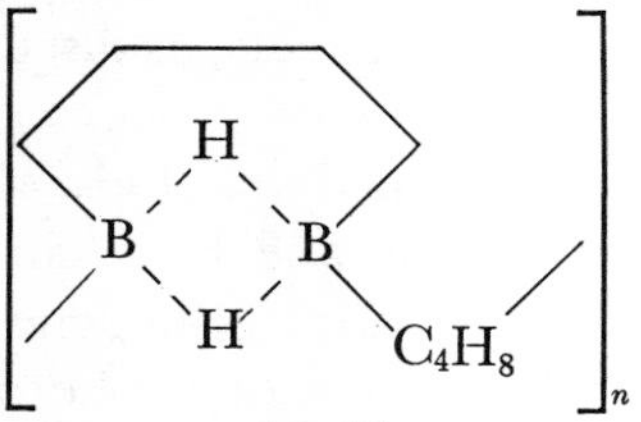

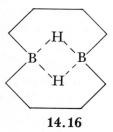

14.15

Distillation of this polymeric material yields an extraordinarily stable dimer, a product that is stable to air and water as well as to olefins. Although it has the empirical composition of a dialkylborane, its behavior does not resemble that of those reactive intermediates. The product has been assigned the structure of 1,6-diboracyclodecane with a highly stable transannular boron-hydrogen bridge[34] (**14.16**).

14.16

The organoboranes have a fascinating structural chemistry that is only now undergoing systematic exploration.

Treatment of acetylenes with the theoretical amount of borane re-

[32] H. C. Brown, E. Negishi, and S. K. Gupta, *J. Amer. Chem. Soc.*, **92**, 2460 (1970).
[33] H. C. Brown, E. Negishi, and P. L. Burke, *J. Amer. Chem. Soc.*, **93**, 3400 (1971).
[34] E. Breuer and H. C. Brown, *J. Amer. Chem. Soc.*, **91**, 4164 (1969).

sults in dihydroboration. The product is predominantly the *gem*-dibora derivatives[29][35] (**14.17**). The precise structure of these *gem*-

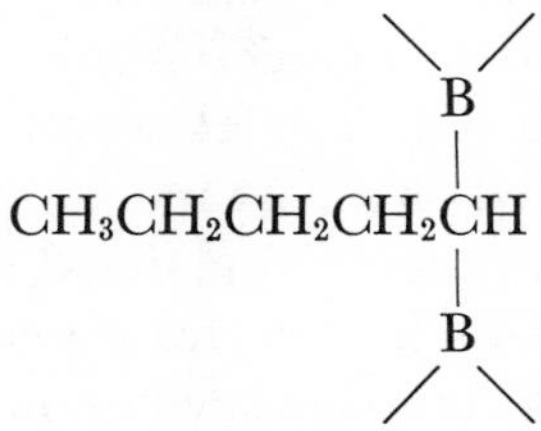

$$CH_3CH_2CH_2CH_2CH \overset{\displaystyle \overset{\diagdown \diagup}{B}}{\underset{\displaystyle \underset{\diagup \diagdown}{B}}{}}$$

14.17

dibora derivatives has not yet been worked out.

Monohydroboration of internal acetylenes can be achieved by using controlled amounts of reagent[29] (14.44).

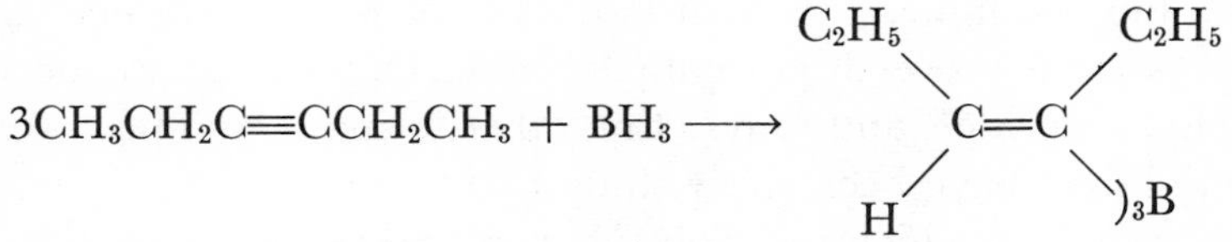

$$3CH_3CH_2C{\equiv}CCH_2CH_3 + BH_3 \longrightarrow \quad \begin{matrix} C_2H_5 & C_2H_5 \\ \diagdown & \diagup \\ C{=}C \\ \diagup & \diagdown \\ H &)_3B \end{matrix} \qquad 14.44$$

However, under these conditions terminal acetylenes undergo preferential dihydroboration (**14.17**) even in the same molar ratio of 3:1.

The use of dialkylboranes greatly simplifies the problem of achieving the monohydroboration of acetylenes to form the desired vinylboranes. This will be discussed in the next chapter.

10. Hydroboration of Functional Derivatives [36-38]

The reactions of diborane with the carbon-carbon double and triple bonds are very fast, much faster in fact than the reaction of diborane with many functional groups (Section XIII-2). An important consequence is that many functional groups can be tolerated in the hydroboration reaction, readily providing organoboranes containing such substituents. For the first time the organic chemist has

[35] G. Zweifel and H. Arzoumanian, *J. Amer. Chem. Soc.*, **89,** 291 (1967).

[36] H. C. Brown and M. K. Unni, *J. Amer. Chem. Soc.*, **90,** 2902 (1968).

[37] H. C. Brown and R. M. Gallivan, *J. Amer. Chem. Soc.*, **90,** 2906 (1968).

[38] H. C. Brown and R. L. Sharp, *J. Amer. Chem. Soc.*, **90,** 2915 (1968).

at his disposal a relatively reactive organometallic with a wide variety of possible substituents.

This is a new development whose full implications have not yet been realized. In the past, organic chemists have relied heavily on organometallics for their syntheses. However, it was almost axiomatic that a synthesis that relied on an organometallic had to be planned to avoid the presence of a reactive functional substituent either in the organometallic itself or in the substrate undergoing reaction. Now we are in the midst of a complete revolution in this area—the ready synthesis and utilization of a wide variety of substituents. Consequently, it is appropriate to review briefly the main features of the hydroboration of functional derivatives.

The presence of relatively inert substituents, such as halogen and ether groupings, should not cause any difficulty in the hydroboration reaction. For example, the hydroboration of *p*-chlorostyrene and *p*-methoxystyrene proceeds normally.[20] Similarly, the aliphatic derivatives, allyl chloride[39] and vinyl ethyl ether,[40] and a number of related derivatives have been successfully hydroborated.[41,42]

Even for cases in which the molecule contains reducible groups, it has been possible to achieve hydroboration. Thus methyl oleate has been converted to an equimolar mixture of 9- and 10-hydroxyoctanoic acids by the hydroboration-oxidation procedure[43] and the methyl ester of 10-undecenoic acid has likewise been converted into the 11-hydroxy derivative.[44] Even derivatives containing the double bond much closer to the ester group are readily converted into the corresponding organoboranes.[45] The high reactivity of the aldehyde group and the free carboxylic acid group toward diborane would doubtless result in a competition for the diborane. However, simple conversion of these groups into the acetal and ester, respectively, circumvents the problem.

[39] M. F. Hawthorne and J. A. Dupont, *J. Amer. Chem. Soc.*, **80,** 5830 (1958).

[40] B. M. Mikhailov and T. A. Shchegoleva, *Izvest. Akad. Nauk. SSSR*, 546 (1959).

[41] H. C. Brown and K. A. Keblys, *J. Amer. Chem. Soc.*, **86,** 1791 (1964).

[42] H. C. Brown and O. J. Cope, *J. Amer. Chem. Soc.*, **86,** 1801 (1964).

[43] S. P. Fore and W. G. Bickford, *J. Org. Chem.*, **24,** 920 (1959).

[44] R. Dulou and Y. Chretien-Bessiere, *Bull. soc. chim. France*, 1362 (1959).

[45] H. C. Brown and K. A. Keblys, *J. Amer. Chem. Soc.*, **86,** 1795 (1964).

The presence of substituents can introduce marked directive influences on the hydroboration reaction. In some cases this may complicate the situation; in others, one can take advantage of it. A systematic study of the hydroboration of 3-butenyl derivatives[36] (**14.18**), 2-butenyl (crotyl) derivatives[37] (**14.19**), 1-butenyl (vinyl) derivatives[38] (**14.20**), and related cyclopentenyl derivatives[46] (**14.21**) provides the basis for an understanding of the directive effects.

$$H_2C{=}CHCH_2CH_2X \qquad H_3CCH{=}CHCH_2X \qquad H_3CCH_2CH{=}CHX$$

$$\textbf{14.18} \qquad\qquad \textbf{14.19} \qquad\qquad \textbf{14.20}$$

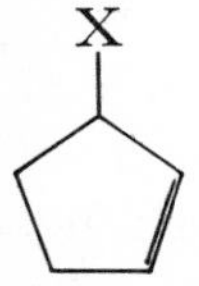

$$\textbf{14.21}$$

Hydroboration of many of the 3-butenyl derivatives proceeds to place as much as 20 percent of the boron in the secondary 3-position. This is undesirable in giving a larger conversion to the minor product. The difficulty is readily overcome by using a suitable dialkylborane, such as disiamylborane, for the hydroboration (Chapter XV). Otherwise, no difficulties were encountered for a wide variety of substituents ($X = OCH_3$, OC_6H_5, OH, O_2CCH_3, Cl, NH_2, SCH_3, and $CO_2C_2H_5$).

In the crotyl derivatives the strong directive influence of the substituents places the boron predominantly ($\smile$ 90 percent) in the 2-position (14.45).

$$CH_3CH{=}CHCH_2X \xrightarrow{\text{HB}} CH_3CH_2\underset{\underset{\diagup\diagdown}{\overset{|}{B}}}{C}HCH_2X \qquad 14.45$$

When X is a good leaving group, such as Cl, elimination occurs to give 1-butene, which then undergoes hydroboration. This side reac-

[46] H. C. Brown and E. F. Knights, *J. Amer. Chem. Soc.*, **90,** 4439 (1968).

tion can be avoided in ethyl ether.[47] Apparently the elimination is catalyzed by the donor properties of tetrahydrofuran.

In the case of the 1-butenyl derivatives, the presence of an alkoxy substituent directs the boron strongly to the 2-position. On the other hand, a chlorine substituent directs the boron to the 1-position. Acetoxy lies in between.

With an appreciation of these phenomena, it is possible to utilize directive effects to produce the desired organoboranes for transfer to carbon or other atoms as required in a specific synthesis.

11. Conclusion

Despite all our progress, we are still in the initial stages of understanding this fascinating reaction. With our increasing grasp of the remarkable versatility of the organoboranes as synthetic intermediates (Chapter XVI), it is evident that present and future findings in this area must somehow be transmitted to new generations of students so that they can take advantage of the rich possibilities. This will be no easy task.

[47] D. J. Pasto and J. Hickman, *J. Amer. Chem. Soc.*, **90,** 4445 (1968).

XV. Hydroboration with Borane Derivatives[1]

1. Advantages of Substituted Boranes

Borane, BH_3, is a trifunctional hydroborating agent of very low steric requirements. Its application in certain special situations can result in experimental difficulties. For example, the hydroboration of terminal acetylenes with borane provides the desired trivinylborane in only very low yield (Section XIV–9). Similarly, the reaction of 1,4-pentadiene with borane yields as major product not the 1,5-diborapentane species anticipated for independent hydroboration of two double bonds, but the 1,4-diborapentane moiety, evidently the result of a cyclic hydroboration (Section XIV–9).

The hydroboration of hindered olefins (Section XIV–7) revealed that it was possible to control the reaction in suitable instances to provide the monofunctional dialkylboranes, such as disiamylborane and dicyclohexylborane, and difunctional monoalkylboranes, such as thexylborane. These reagents offered promise of solving some of problems just mentioned. A study was therefore undertaken; it led to simple solutions of many of these problems.

2. Selective Hydroboration with Disiamylborane [2]

The observation that the hydroboration of 2-methyl-2-butene proceeds rapidly to the dialkylborane stage (15.1), but only very slowly

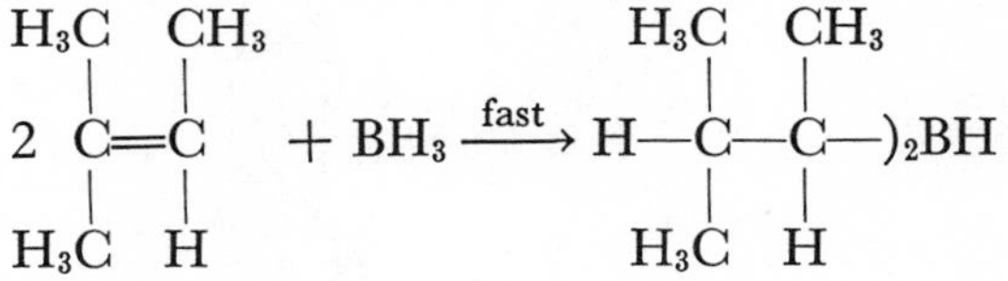

$$2 \begin{array}{c} H_3C \quad CH_3 \\ | \quad | \\ C\!=\!C \\ | \quad | \\ H_3C \quad H \end{array} + BH_3 \xrightarrow{\text{fast}} \begin{array}{c} H_3C \quad CH_3 \\ | \quad | \\ H\!-\!C\!-\!C\!-\!)_2BH \\ | \quad | \\ H_3C \quad H \end{array} \quad 15.1$$

[1] H. C. Brown, *Hydroboration* (W. A. Benjamin, New York, 1962).
[2] H. C. Brown and G. Zweifel, *J. Amer. Chem. Soc.*, **82,** 3222 (1960).

to the trialkylborane end product (15.2), indicates that the last stage must be strongly influenced by the large steric requirements of both the reagent and the olefin.

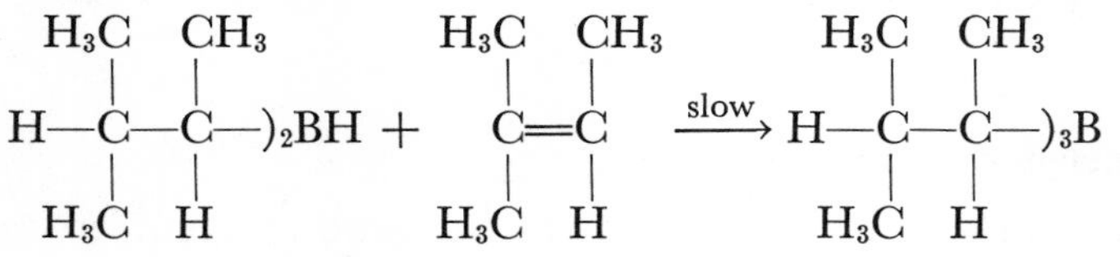

$$
\underset{\substack{| \\ H_3C \quad H}}{\overset{\substack{H_3C \quad CH_3 \\ |}}{H-C-C-}})_2BH \;+\; \underset{\substack{| \\ H_3C \quad H}}{\overset{\substack{H_3C \quad CH_3 \\ |}}{C=C}} \xrightarrow{\text{slow}} \underset{\substack{| \\ H_3C \quad H}}{\overset{\substack{H_3C \quad CH_3 \\ |}}{H-C-C-}})_3B \qquad 15.2
$$

This suggested that the reagent might possess an enhanced sensitivity to the steric requirements of the substituents on the double bond and thereby exert a steric influence on the direction of hydroboration and on the relative reactivities of various olefin structures.

Indeed, 1-hexene reacts readily with the reagent, and oxidation of the product yields 1-hexanol in an isomeric purity of at least 99 percent (**15.1**), whereas diborane yields 6 percent of the secondary alcohol. Similarly, styrene is converted by this procedure into β-phenylethanol of 98 percent purity (**15.1**) in contrast to the 81:19 distribution realized with diborane. Finally, *cis*-4-methyl-2-pentene undergoes attachment of the boron atom to the less hindered position, providing the alcohol 4-methyl-2-pentanol in isomeric purity of 97 percent (**15.3**).

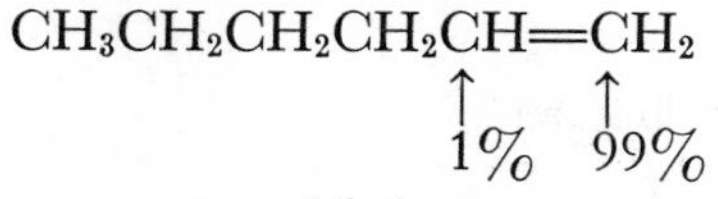

$$
\underset{\substack{\uparrow \quad \uparrow \\ 1\% \quad 99\%}}{CH_3CH_2CH_2CH_2CH=CH_2}
$$

15.1

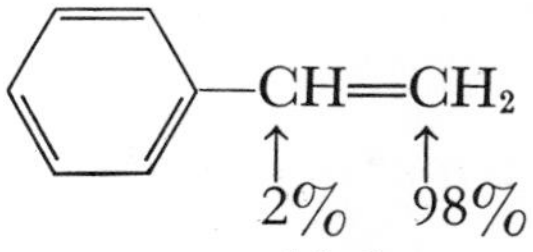

$$
\text{C}_6H_5\text{—}\underset{\substack{\uparrow \quad \uparrow \\ 2\% \quad 98\%}}{CH=CH_2}
$$

15.2

$$
\underset{\substack{| \quad \uparrow \quad \uparrow \quad | \\ H \quad 3\% \quad 97\% \quad H}}{\overset{(CH_3)_2CH \qquad CH_3}{C=C}}
$$

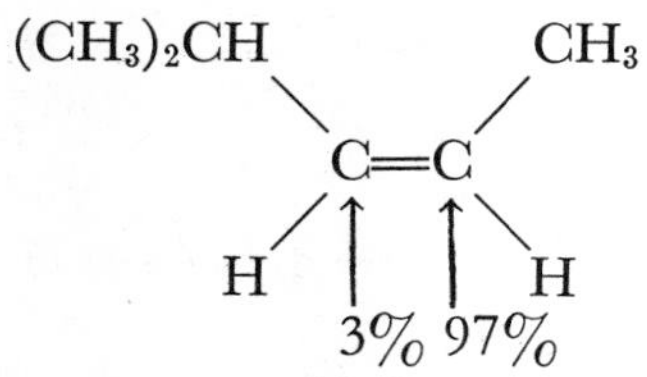

15.3

A major development along this line is the recent report that treatment of Δ¹-cholestene with disiamylborane results in the predominant formation of cholestan-2α-ol, in contrast to the nearly 50:50 mixture of cholestan-1α-ol and cholestan-2α-ol realized with diborane[3] (15.3).

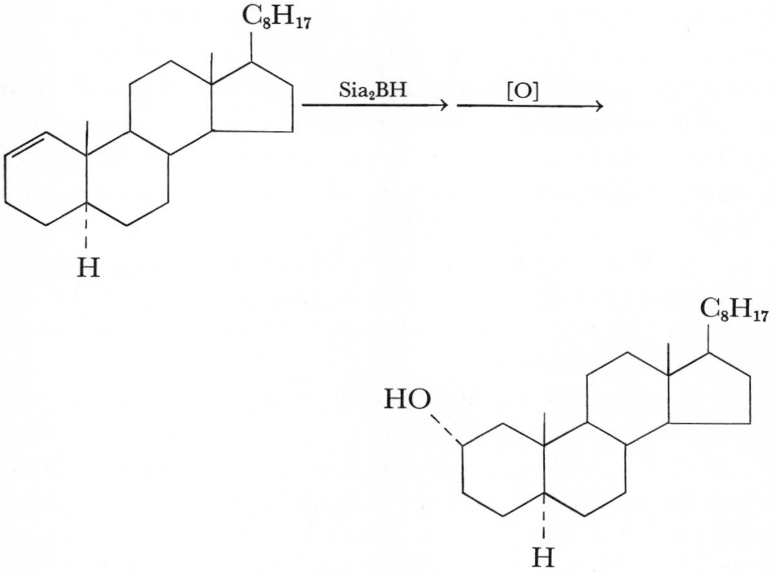

15.3

These studies revealed major differences in the rates of reaction of the reagent with different olefins. Thus the reaction of the reagent with 1-hexene and 2-methyl-1-pentene is very rapid at 0°, whereas the reactions of internal olefins are considerably slower, with cyclopentene reacting faster than *trans*-2-hexene and the latter reacting considerably faster than cyclohexene. *Cis*-2-hexene reacts considerably faster than the *trans* isomer. Trisubstituted olefins, such as 2-methyl-2-butene and 1-methylcyclohexene, react very slowly.

The result of these qualitative experiments may be expressed in the following series:[4] 1-hexene ≥ 3-methyl-1-butene > 2-methyl-1-butene > 3,3-dimethyl-1-butene > *cis*-2-hexene ≥ cyclopentene > *trans*-2-hexene > *trans*-4-methyl-2-pentene > cyclohexene ≥

---

[3] F. Sondheimer and M. Nussim, *J. Org. Chem.*, **26,** 630 (1961).

[4] H. C. Brown and G. Zweifel, *J. Amer. Chem. Soc.*, **83,** 1241 (1961).

1-methylcyclopentene $>$ 2-methyl-2-butene $>$ 1-methylcyclohexene $\geq$ 2,3-dimethyl-2-butene.

The observed differences in reactivity are quite large and can be utilized for the selective hydroboration of a more reactive olefin in the presence of a less reactive structure. Thus, treatment of a mixture of 1-pentene and 2-pentene with sufficient reagent to react with the more reactive terminal olefin yields the 2-pentene free of the 1-isomer. In the same way, the 1-hexene, in a mixture with cyclohexene, can be selectively hydroborated to yield essentially pure cyclohexene.

It is also possible to react selectively cyclopentene in a mixture with cyclohexene. Finally, it is possible to take advantage of the differences in reactivity to remove the more reactive *cis* olefin from a mixture of *cis* and *trans* isomers.

It is apparent from these results that the reagent should permit the selective hydroboration of complex molecules containing two or more reactive centers. Some applications of this kind are presented later.

Although the reaction of diborane with olefins is too fast for measurement, it proved possible to make a kinetic study of the reaction of disiamylborane with a number of olefins.[5] The reaction in tetrahydrofuran at 0° is second-order, first-order in disiamylborane dimer, and first-order in olefin.

The rate constant for cyclopentene, $k_2 = 14.0 \times 10^{-4}$ liter mole^{-1} sec^{-1}, decreases 100-fold for cyclohexene, $k_2 = 0.134 \times 10^{-4}$ liter mole^{-1} sec^{-1}, but rises sharply again by a factor of 500 for cycloheptene, $k_2 = 72 \times 10^{-4}$ liter mole^{-1} sec^{-1}. The rate constant for *cis*-2-butene, with $k_2 = 23 \times 10^{-4}$, is greater than that of *trans*-2-butene, $k_2 = 3.8 \times 10^{-4}$, by a factor of 6.

The results suggest that strain in the olefin system greatly facilitates the reaction. Thus both cyclopentene and cycloheptene are strained olefins relative to cyclohexene. Similarly, *trans*-2-butene is thermodynamically more stable than the *cis* isomer.

Finally, the second-order kinetics appear to require a transition state composed of one molecule of the dimer and one molecule of olefin (15.4).

[5] H. C. Brown and A. W. Moerikofer, *J. Amer. Chem. Soc.*, **83**, 3417 (1961).

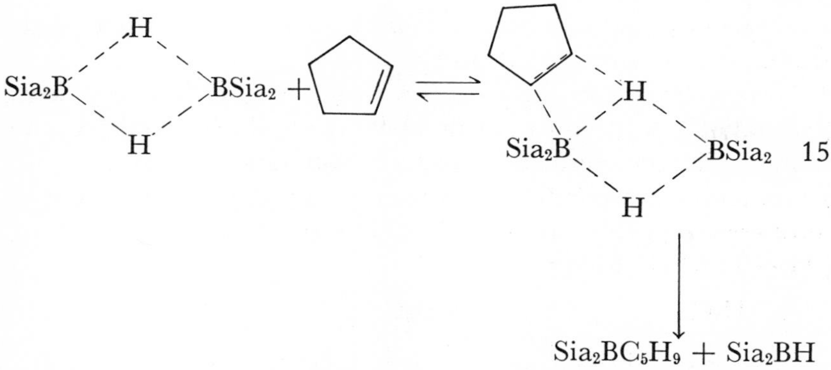

$$Sia_2BC_5H_9 + Sia_2BH$$

The reaction presumably proceeds with the formation of the product and one molecule of monomer, which should react with a second molecule of olefin in a rapid second step, or undergo dimerization to *sym*-tetrasiamyldiborane.

The ether solvent probably facilitates the reaction by coordinating with disiamylborane monomer, the "leaving group" in this reaction.

As pointed out earlier, disiamylborane is only one of a number of dialkylboranes readily synthesized via hydroboration. The utility of these derivatives for selective hydroboration was examined.[6] The results with diisopinocampheylborane are of especial interest and importance and will, therefore, be considered in some detail.

3. Asymmetric Hydroboration with Diisopinocampheylborane [7, 8]

As pointed out earlier, the hydroboration of α-pinene proceeds readily to the formation of diisopinocampheylborane (15.5).

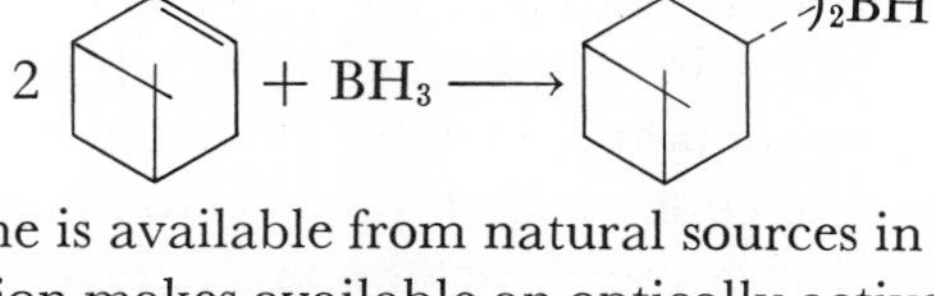

Since α-pinene is available from natural sources in optically active form, this reaction makes available an optically active dialkylborane.

[6] H. C. Brown, N. R. Ayyangar, and G. Zweifel, *J. Amer. Chem. Soc.*, **85,** 2072 (1963).

[7] G. Zweifel and H. C. Brown, *J. Amer. Chem. Soc.*, **86,** 393 (1964).

[8] H. C. Brown, N. R. Ayyangar, and G. Zweifel, *J. Amer. Chem. Soc.*, **86,** 397 (1964).

This reagent exhibits a remarkable ability to achieve the asymmetric hydroboration of suitable olefins.[8-10]

For example, cis-2-butene reacts with the reagent from α-pinene, $[\alpha]^{20}_D + 47.6°$, to produce an organoborane which, oxidized with alkaline hydrogen peroxide, produces 2-butanol, $[\alpha]^{20}_D - 11.8°$.[8] This rotation corresponds to an optical purity of 87 percent. Use of levorotatory α-pinene produces 2-butanol of the opposite rotation, $[\alpha]^{20}_D + 11.8°$ (15.6).[8]

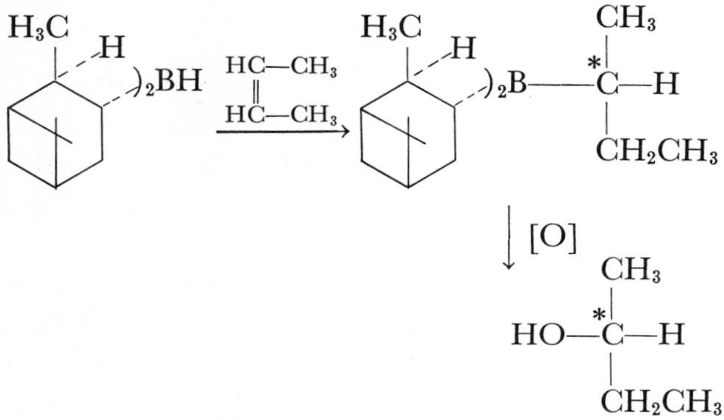

15.6

Similarly, cis-3-hexene was converted in 81 percent yield to 3-hexanol, $[\alpha]^{20}_D - 6.5°$, indicating an optical purity of 91 percent. Application of the procedure to norbornene produces exo-norborneol in a yield of 62 percent. The observed rotation, $[\alpha]^{20}_D - 2.0°$, also indicates a high optical purity.

These results establish that a boron atom at the asymmetric center, RR′C*HB$\langle$, is capable of maintaining asymmetry without significant racemization over periods of several hours. The unusually high optical purities realized in this asymmetric hydroboration and the ease with which organoboranes may be converted into other derivatives with retention of configuration promise to make this approach to optically active derivatives a useful one for the synthetic chemist.

In the three cases mentioned, the alcohols from dextrorotatory

[9] H. C. Brown, N. R. Ayyangar, and G. Zweifel, *J. Amer. Chem. Soc.*, **86,** 1071 (1964).

[10] G. Zweifel, N. R. Ayyangar, T. Munekata, and H. C. Brown, *J. Amer. Chem. Soc.*, **86,** 1076 (1964).

α-pinene have configurations that can be correlated with the structure of the reagent. Consequently, the procedure also promises to be valuable as a tool to establish the absolute configuration of alcohols and other derivatives.[8]

4. Hydroboration of Acetylenes with Dialkylboranes [11]

As pointed out earlier, treatment of an internal acetylene, such as 3-hexyne, with the calculated quantity of borane provides the desired trivinylborane in reasonable yield. However, the same procedure is unsuccessful with terminal acetylenes, such as 1-hexyne. Here double hydroboration to give the 1,1-dibora derivative occurs preferentially (Section XIV–9). For such hydroborations of acetylenes to form the vinylbora derivatives, the dialkylboranes have major advantages.[11]

Thus the hydroboration of 3-hexyne with disiamylborane or dicyclohexylborane goes quantitatively to the desired intermediate (15.7).

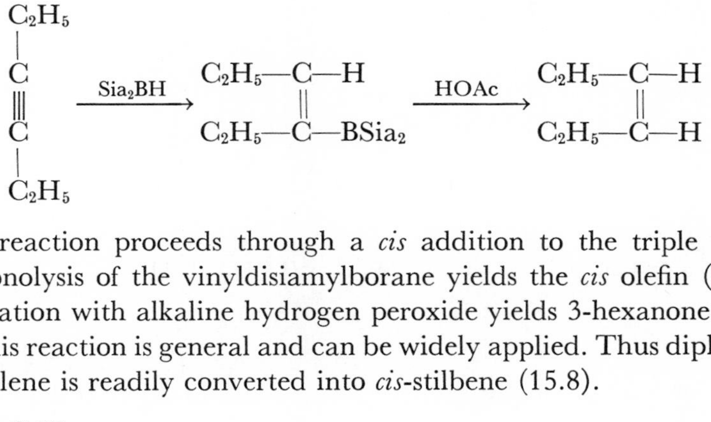

$$
\begin{array}{c}
C_2H_5 \\
| \\
C \\
\|\|\| \\
C \\
| \\
C_2H_5
\end{array}
\xrightarrow{\ Sia_2BH\ }
\begin{array}{c}
C_2H_5-C-H \\
\| \\
C_2H_5-C-BSia_2
\end{array}
\xrightarrow{\ HOAc\ }
\begin{array}{c}
C_2H_5-C-H \\
\| \\
C_2H_5-C-H
\end{array}
\qquad 15.7
$$

The reaction proceeds through a *cis* addition to the triple bond. Protonolysis of the vinyldisiamylborane yields the *cis* olefin (15.7). Oxidation with alkaline hydrogen peroxide yields 3-hexanone.

This reaction is general and can be widely applied. Thus diphenylacetylene is readily converted into *cis*-stilbene (15.8).

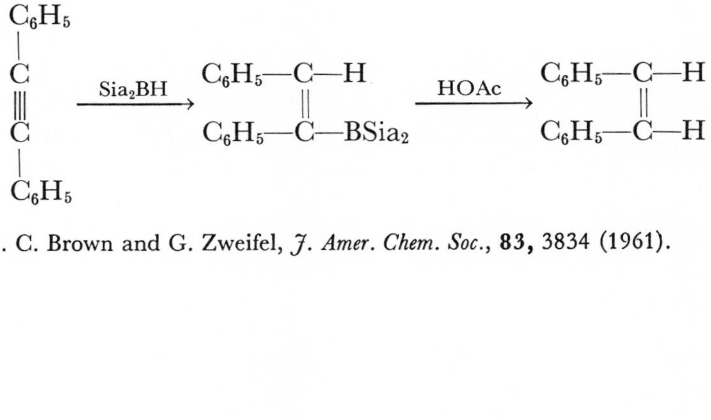

$$
\begin{array}{c}
C_6H_5 \\
| \\
C \\
\|\|\| \\
C \\
| \\
C_6H_5
\end{array}
\xrightarrow{\ Sia_2BH\ }
\begin{array}{c}
C_6H_5-C-H \\
\| \\
C_6H_5-C-BSia_2
\end{array}
\xrightarrow{\ HOAc\ }
\begin{array}{c}
C_6H_5-C-H \\
\| \\
C_6H_5-C-H
\end{array}
\qquad 15.8
$$

[11] H. C. Brown and G. Zweifel, *J. Amer. Chem. Soc.*, **83**, 3834 (1961).

Hydroboration

The use of dialkylboranes is especially valuable for terminal acetylenes. Here either disiamylborane or dicyclohexylborane adds rapidly and cleanly to give a quantitative yield of the desired vinyl derivative (15.9).

$$n\text{-}C_4H_9C\equiv CH + Sia_2BH \longrightarrow n\text{-}C_4H_9\underset{\underset{H}{|}\quad\underset{BSia_2}{|}}{C}=CH \qquad 15.9$$

Protonolysis with acetic acid produces the 1-alkene. Oxidation with alkaline hydrogen peroxide produces the aldehyde (15.10).

$$n\text{-}C_4H_9\underset{\underset{H}{|}\quad\underset{BSia_2}{|}}{C}=CH \xrightarrow{\text{[O]}} n\text{-}C_4H_9CH_2CHO \qquad 15.10$$

Thus this procedure provides a convenient means of achieving the anti-Markovnikov hydration of terminal acetylenes.

A second addition of disiamylborane is so slow as to be impractical. However, two equivalents of dicyclohexylborane can be added to give the 1,1-dibora derivative (15.11).

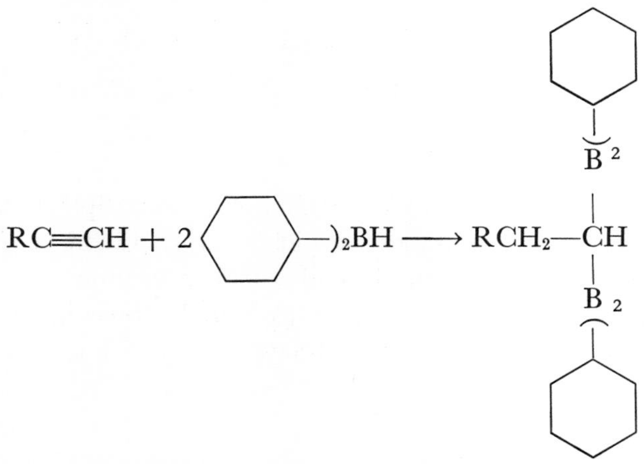

$$RC\equiv CH + 2\left(\bigcirc\!-\right)_2BH \longrightarrow RCH_2\!-\!CH \qquad 15.11$$

The conclusion that the dihydroboration of acetylenes proceeds to place two boron atoms preferentially on the same carbon atom has been questioned.[12] However, a careful reexamination of the dihydro-

[12] D. J. Pasto, *J. Amer. Chem. Soc.*, **86**, 3039 (1964).

boration of 1-hexyne using deuterium as a tracer has reaffirmed the original conclusion.[13]

These vinylboranes evidently have rich possibilities for organic synthesis. For example, the hydroboration of substituted propargyl chlorides provides a convenient route to terminal allenes[14] (15.12).

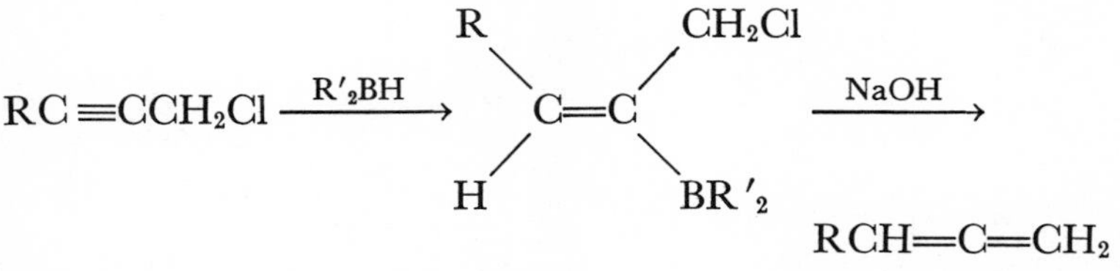

$$RC{\equiv}CCH_2Cl \xrightarrow{R'_2BH} \underset{H}{\overset{R}{\diagdown}}C=C\underset{BR'_2}{\overset{CH_2Cl}{\diagup}} \xrightarrow{NaOH}$$

$$RCH{=}C{=}CH_2 \quad 15.12$$

Moreover, these intermediates can be used for the stereoselective synthesis of substituted alkenes[15] (15.13) and conjugated dienes.[16]

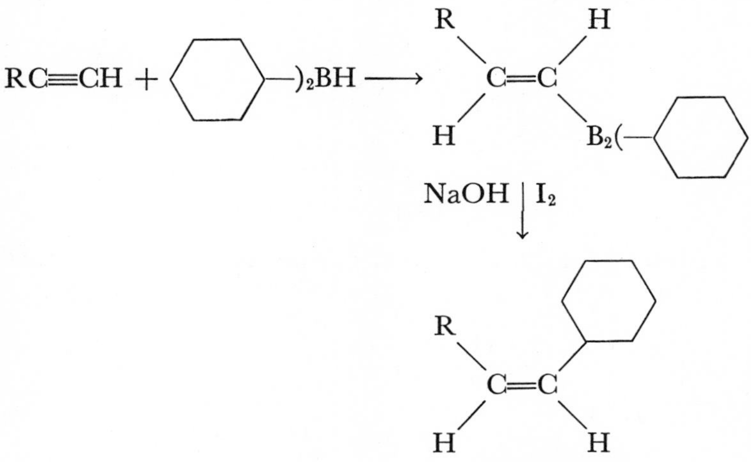

cis (99% pure)

Attention is also called to a study of the selective hydroboration of conjugated diynes with dialkylboranes, providing a convenient route

---

[13] G. Zweifel and H. Arzoumanian, *J. Amer. Chem. Soc.*, **89**, 291 (1967).

[14] G. Zweifel, A. Horng, J. T. Snow, *J. Amer. Chem. Soc.*, **89**, 291 (1967).

[15] G. Zweifel, H. Arzoumanian, and C. C. Whitney, *J. Amer. Chem. Soc.*, **89**, 3652 (1967).

[16] G. Zweifel, N. L. Polston, and C. C. Whitney, *J. Amer. Chem. Soc.*, **89**, 6243 (1967).

to conjugated *cis*-enynes, α,β-acetylenic ketones, and *cis,cis*-dienes[17] (15.14).

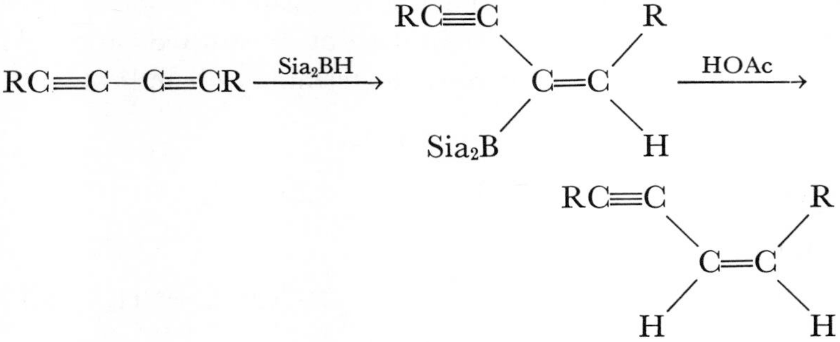
RC≡C—C≡CR
Sia₂BH
RC≡C
R
C=C
Sia₂B
H
HOAc
RC≡C
R
C=C
H
H

15.14

Finally, it should be mentioned that catechol reacts with borane to produce a reagent (**15.4**) with useful properties.[18] This reagent

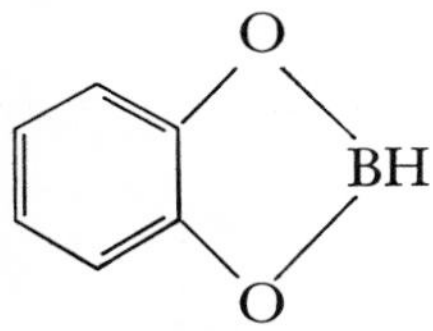
O
BH
O

15.4

provides a simple route to alkyl- and vinylboronic acids and esters (15.15).

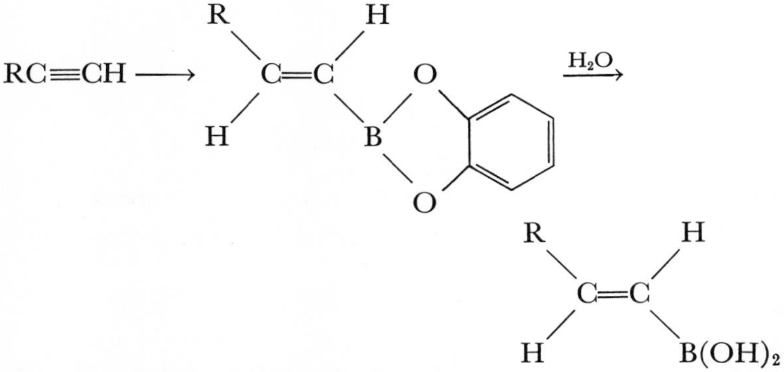
RC≡CH
R
H
C=C
H
B
O
O
H₂O
R
H
C=C
H
B(OH)₂

15.15

[17] G. Zweifel and N. L. Polston, *J. Amer. Chem. Soc.*, **92,** 4068 (1970).

[18] H. C. Brown and S. K. Gupta, *J. Amer. Chem. Soc.*, **93,** 1816 (1971), and research in progress.

5. Hydroboration of Dienes with Disiamylborane [4, 19]

Disiamylborane, as a monofunctional hydroborating agent with large steric requirements, solves many of the problems previously noted in the hydroboration of dienes. Thus, 1,3-butadiene can be converted into the 1,4-butanediol in higher isomeric purity (10 percent 1,3-; 90 percent 1,4-). Similarly, 1,5-hexadiene forms 93 percent of 1,6-hexanediol with only 7 percent of isomeric impurities. Clearly the cyclization reaction observed with borane (Section XIV–9) is avoided.

The monohydroboration of conjugated aliphatic dienes is a difficult problem. The conjugated dienes are less reactive than the corresponding olefins. Consequently, the monohydroboration of such a diene converts it into an unsaturated borane that is far more reactive toward the hydroborating agent than the original diene. As a result, the yield of monohydroborated product is low.

The situation is less difficult in the case of nonconjugated dienes. Thus 1,5-hexadiene has been converted into 5-hexen-1-ol (15.16) and bicycloheptadiene has been converted into *exo*-5-norborneol in reasonable yield (15.17).

$$\begin{array}{ll}
\begin{array}{l} CH_2-CH\!=\!CH_2 \\ | \\ CH_2-CH\!=\!CH_2 \end{array}
\xrightarrow{\;HB\;}\xrightarrow{\;[O]\;}
\begin{array}{l} CH_2-CH\!=\!CH_2 \\ | \\ CH_2-CH_2-CH_2OH \end{array}
& 15.16
\end{array}$$

15.17

Disiamylborane would appear to have major advantages in the monohydroboration of dienes with two reaction centers of considerably different reactivities. Thus, both 4-vinylcyclohexene (15.18) and *d*-limonene (15.19) are readily converted into unsaturated alcohols by this reagent, with attack at the more reactive exocyclic double bond.

Even in the case of conjugated dienes, it has proved possible to achieve a good yield of the unsaturated alcohol by taking advantage

[19] G. Zweifel, K. Nagase, and H. C. Brown, *J. Amer. Chem. Soc.*, **84**, 190 (1962).

15.18

15.19

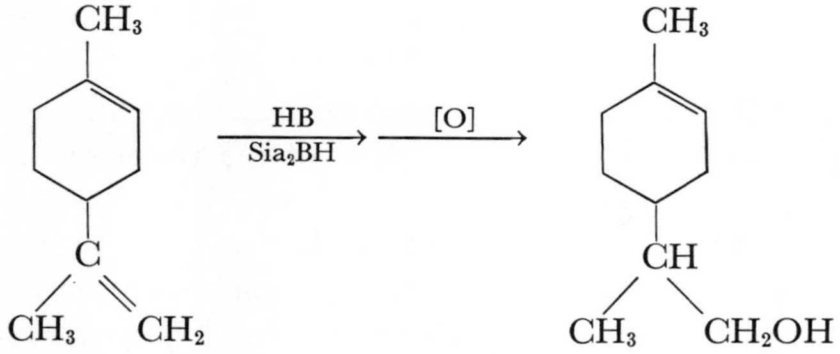

of the sensitivity of disiamylborane toward different olefinic structures (15.20).

$$CH_3CH{=}CH{-}CH{=}CH_2 \xrightarrow[Sia_2BH]{HB} \xrightarrow{[O]} CH_3CH{=}CHCH_2CH_2OH \qquad 15.20$$

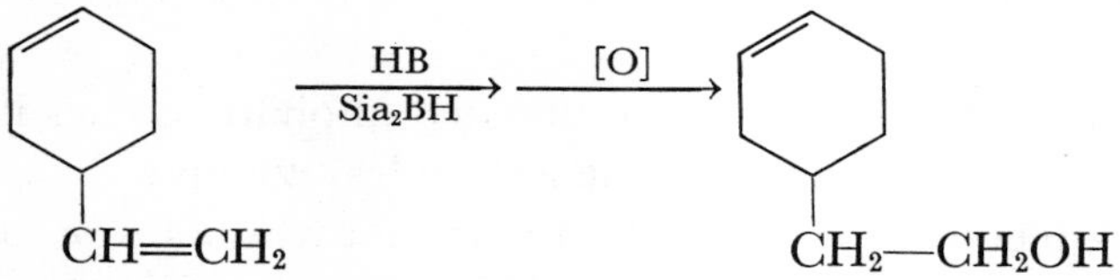

Interesting and useful results have been realized in the monohydroboration of terpenes, such as myrcene[20] (15.21) and caryophyllene,[21] with this reagent.

15.21

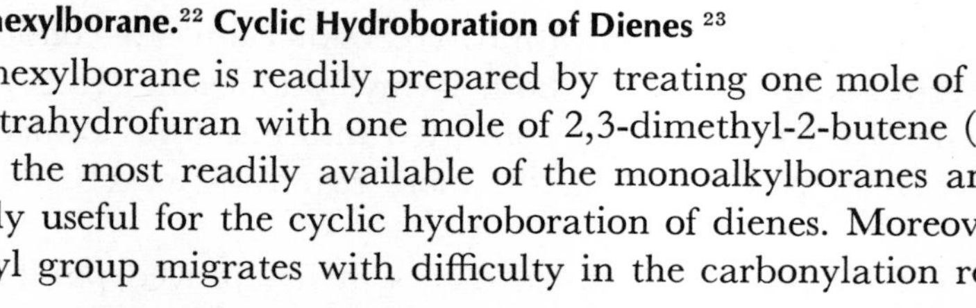

6. Thexylborane.[22] Cyclic Hydroboration of Dienes [23]

Thexylborane is readily prepared by treating one mole of borane in tetrahydrofuran with one mole of 2,3-dimethyl-2-butene (15.22). It is the most readily available of the monoalkylboranes and it is highly useful for the cyclic hydroboration of dienes. Moreover, the thexyl group migrates with difficulty in the carbonylation reaction

[20] H. C. Brown, K. P. Singh, and B. J. Garner, *J. Organometal. Chem.*, **1**, 2 (1963).

[21] Unpublished studies with K. R. Varma, K. P. Singh, and S. P. Acharya.

[22] G. Zweifel and H. C. Brown, *J. Amer. Chem. Soc.*, **85**, 2066 (1963).

[23] H. C. Brown and C. D. Pfaffenberger, *J. Amer. Chem. Soc.*, **89**, 5475 (1967).

$$
\begin{array}{c}
\underset{\substack{| \\ H_3C}}{\overset{\substack{H_3C \\ |}}{C}}=\underset{\substack{| \\ CH_3}}{\overset{\substack{CH_3 \\ |}}{C}} + BH_3 \longrightarrow H-\underset{\substack{| \\ H_3C}}{\overset{\substack{H_3C \\ |}}{C}}-\underset{\substack{| \\ CH_3}}{\overset{\substack{CH_3 \\ |}}{C}}-BH_2 \qquad 15.22
\end{array}
$$

(Chapter XVII). Consequently, thexylborane makes possible many interesting syntheses via this reaction.

Although this monoalkylborane is a hindered derivative, it will hydroborate two additional molecules of olefin, provided they are not too hindered. The hydroboration can be achieved step-wise, so that one can prepare an organoborane containing three different groups (15.23).

$$
\begin{array}{l}
\raisebox{0pt}{$\displaystyle \overline{}\text{--}BH_2$} \xrightarrow[\text{CH}_2\text{CH(CH}_3)_2]{(\text{CH}_3)_2\text{C}=\text{CH}_2} \\[2em]
\qquad \overline{}\text{--}B\!\!\begin{array}{l} \diagup \\ \diagdown \\ H \end{array} \xrightarrow{\text{H}_2\text{C}=\text{CHCH}_2\text{CO}_2\text{C}_2\text{H}_5} \qquad 15.23 \\[3em]
\qquad\qquad\qquad \overline{}\text{--}B\!\!\begin{array}{l} \diagup \text{CH}_2\text{CH(CH}_3)_2 \\ \diagdown \text{CH}_2\text{CH}_2\text{CH}_2\text{CO}_2\text{C}_2\text{H}_5 \end{array}
\end{array}
$$

In this procedure it is desirable to add the more hindered olefin first, followed by the less hindered olefin. To illustrate the utility of this feature, carbonylation involves the two alkyl groups attached to boron, other than the thexyl group. Thus this provides a general synthesis of ketones from almost any two olefins[24] (15.24).

$$
\begin{array}{l}
(\text{CH}_3)_2\text{C}=\text{CH}_2 \\
\qquad\qquad\qquad + \text{CO} \longrightarrow \\
\text{C}_2\text{H}_5\text{O}_2\text{CCH}_2\text{CH}=\text{CH}_2 \\
\qquad\qquad (\text{CH}_3)_2\text{CHCH}_2\underset{\substack{\| \\ O}}{C}\text{CH}_2\text{CH}_2\text{CH}_2\text{CO}_2\text{C}_2\text{H}_5 \qquad 15.24
\end{array}
$$

Thexylborane is especially valuable for achieving the cyclic hydroboration of dienes[25] (15.25, 15.26, 15.27).

[24] H. C. Brown and E. Negishi, *J. Amer. Chem. Soc.*, **89**, 5285 (1967).
[25] H. C. Brown and E. Negishi, *J. Amer. Chem. Soc.*, **89**, 5477 (1967).

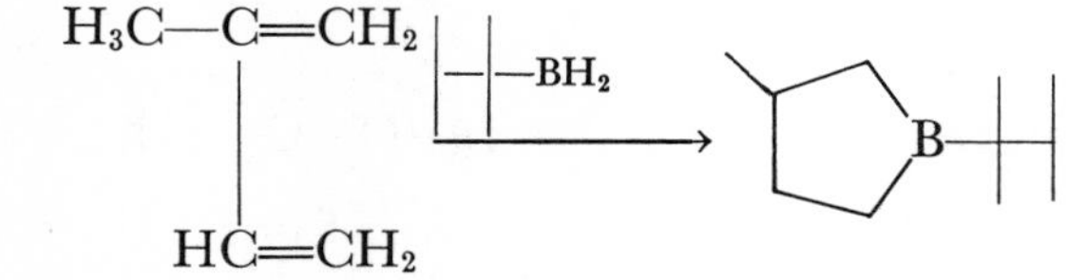

Hydroboration

$$H_3C-C=CH_2 \quad | \overline{\quad}-BH_2 \longrightarrow \quad (15.25)$$

$$HC=CH_2$$

15.25

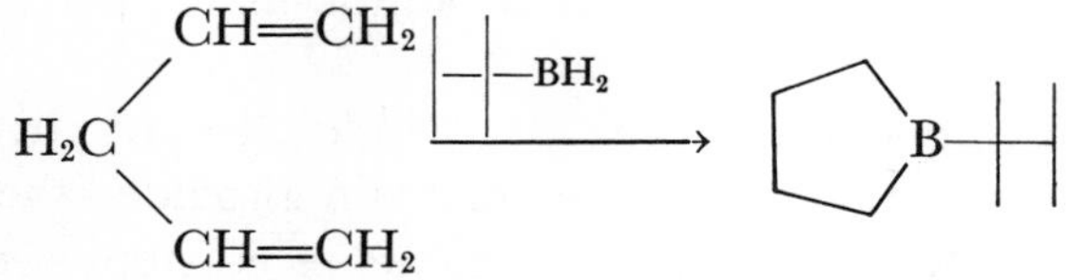

15.26

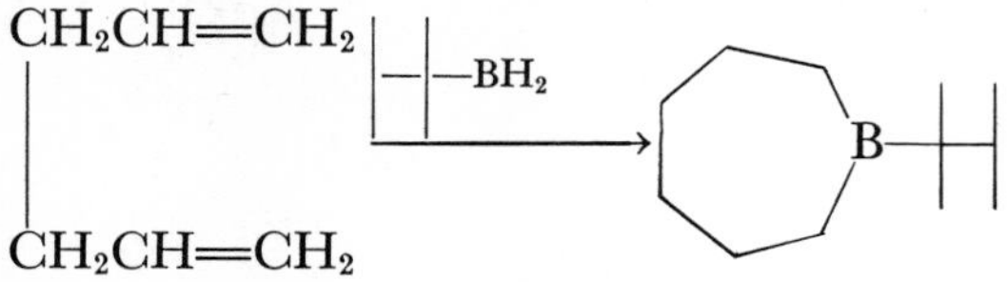

15.27

The utility of this approach to cyclic hydroboration is illustrated by the stereospecific conversion of limonene into the pure *cis*-diol (15.28) or into pure carvomenthol[21] (15.29).

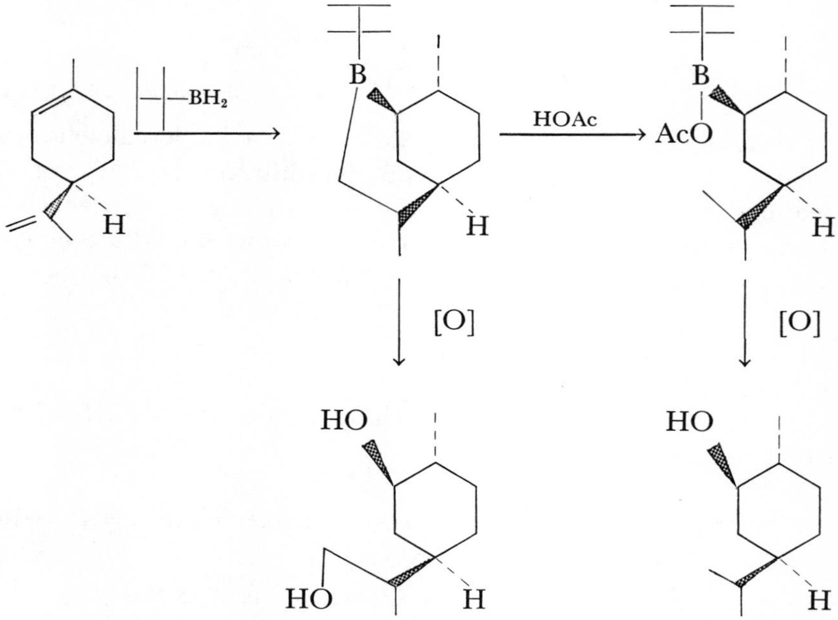

15.28

15.29

7. 9-Borabicyclo[3.3.1]Nonane [26] (9-BBN) and Other Reagents

The reaction of 1,5-cyclooctadiene with borane in tetrahydrofuran can be controlled to provide the bicyclic borane, 9-borabicyclo-[3.3.1]nonane (15.30).

$$+ BH_3 \longrightarrow \qquad\qquad 15.30$$

(dimer)

This material, termed 9-BBN for convenience, possesses remarkable thermal stability. Moreover, it exhibits remarkable air stability for a dialkylborane. Yet it hydroborates olefins readily to produce the corresponding 9-alkyl-9-BBN (15.31).

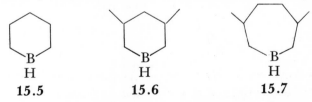

$$CH_3 + 9\text{-BBN} \longrightarrow H_3C \qquad\qquad 15.31$$

Many of the new reactions of organoboranes (Chapter XVI) use only one of the three alkyl groups on the trialkylborane. For many of these reactions the use of 9-BBN as the hydroborating agent solves the problem: only the 9-R group participates in these reactions.

Methods have recently been developed for the synthesis of bora-cyclohexane[27] (**15.5**) and the related derivatives[28] (**15.6** and **15.7**).

15.5	15.6	15.7
B	B	B
H	H	H

[26] E. F. Knights and H. C. Brown, *J. Amer. Chem. Soc.*, **90,** 5280, 5281, 5283 (1968).

[27] H. C. Brown and E. Negishi, *J. Organometal. Chem.*, **27,** C67 (1971).

[28] *Ibid.*, **28,** C1 (1971).

Hydroboration

These reagents also possess interesting properties and are finding valuable application.

Catechol borane, although not strictly a dialkylborane, readily reacts with alkenes and alkynes to produce the corresponding esters of the boronic acid.[18] This provides a simple convenient route to the alkyl- (15.32) and vinylboronic acids (15.15).

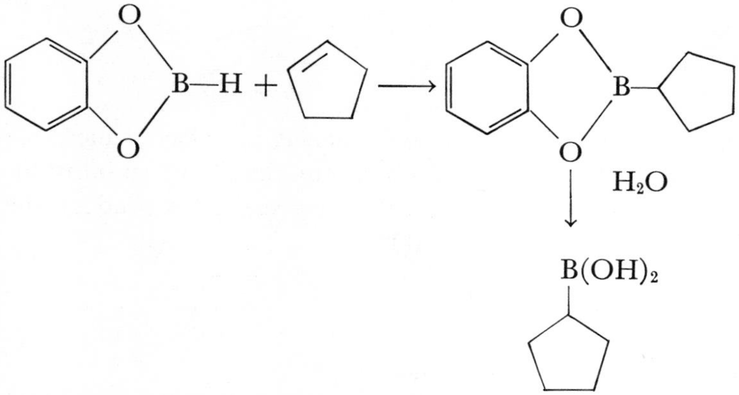

$$15.32$$

Finally, the treatment of boronic and borinic esters with lithium aluminum hydride in the presence of olefins[29] or with aluminum hydride[30] provides convenient routes to "mixed" organoboranes, probably through the intermediacy of the corresponding dialkylborane (15.33).

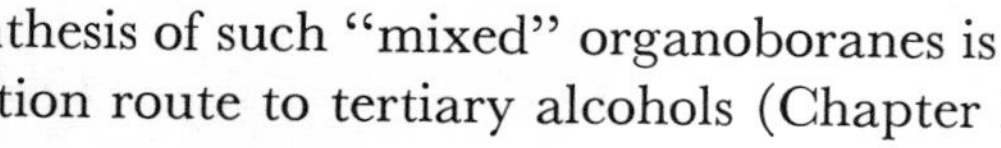

$$15.33$$

The ready synthesis of such "mixed" organoboranes is important for the carbonylation route to tertiary alcohols (Chapter XVII).

[29] H. C. Brown, E. Negishi, and S. K. Gupta, *J. Amer. Chem. Soc.*, **92**, 6648 (1970).
[30] H. C. Brown and S. K. Gupta, *J. Amer. Chem. Soc.*, **93**, 1818 (1971).

8. Conclusion

The availability of both monofunctional and bifunctional hydroborating reagents adds new dimensions to a reaction already remarkable for its versatility. It opens up a major new area of structural chemistry of the organoboranes, an area now only in its infancy.

We are now able to synthesize an extraordinarily wide variety of organoboranes. In the next chapter we consider the reactions to which these organoboranes may be subjected.

PART SIX

ORGANOBORANES

XVI. The Versatile Organoboranes[1,2]

1. Background

For many years the only practical route to the organoboranes involved the reaction of an organometallic, generally the organozinc[3] or the organomagnesium[4] derivative, with the boron ester or halide. Since the synthesis of the organoborane involved the prior formation of an organometallic, there was little incentive to explore the possible utility of the organoboranes, another class of organometallics, as intermediates in organic synthesis. Consequently, until recently, little information was available on reactions of interest to the chemist active in synthesis.[5]

For example, it was known that the aliphatic organoboranes are typical nonpolar substances, which in their physical properties resemble closely the corresponding hydrocarbons. They are relatively stable to water, but are sensitive to oxygen (16.1). They react sluggishly with the halogens and with the halogen acids (16.1), but the reactions did not appear to be of synthetic value.

$$R_3B + O_2 \xrightarrow{H_2O} R_2BOR$$
$$R_3B + O_2 \xrightarrow{dry} RB(OR)_2$$
$$R_3B + Br_2 \longrightarrow R_2BBr + RBr \qquad 16.1$$
$$R_3B + HBr \longrightarrow R_2BBr + RH$$

The organoboranes are coordinatively unsatisfied. They readily form addition compounds with ammonia and the amines (16.2).

$$R_3B + NR_3 \rightleftharpoons R_3B{:}NR_3 \qquad 16.2$$

---

[1] H. C. Brown, *Hydroboration* (W. A. Benjamin, New York, 1962).

[2] H. C. Brown, "Moderni Sviluppi Delle Sintesi Organica," *Conferenze X, Corso Estivo di Chimica, Academia Nazionale dei Lincei* (Rome, 1968), pp. 73–103.

[3] E. Frankland and B. F. Duppa, *Proc. Roy. Soc.* (*London*), **10**, 568 (1859).

[4] E. Krause and R. Nitsche, *Ber.*, **54**, 2784 (1921).

[5] M. F. Lappert, *Chem. Revs.*, **56**, 959 (1956).

Although this reaction has proved of major value in the investigation of steric effects (Chapters V, VI), it apparently also possessed no significance for organic synthesis.

Indeed, only the ready oxidation of organoboranes with alkaline hydrogen peroxide, briefly mentioned by Johnson and Van Campen[6] (16.3), appeared to offer utility to the synthetic chemist.

$$R_3B + 3H_2O_2 + NaOH \longrightarrow 3ROH + NaB(OH)_4 \qquad 16.3$$

Now that the organoboranes were readily available from olefins without an organometallic as intermediate, it appeared desirable to initiate a research program involving systematic exploration of their chemistry. This exploration has proven remarkably fruitful. Many new reactions of major significance to synthetic chemistry have been and are being discovered. This chapter will review some of the more promising developments in this area. There is little question but that the organoboranes are the most versatile organometallics now available for organic synthesis.

2. Isomerization of Organoboranes [7, 8]

At moderate temperatures the organoboranes undergo a facile isomerization that proceeds to place the boron atom at the least hindered position of the alkyl groups (16.4).

$$16.4$$

[6] J. R. Johnson and M. G. Van Campen, Jr., *J. Amer. Chem. Soc.*, **60**, 121 (1938).

[7] H. C. Brown and B. C. Subba Rao, *J. Amer. Chem. Soc.*, **81**, 6434 (1959).

[8] H. C. Brown and G. Zweifel, *J. Amer. Chem. Soc.*, **88**, 1433 (1966).

This makes possible a number of very valuable syntheses, not otherwise practicable. For example, the following syntheses may be readily achieved in essentially quantitative yield by the use of hydroboration and isomerization[8] (16.5).

$$
\begin{array}{c}
\text{C} \\
| \\
\text{C} \\
| \\
\text{C—C—C—C—C} \\
| \\
\text{OH}
\end{array}
\xrightarrow{-\text{H}_2\text{O}}
\begin{array}{c}
\text{C} \\
| \\
\text{C} \\
| \\
\text{C—C—C=C—C}
\end{array}
$$

$$\Big\downarrow \text{HB}$$

$$
\begin{array}{c}
\text{C} \\
| \\
\text{C} \\
| \\
\text{C—C—C—C—C} \\
| \\
\text{O} \\
\text{H}
\end{array}
\xleftarrow{[\text{O}]}
\begin{array}{c}
\text{C} \\
| \\
\text{C} \\
| \\
\text{C—C—C—C—C} \\
| \\
\text{B} \\
\diagup\diagdown
\end{array}
\qquad 16.5
$$

$$\Big\downarrow \Delta$$

$$
\begin{array}{c}
\text{C} \\
| \\
\text{C} \\
| \\
\text{C—C—C—C—C} \\
| \\
\text{O} \\
\text{H}
\end{array}
\xleftarrow{[\text{O}]}
\begin{array}{c}
\text{C} \\
| \\
\text{C} \\
| \\
\text{C—C—C—C—C} \\
| \\
\text{B} \\
\diagup\diagdown
\end{array}
$$

It is also possible to move the boron atom out of the ring and into the side chain[9] (16.6).

This ready isomerization of organoboranes under mild conditions makes possible a number of interesting syntheses of desired compounds from alternative, more readily available intermediates, such

[9] H. C. Brown and G. Zweifel, *J. Amer. Chem. Soc.*, **89,** 561 (1967).

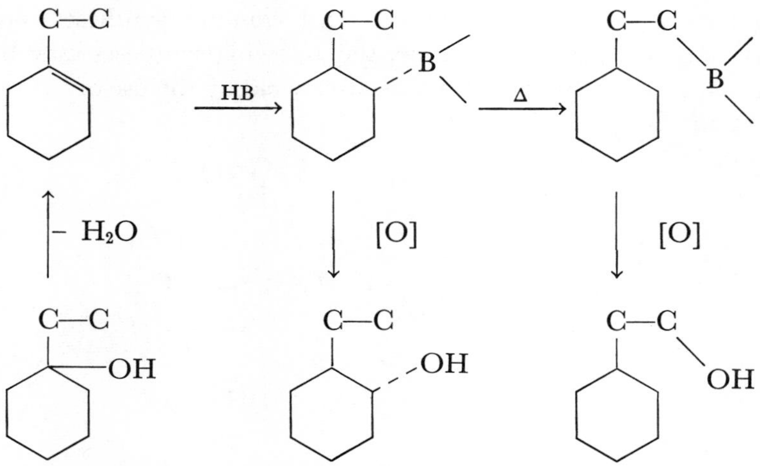

Organoboranes

$$16.6$$

as the synthesis of the exocyclic alcohol, 2-cyclohexylethanol, from the readily available endocyclic olefin, 1-ethylcyclohexene, illustrated above.

3. Displacement Reactions of Organoboranes

The mechanism of the isomerization reaction appears to involve a partial dissociation of the organoborane into olefin and the boron-hydrogen bond, followed by readition. The process occurs repeatedly, until the boron atoms end up at the least hindered position of the molecule, thereby yielding the most stable of the organoboranes derivable from the particular alkyl groups used[7] (16.7a and 16.7b).

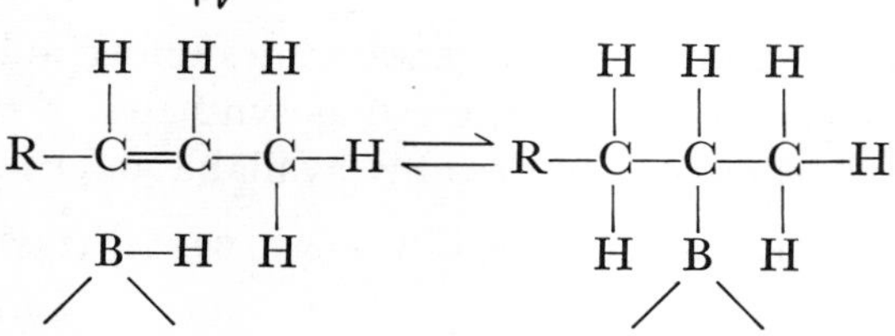

$$16.7a$$

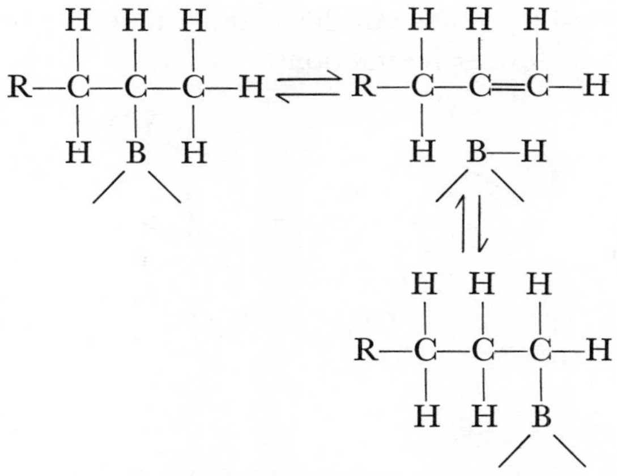

$$
\underset{\substack{| \quad | \quad | \\ H \quad B \quad H \\ \diagup \diagdown}}{R-\overset{\overset{\displaystyle H \ \ H \ \ H}{|}}{C}-\overset{|}{C}-\overset{|}{C}-H} \ \rightleftharpoons \ \underset{\substack{| \quad | \\ H \quad B-H \\ \diagup \diagdown}}{R-\overset{\overset{\displaystyle H \ \ H \ \ H}{|}}{C}-\overset{|}{C}=\overset{|}{C}-H}
$$

16.7b

$$
\underset{\substack{| \quad | \quad | \\ H \quad H \quad B \\ \diagup \diagdown}}{R-\overset{\overset{\displaystyle H \ \ H \ \ H}{|}}{C}-\overset{|}{C}-\overset{|}{C}-H}
$$

It follows from this mechanism that the presence of another olefin, of equal or greater reactivity, will cause the original olefin to be displaced from the organoborane.[10,11] The combination of hydroboration, isomerization, and displacement makes possible the contra-thermodynamic isomerization of olefins[12,13] (16.8).

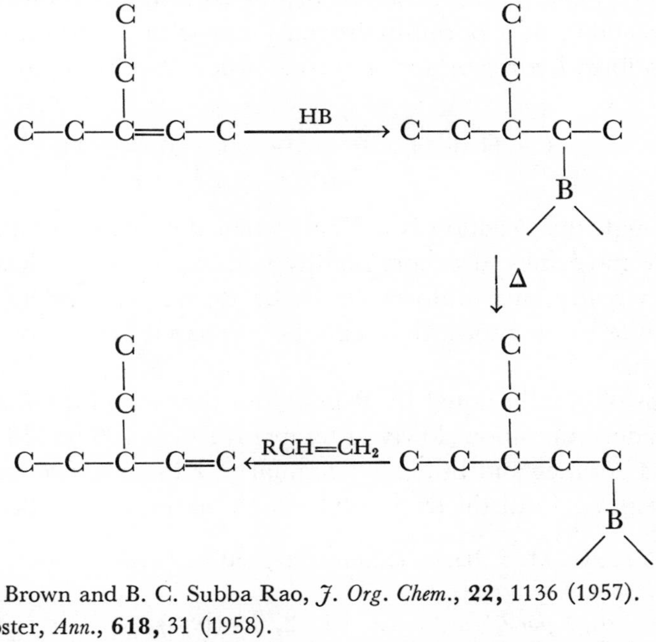

$$
\underset{\substack{| \\ C \\ | \\ C}}{C-C-C}=C-C \ \xrightarrow{\text{HB}} \ \underset{\substack{| \\ C \\ | \\ C \\ \\ B \\ \diagup\diagdown}}{C-C-C-C-C}
$$

16.8

$$
\Big\downarrow \Delta
$$

$$
\underset{\substack{| \\ C \\ | \\ C}}{C-C-C}-C=C \ \xleftarrow{\text{RCH}=\text{CH}_2} \ \underset{\substack{| \\ C \\ | \\ C \\ \\ B \\ \diagup\diagdown}}{C-C-C-C-C}
$$

[10] H. C. Brown and B. C. Subba Rao, *J. Org. Chem.*, **22,** 1136 (1957).
[11] R. Köster, *Ann.*, **618,** 31 (1958).
[12] H. C. Brown and M. V. Bhatt, *J. Amer. Chem. Soc.*, **82,** 2074 (1960).
[13] H. C. Brown and M. V. Bhatt, *J. Amer. Chem. Soc.*, **88,** 1440 (1966).

It is also possible to move in this manner a double bond from the endocyclic to the exocyclic position[14] (16.9).

$$16.9$$

4. Cyclization

Besides isomerization, the organoboranes undergo other interesting reactions under the influence of heat.[15] The most promising is the ready reaction of a boron-hydrogen bond with a carbon-hydrogen bond to form hydrogen and a boron-carbon bond (16.10).

$$16.10$$

Although this reaction is difficult when the two groupings are in separate molecules, it occurs readily within a single molecule, providing a ready entry into cyclic boron derivatives and to the corresponding compounds that can be prepared from boron intermediates.

It was originally noted by Rosenblum that tri-n-butylborane undergoes decomposition slowly at temperatures of 100 to 130° to form n-butyldiboranes and butenes.[16] Similarly, Köster has reported that in attempting to distill tri-n-decylborane he obtained 1-decene and

---

[14] H. C. Brown, M. V. Bhatt, T. Munekata, and G. Zweifel, *J. Amer. Chem. Soc.*, **89**, 567 (1967).

[15] R. Köster, *Angew. Chem. Internat. Edit.*, **3**, 174 (1964).

[16] L. Rosenblum, *J. Amer. Chem. Soc.*, **77**, 5016 (1955).

n-decyldiboranes,[11] and Ashby has noted a smooth decomposition of tri-n-octylborane, upon attempted distillation at 169° at 0.2 mm, into *trans*-2-octene and tetra-n-octyldiborane[17] (16.11).

$$2(n-C_4H_9)_3B \xrightarrow{\Delta} 2\ n-C_4H_8 + [(n-C_4H_9)_2BH]_2 \quad 16.11$$

These results are all consistent in indicating a decomposition of the organoborane at moderate temperatures into the dialkylborane (dimer) and olefin. However, there appears to be some question as to the structure of the olefin produced in the reaction.

A somewhat different reaction was observed by Winternitz for tri-n-pentylborane.[18] Maintained at its boiling point, it evolved hydrogen and *trans*-2-pentene in equimolar amounts, and was transformed into a new product, identified as a cyclic derivative, 1-n-pentyl-2-methylboracyclopentane (16.12).

$$(n-C_5H_{11})_3\overset{.}{B} \xrightarrow{\Delta} n-C_5H_{10} + H_2 + \begin{array}{c} H_2C\!-\!\!-\!\!-\!CH_2 \\ |\qquad\quad| \\ H_2C\qquad CH\!-\!CH_3 \\ \diagdown\ \ \diagup \\ B \\ | \\ n\!-\!C_5H_{11} \end{array} \quad 16.12$$

It was also reported that tri-n-hexylborane reacts considerably slower, but likewise yields hydrogen, an olefin, and the related cyclic compound[18] (16.13).

$$(n-C_6H_{13})_3B \xrightarrow{\Delta} n-C_6H_{12} + H_2 + \begin{array}{c} CH_2 \\ \diagup\quad\ \diagdown \\ H_2C\qquad\ CH_2 \\ |\qquad\qquad| \\ H_2C\qquad\ CH\!-\!CH_3 \\ \diagdown\quad\ \diagup \\ B \\ | \\ n\!-\!C_6H_{13} \end{array} \quad 16.13$$

[17] E. C. Ashby, *J. Amer. Chem. Soc.*, **81**, 4791 (1959).

[18] P. F. Winternitz and A. A. Carotti, *J. Amer. Chem. Soc.*, **82**, 2430 (1960).

This reaction presumably proceeds through an initial dissociation to the dialkylborane, followed by closure of the ring (16.14).

$$(n-C_5H_{11})_3B \xrightarrow{\Delta} (n-C_5H_{11})_2BH + n-C_5H_{10}$$

$$\xrightarrow{\Delta} + H_2 + \qquad \qquad \text{16.14}$$

One puzzling aspect of this experiment calls for clarification. Our own study of related systems indicates that primary carbon-hydrogen bonds participate more readily than secondary.[19] This could account for the difference in rate between the *n*-pentyl and *n*-hexylboranes. Saegebarth has reported that in the isomerization of cyclic organoboranes derived from dienes the reaction proceeds preferentially to the formation of six-membered ring derivatives.[20] On this basis, the product from the cyclization of tri-*n*-pentylborane would be expected to be predominantly the 1-*n*-pentylboracyclohexane derivative (**16.1**).

16.1

In the hydroboration of *trans*-di-*t*-butylethylene, Logan and Flautt observed that the hydroboration ceases at the monoalkylborane stage. The product could be isolated, but lost hydrogen readily above 100° to form a cyclic organoborane. Oxidation of the latter with

[19] H. C. Brown, K. J. Murray, H. Muller, and G. Zweifel, *J. Amer. Chem. Soc.*, **88**, 1443 (1966).

[20] K. A. Saegebarth, *J. Amer. Chem. Soc.*, **82**, 2081 (1960).

alkaline hydrogen peroxide provided the glycol, 2,2,5,5-tetramethyl-1,4-hexanediol[21] (16.15).

This reaction has been explored as a general synthesis of glycols from olefins.[19,22] Thus the hydroboration of 2,4,4-trimethyl-1-pentene to the dialkylborane stage, followed by a short period of heating at 200 to 220°, permits the synthesis of 2,4,4-trimethyl-1,5-pentanediol in 80 percent yield (16.16).

The use of thexylborane has many advantages for such syntheses[19] (16.17).

By heating tri-n-butylborane in an autoclave at 300°, Köster was able to achieve cyclization in spite of the fact that this requires the formation of the apparently less favored 5-membered ring. Oxidation then afforded 1,4-butanediol in 80 percent yield[22] (16.18).

Because of the many possibilities for isomerization, the products from higher boron alkyls, such as tris(2-methyl-1-pentyl)borane, are mixtures and the reaction is less favorable for the synthesis of pure compounds.

[21] T. J. Logan and T. J. Flautt, *J. Amer. Chem. Soc.*, **82**, 3446 (1960).
[22] R. Köster and G. Rotermund, *Angew. Chem.*, **72**, 138 (1960).

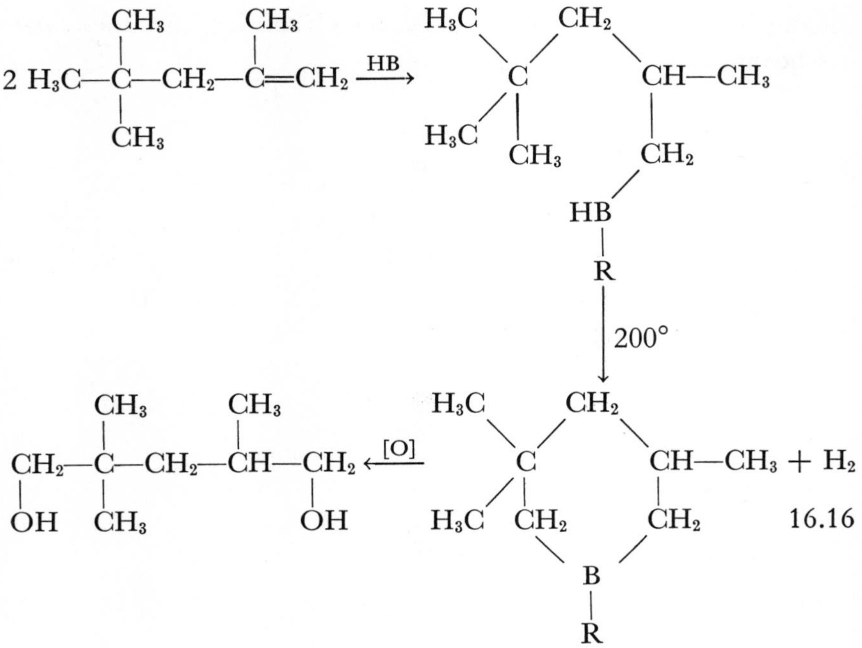

Organoboranes

$$2 \; H_3C-\underset{\underset{CH_3}{|}}{\overset{\overset{CH_3}{|}}{C}}-CH_2-\underset{\overset{|}{CH_3}}{C}=CH_2 \xrightarrow{\text{HB}}$$

$$H_3C-\underset{\underset{CH_3}{|}}{\overset{\overset{CH_3}{|}}{C}}-CH_2-\underset{H}{\overset{|}{C}}H-CH_3$$ with HB–R group, 200°

$$\underset{\overset{|}{OH}}{CH_2}-\underset{\overset{|}{CH_3}}{\overset{\overset{CH_3}{|}}{C}}-CH_2-\underset{}{C}H-\underset{\overset{|}{OH}}{CH_2} \xleftarrow{\text{[O]}}$$

ring structure with B–R $+ H_2$

16.16

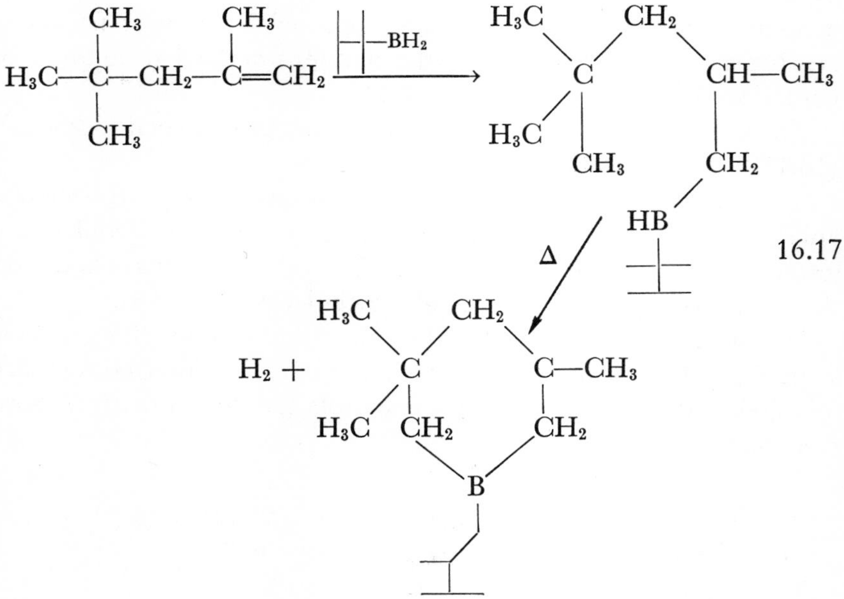

$$H_3C-\underset{\underset{CH_3}{|}}{\overset{\overset{CH_3}{|}}{C}}-CH_2-\underset{\overset{|}{CH_3}}{C}=CH_2 \xrightarrow{\quad\text{BH}_2}$$

ring structures with HB, Δ

16.17

$$H_2 +$$ ring structure with B

$$(n{-}C_4H_9)_3B \xrightarrow{300°} \begin{array}{c} H_2C{-}\!\!-\!\!-CH_2 \\ | \qquad\qquad | \\ H_2C \qquad CH_2 \\ \diagdown \quad \diagup \\ B \\ | \\ n{-}C_4H_9 \end{array} + H_2 + C_4H_8$$

$$\Big\downarrow [O] \qquad\qquad\qquad 16.18$$

$$\begin{array}{ccc} CH_2CH_2CH_2CH_2 & + & CH_3CH_2CH_2CH_2 \\ | \qquad\qquad | & & | \\ OH \qquad\quad OH & & OH \end{array}$$

A much simpler, related reaction occurs with the aromatic derivatives[23] (16.19).

$$+ \qquad\qquad + H_2 \quad 16.19$$

The noncyclic alkyl groups, R, can be removed by hydrogenation at 160° to provide products, such as boraindane (**16.2**), 3-methylboraindane (**16.3**), and boratetralin (**16.4**), dimers in solution.

[23] R. Köster and K. Reinert, *Angew. Chem.*, **71**, 521 (1959).

$$\begin{bmatrix} \text{16.2} \end{bmatrix} \quad \begin{bmatrix} \text{16.3} \end{bmatrix} \quad \begin{bmatrix} \text{16.4} \end{bmatrix}$$

Finally, Köster reports that tri-*n*-octylborane (16.20) and tri-*n*-nonylborane (16.21) yield bicyclic products in addition to the simple monocyclic derivatives of the type realized with tri-*n*-pentylborane and tri-*n*-hexylborane.[15] The conditions are rigorous, temperatures between 250 and 350° being employed.

$$(n\text{-}C_8H_{17})_3B \longrightarrow \quad + 2C_8H_{16} + 2H_2 \qquad 16.20$$

$$(n\text{-}C_9H_{19})_3B \longrightarrow$$

8%   ~80%   ~12%

16.21

$$\underset{\displaystyle OH}{CH_2CH_2CH_2CH_2}\underset{\displaystyle OH}{CHCH_2CH_2CH_2CH_2}\underset{\displaystyle OH}{}$$

CH₂CH₂CH₂CH₂CHCH₂CH₂CH₂CH₂
   | | |
  OH   OH   OH

It is possible to achieve the synthesis of such cyclic boron derivatives (16.22, 16.23) under much milder conditions via the hydroboration reaction[24] (Section XV–6).

[24] H. C. Brown and C. D. Pfaffenberger, *J. Amer. Chem. Soc.*, **89,** 5475 (1967).

$$H_3C-C=CH_2 \quad + H_2B- \quad \longrightarrow \quad H_3C-CH-CH_2 \quad B- \quad 16.22$$
$$CH=CH_2 \quad\quad\quad\quad CH_2-CH_2$$

$$CH_2CH=CH_2 \quad + H_2B- \quad \longrightarrow \quad CH_2CH_2CH_2 \quad B- \quad 16.23$$
$$CH_2CH=CH_2 \quad\quad\quad\quad CH_2CH_2CH_2$$

Similarly, 9-boradecalin can be obtained under relatively mild conditions from the 1,4,8-nonatriene[25] (16.24).

$$CH_2=CHCH_2CH=CHCH_2CH_2CH=CH_2 \xrightarrow[\Delta]{HB}$$

16.24

5. Protonolysis

The trialkylboranes are remarkably stable toward water. Ulmschneider and Goubeau report that the hydrolysis of trimethylborane with an equivalent of water for 7 hours at 180° affords a 69 percent yield of dimethylborinic acid.[26] The reaction with hydrogen sulfide is even more sluggish: a temperature of 280° is required to achieve conversion to boron sulfide. The reactions with alcohols and phenols are also slow. Very high temperatures were required to achieve the indicated reaction with ethylene glycol (16.25) and catechol (16.26).

$$CH_2OH \quad + B(CH_3)_3 \xrightarrow{340°} \quad CH_2-O \quad B-CH_3 + 2CH_4 \quad 16.25$$
$$CH_2OH \quad\quad\quad\quad\quad\quad CH_2-O$$

[25] H. C. Brown and E. Negishi, *J. Amer. Chem. Soc.*, **91**, 1224 (1969).
[26] D. Ulmschneider and J. Goubeau, *Ber.*, **90**, 2733 (1957).

Organoboranes

$$+ B(CH_3)_3 \xrightarrow{290°}$$

$$B\text{—}CH_3 + 2CH_4 \quad 16.26$$

The addition of alkali to water stabilizes the trialkylboron toward hydrolysis.[27] Treatment with concentrated mineral acids facilitates the hydrolysis. However, even heating under reflux with such acids brings about the loss of only one alkyl group. Thus, Johnson *et al.* report that in heating tri-*n*-butylborane with 48 percent hydrobromic acid under reflux a quantitative yield of di-*n*-butylboronic acid was obtained in 1 hour[28] (16.27).

$$(n\text{-}C_4H_9)_3B + HBr \longrightarrow (n\text{-}C_4H_9)_2BBr + n\text{-}C_4H_{10} \quad 16.27$$
$$(n\text{-}C_4H_9)_2BBr + H_2O \longrightarrow (n\text{-}C_4H_9)_2BOH + HBr$$

Somewhat unexpectedly, the organoboranes have proved to be more susceptible to attack by carboxylic acids than by mineral acids. Thus, it has been observed that under relatively mild conditions triethylborane can be converted into diethylboron acetate and ethane.[29,30]

$$(C_2H_5)_3B + CH_3CO_2H \xrightarrow{100°} (C_2H_5)_2BOCOCH_3 + C_2H_6 \quad 16.28$$

A detailed study of the action of carboxylic acids on organoboranes has revealed that two of the three groups can be removed by excess anhydrous acid at room temperature, and all three groups can generally be removed by refluxing the organoborane in diglyme solution

[27] H. C. Brown and N. C. Hebert, unpublished observations.

[28] J. R. Johnson, H. R. Snyder, and M. G. Van Campen, Jr., *J. Amer. Chem. Soc.*, **60**, 115 (1938).

[29] H. Meerwein, G. Hinz, H. Majert, and H. Sönke, *J. prakt. Chem.*, **147**, 226 (1936).

[30] J. Goubeau, R. Epple, D. Ulmschneider, and H. Lehmann, *Angew. Chem.*, **67**, 710 (1955).

with a moderate excess of propionic acid for 2 to 3 hours.[31] Conse-quently, hydroboration of olefins in diglyme, followed by refluxing with propionic acid, offers a convenient noncatalytic procedure for the hydrogenation of double bonds (16.29).

$$3RCH=CH_2 \xrightarrow{HB} 3(RCH_2CH_2)_3B \xrightarrow[\Delta]{C_2H_5CO_2H} 3RCH_2CH_3 \quad 16.29$$

Secondary alkyl groups undergo protonolysis less readily than primary. Consequently, in hydrogenating internal olefins, especially those containing bulky groups, it is preferable that the boron atom be transferred to the terminal position by heating under reflux prior to the addition of the acid. Moreover, use of the less volatile solvent, triglyme, permits completion of the protonation stage in $\frac{1}{2}$ to 1 hour.

The combined hydroboration-protonolysis procedure has been ap-plied to a considerable number of representative olefins, such as 1-hexene, 2-hexene, 1-octene, 2,4,4-trimethyl-1-pentene, 2,4,4-trimethyl-2-pentene, cyclopentene, and cyclohexene. Yields of 80 to 90 percent of the hydrogenated product are realized.

Since olefins containing active sulfur, chlorine, and nitrogen sub-stituents readily undergo hydroboration, this procedure opens up the possibility of hydrogenating olefinic derivatives containing such groups. In this way allylmethylsulfide is converted into *n*-propyl-methylsulfide in a yield of 78 percent.

The reaction appears to proceed with retention of configuration. Thus tri-*exo*-norbornylborane undergoes deuterolysis to yield *exo*-deuteronorbornane[32] (16.30).

16.30

The complete protonolysis of the alkylboranes has been applied for the quantitative analysis of these derivatives.[33]

[31] H. C. Brown and K. Murray, *J. Amer. Chem. Soc.*, **81**, 4108 (1959).

[32] H. C. Brown and K. J. Murray, *J. Org. Chem.*, **26**, 631 (1961).

[33] J. Crighton, A. K. Holliday, A. G. Massey, and N. R. Thompson, *Chem. and Ind.* (London), 347 (1960).

Organoboranes

Organoboranes of the vinyl type, readily obtained by the hydroboration of acetylenes (Section XV–4), undergo protonolysis much more readily. These compounds undergo complete protonolysis with acetic acid at 0°, and the reaction provides a convenient means of converting acetylenes into *cis* olefins of high purity[34] (16.31).

$$
\begin{array}{c}
R \\
| \\
C \\
||| \\
C \\
| \\
R
\end{array}
\xrightarrow{\text{HB}}
\begin{array}{c}
R\!-\!C\!-\!H \\
|| \\
R\!-\!C\!-\!B
\end{array}
\!\!\diagup\!\!
\xrightarrow[0^\circ]{\text{CH}_3\text{CO}_2\text{H}}
\begin{array}{c}
R\!-\!C\!-\!H \\
|| \\
R\!-\!C\!-\!H
\end{array}
\qquad 16.31
$$

Here, also, it has been demonstrated that carboxylic acids are especially effective. The protonolysis fails to proceed under these conditions with strong mineral acids, such as hydrochloric or sulfuric acid.[35]

It would appear that the unique effectiveness of carboxylic acids must derive from the structure of the carboxylic acid group. Presumably the trialkylborane is capable of coordinating at the oxygen of the carbonyl group, placing the boron-carbon bond in position for an intramolecular attack by the proton (16.32).

$$
\text{CH}_3\text{C}
\begin{array}{c}
\diagup\text{O} \\
\diagdown\text{O}\!-\!\text{H}
\end{array}
\xrightarrow{\text{BR}_3}
\text{CH}_3\text{C}
\begin{array}{c}
\diagup\text{O:B}\!\!\begin{smallmatrix} R \\ \diagup \\ \diagdown R \end{smallmatrix} \\
\diagdown\text{O}\!-\!\text{H}
\end{array}
$$

$$
\downarrow \qquad\qquad 16.32
$$

$$
\text{CH}_3\text{C}
\begin{array}{c}
\diagup\text{O}\!-\!\text{BR}_2 \\
\diagdown\diagdown\text{O}
\end{array}
\quad + \text{RH}
$$

[34] H. C. Brown and G. Zweifel, *J. Amer. Chem. Soc.*, **81,** 1512 (1959).
[35] H. C. Brown and D. Bowman, unpublished observations.

The increasing difficulty in each successive stage presumably arises in large part from the decreased ability of the dialkylboron acetate and the monoalkylboron· diacetate to accept an electron-pair and become coordinated to the carbonyl group.

6. Halogenolysis[36, 37]

The rupture of the carbon-boron bond by direct reaction of halogens with the trialkylboranes has proven to be surprisingly difficult. Thus bromine reacts only relatively slowly with tri-*n*-butylborane in the dark in the absence of a solvent to form *n*-butyl bromide, di-*n*-butylbromoborane, *n*-butane, and hydrogen bromide.[38] The reaction is greatly facilitated by the use of methylene chloride as a solvent and provides a convenient procedure for the *anti*-Markovnikov hydrobromination of olefins[39] (16.33).

$$[(CH_3)_2CHCH_2]_3B + Br_2 \xrightarrow[\text{dark}]{\underset{CH_2Cl_2}{25°}} (CH_3)_2CHCH_2Br + [(CH_3)_2CHCH_2]_2BBr \quad 16.33$$

The use of B-R-9-BBN derivatives (Section XV–7) provides a higher utilization of the alkyl group in many cases, and a good yield of the *anti*-Markovnikov bromide from the olefin[40] (16.34).

Investigation revealed an unexpected feature—the reaction does not involve simple rupture of the carbon-boron bond by bromine. Instead, the reaction proceeds through the fast α-bromination of the organoborane and subsequent reaction of the intermediate with the hydrogen bromide[30,41] (16.35).

[36] H. C. Brown, M. W. Rathke, and M. M. Rogić, *J. Amer. Chem. Soc.*, **90,** 5038 (1968).

[37] H. C. Brown and C. F. Lane, *J. Amer. Chem. Soc.*, **92,** 6660 (1970).

[38] J. R. Johnson, H. R. Snyder, and M. G. Van Campen, Jr., *J. Amer. Chem. Soc.*, **60,** 115 (1938).

[39] C. F. Lane and H. C. Brown, *J. Amer. Chem. Soc.*, **92,** 7212 (1970).

[40] C. F. Lane and H. C. Brown, *J. Organometal. Chem.*, **26,** C51 (1971).

[41] The substitution in the *alpha* position of the organoborane is greatly facilitated by light. The α-bromo derivatives are readily transformed into new carbon structure by treatment with base. Consequently, this development provides a promising new route for the synthesis of carbon structures (Chapter XVIII). C. F. Lane and H. C. Brown, *J. Amer. Chem. Soc.*, **93,** 1025 (1971).

Organoboranes

$$R_2BCH_2CH(CH_3)_2 + Br_2 \xrightarrow{\text{fast}} R_2BCHCH(CH_3)_2 + HBr$$

with Br below the C, and

$$\downarrow \text{slow} \qquad\qquad 16.35$$

$$R_2BBr + (CH_3)_2CHCH_2Br$$

These difficulties in the brominolysis of organoboranes are largely overcome by treating the organoborane with bromine at 0° in the presence of sodium methoxide in methanol[37] (16.36).

$$(CH_3CH_2CH_2\overset{\overset{\displaystyle CH_3}{|}}{C}HCH_2)_3B + 3Br_2 + 4NaOCH_3 \longrightarrow$$

$$3CH_3CH_2CH_2\overset{\overset{\displaystyle CH_3}{|}}{C}HCH_2Br + 3NaBr + NaB(OCH_3)_4 \qquad 16.36$$
$$(99\%)$$

Similarly, the reaction of iodine with organoboranes is exceedingly sluggish.[42] However, the addition of sodium hydroxide in methanol

[42] L. H. Long and D. Dollimore, *J. Chem. Soc.*, 3902, 3906 (1953).

to the organoborane and iodine brings about a rapid reaction[36] (16.37).

$$(RCH_3CH_2)_3B + 2I_2 + 2NaOH \longrightarrow 2RCH_2CH_2I + 2NaI$$
$$+ RCH_2CH_2B(OH)_2 \quad 16.37$$

Only two of three primary alkyl groups react, and the situation is even less favorable for secondary alkyl groups. Fortunately, the use of disiamylborane circumvents this difficulty (16.38).

$$RCH{=}CH_2 + Sia_2BH \longrightarrow RCH_2CH_2BSia_2 \qquad 16.38$$
$$RCH_2CH_2BSia_2 + I_2 + NaOH \longrightarrow RCH_2CH_2I + Sia_2BOH + NaI$$

The reaction appears to be widely applicable, as shown by the following representative transformations (16.39–16.41).

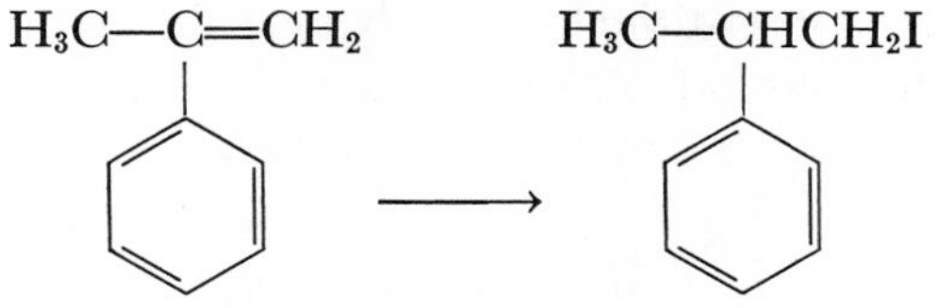

16.39

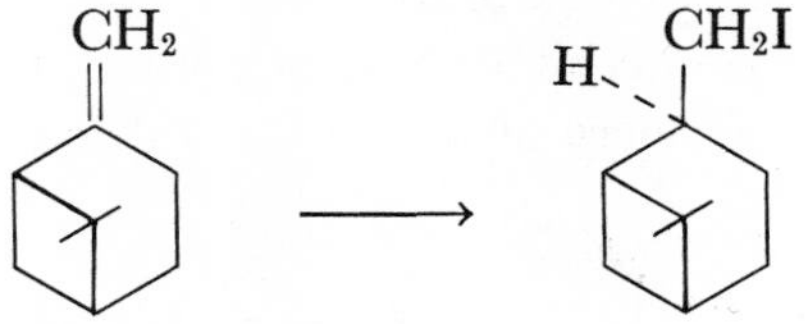

16.40

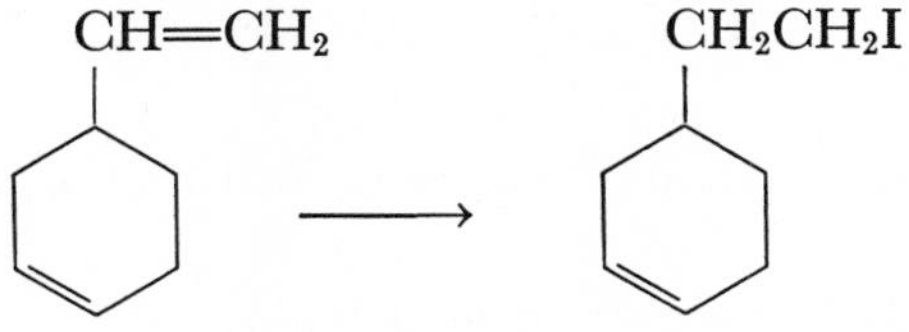

16.41

Consequently the treatment of the organoboranes with bromine or iodine under alkaline conditions permits the halogenolysis of the organoboranes under mild conditions. This permits the synthesis in many instances of the desired organic halide while taking advantage of the high selectivity and stereospecificity of the hydroboration reaction.

7. Oxidation—Oxygen [43]

It has long been known that organoboranes, both aliphatic and aromatic, are highly reactive toward oxygen.[44] Considerable attention has been paid to the mechanism of the autoxidation reaction.[45] However, the reaction has shown no promise for synthetic purposes.[46,47] Sufficient oxygen is readily absorbed to oxidize all three carbon-boron bonds, but the reaction is reported to be complex. Long reaction times and elevated temperatures were used for the oxidations in an unsuccessful attempt to achieve high conversion to the alcohol.

It is quite clear that the initial product of such oxidations is the peroxide[45] (16.42).

$$R_3B + O_2 \longrightarrow R_2BO_2R \qquad\qquad 16.42$$

It appeared possible that the low yields were the result of the thermal instability of the peroxide as oxygen was forced in to achieve complete conversion of the carbon-boron bonds.[46] Consequently, we utilized gentle conditions (0°) to introduce the theoretical amount of oxygen (1.5 moles of O_2 per mole of R_3B) into a solution of the organoborane in tetrahydrofuran.[48] The reaction product, possibly $RB(O_2R)OR$, was treated with aqueous alkali. Essentially quantitative yields of alcohols were obtained (16.43).

$$16.43$$

98%

[43] H. C. Brown, M. M. Midland, and G. W. Kabalka, *J. Amer. Chem. Soc.*, **93,** 1024 (1971).

[44] Certain triarylboranes with large sterically demanding groups, such as trimesitylborane, are stable to oxygen and can be handled in air without special precautions. H. C. Brown and V. H. Dodson, *J. Amer. Chem. Soc.*, **79,** 2302 (1957).

[45] A. G. Davies and B. P. Roberts, *J. Chem. Soc.* **B,** 311 (1969).

[46] S. B. Mirviss, *J. Amer. Chem. Soc.*, **83,** 3051 (1961).

[47] S. B. Mirviss, *J. Org. Chem.*, **32,** 1713 (1967).

[48] We used the small automatic hydrogenator, C. A. Brown and H. C. Brown, *J. Amer. Chem. Soc.*, **84,** 3892 (1962), modified to generate oxygen.[43]

A disadvantage of this procedure is that it does not possess the stereospecificity of the oxidation with alkaline hydrogen peroxide (Section XVI–8). For example, the latter procedure gives *exo*-norborneol in an isomeric purity of 99.5 percent. Similarly, the alcohol from 1-methylcyclopentene is essentially pure *trans*-2-methyl-cyclopentanol. The oxidation with oxygen gas produces a much larger amount of the alternate isomer (16.44, 16.45).

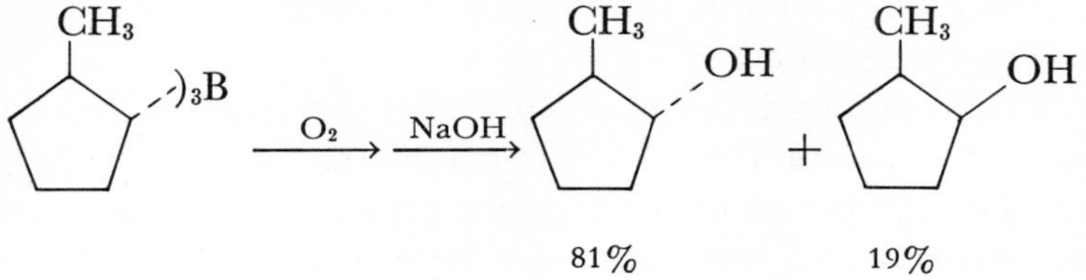

16.44

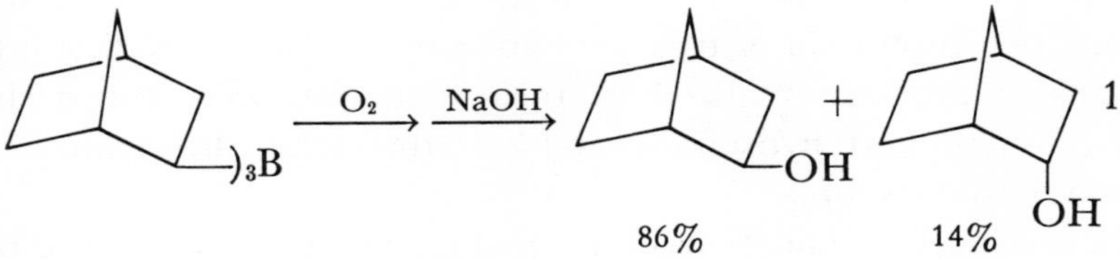

16.45

The formation of the alternate isomers is presumably a result of the free-radical chain reaction involved in the uptake of oxygen by the organoborane[45] (16.46).

$$R_3B + O_2 \longrightarrow R \cdot$$
$$R \cdot + O_2 \longrightarrow RO_2 \cdot \qquad\qquad 16.46$$
$$RO_2 \cdot + R_3B \longrightarrow RO_2BR_2 + R \cdot$$

In cases where such stereochemical properties are not involved, simple oxidation with oxygen can be utilized to achieve an essentially quantitative conversion of the organoborane to the corresponding alcohol. Alternatively, oxidation by alkaline hydrogen peroxide (Section XVI–8) does not involve free-radical intermediates and is very widely applicable.

8. Oxidation—Alkaline Hydrogen Peroxide

It was early demonstrated by Johnson and Van Campen[6] that perbenzoic acid in chloroform solution at 0° reacts quantitatively with tri-*n*-butylborane, and the reaction can be utilized for the quantitative analysis of organoboranes. A reaction of benzoyl peroxide was observed also, but the results were reported to be erratic. In this paper it was stated that aqueous hydrogen peroxide, in the presence of dilute alkalies, effects a complete dealkylation of tri-*n*-butylborane with the formation of boric acid and *n*-butyl alcohol. The reaction was suggested as the basis of a convenient method for the determination of boron in organoboron compounds. Unfortunately, no experimental details were reported.

The reaction was developed as an analytical procedure by Belcher, Gibbons, and Sykes in 1952.[49] Their conditions were exceedingly vigorous. They recommended treating the organoborane with excess hydrogen peroxide and concentrated sodium hydroxide, under reflux. Essentially these conditions were utilized by Winternitz and Carotti[18] in their study of the boron heterocycles formed in the thermal decomposition of tri-*n*-pentylborane and tri-*n*-hexylborane. The major products realized in their oxidation were the cyclic ethers, 2-methyltetrahydrofuran and 2-methyltetrahydropyran, and not the simple glycols.

Our own early applications of this method were also influenced by the vigorous conditions recommended for the analytical procedure.[50,51] However, with continued use of this reaction it was observed that the milder the conditions, the better the yield. In view of the growing importance of this reaction in synthetic chemistry,[52] it appeared desirable to undertake a detailed study of the reaction of alkaline hydrogen peroxide with trialkylboranes[53] (16.47).

$$R_3B + 3H_2O_2 + NaOH \longrightarrow 3ROH + NaB(OH)_4 \qquad 16.47$$

It was observed that the treatment of the organoborane from the usual hydroboration of 50 mmoles of 1-hexene in 40 ml of diglyme

[49] R. Belcher, D. Gibbons, and A. Sykes, *Mikrochim. Acta*, **40,** 76 (1952).

[50] H. C. Brown and B. C. Subba Rao, *J. Amer. Chem. Soc.*, **78,** 5694 (1956).

[51] H. C. Brown and B. C. Subba Rao, *J. Org. Chem.*, **22,** 1136 (1957).

[52] G. Zweifel and H. C. Brown, "Hydration of Olefins, Dienes, and Acetylenes via Hydroboration," *Organic Reactions*, **13,** 1 (1964) (John Wiley, New York).

[53] H. C. Brown, C. Snyder, B. C. Subba Rao, and G. Zweifel, manuscript in preparation.

with 15 mmoles of sodium hydroxide (5 ml of a 3 M solution), followed by the slow addition (rapid vigorous reaction!) of 60 mmoles of hydrogen peroxide, 20 percent excess (6.0 ml of 30 percent hydrogen peroxide) at 25 to 30° led to essentially instantaneous oxidation, with yields of 94 to 97 percent of alcohol based on the original olefin utilized. The reaction proceeded satisfactorily even at 0°, with a yield of 89 percent.

The three solvents commonly utilized for hydroborations are diglyme, tetrahydrofuran, and ethyl ether. It was observed that the oxidation proceeded with equal ease in tetrahydrofuran. However, the reaction in ethyl ether was more sluggish. Both diglyme and tetrahydrofuran are partially miscible with the aqueous phase in these oxidations, whereas ethyl ether is not. However, the simple addition of ethanol as a co-solvent avoided the difficulty in the ethyl ether system and brought about an increase in the yield to 98 percent.

Wide variations in the structure of the olefin can be tolerated in this reaction. The oxidation appears to be remarkably free of the steric requirements of the alkyl group in the organoborane, as indicated by the essentially quantitative conversions realized with trinorbornylborane (16.48), diisopinocampheylborane (16.49), and similar organoboranes containing large bulky alkyl groups (16.50).

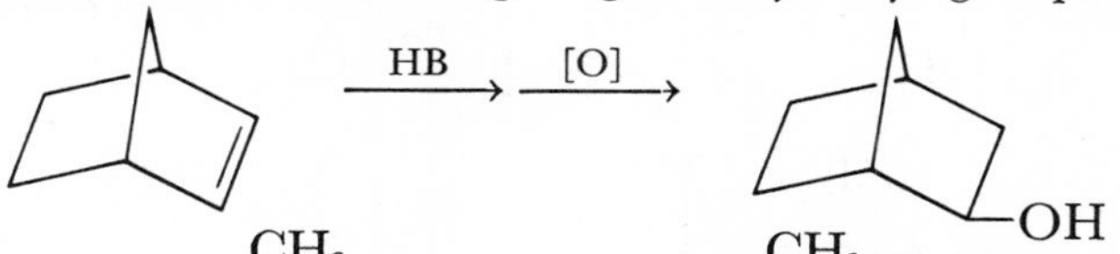

16.48

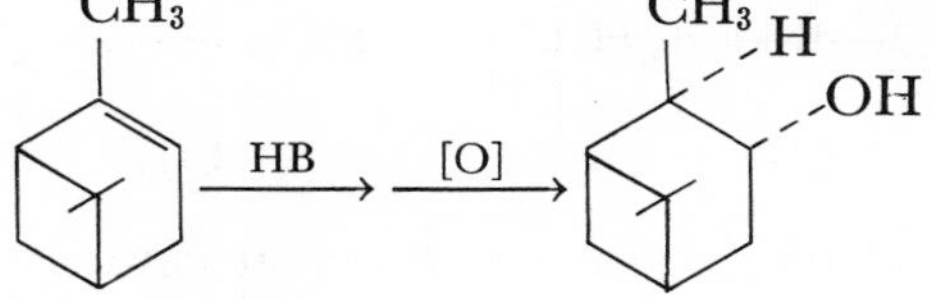

16.49

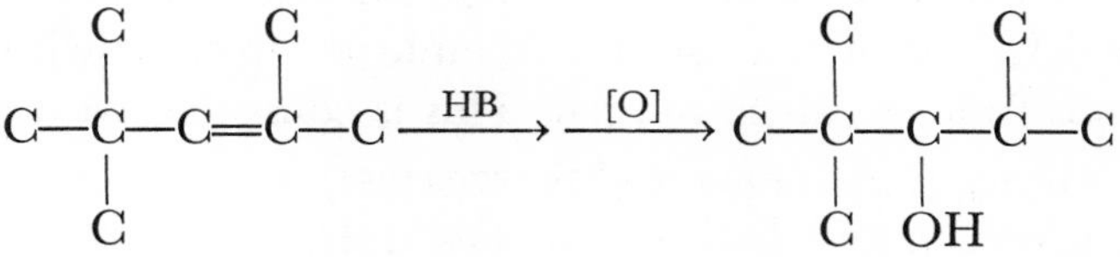

16.50

It is fortunate that the reaction appears to be almost specific for the boron-carbon linkage and is largely insensitive to other functional

groupings. This was demonstrated by carrying out the oxidation of
tri-*n*-hexylborane above in the presence of added reagents, such as
1-hexene, 3-hexyne, 1-hexyne, 1,3-cyclohexadiene, isobutyraldehyde,
methylethylketone, *n*-butyl bromide, and acetonitrile. All but isobutyraldehyde were demonstrated to be present in essentially unchanged form and concentration at the end of the oxidation. In the
case of isobutyraldehyde, the recovery was only 83 percent, evidently
the result of some condensation under the alkaline conditions. However, by a more careful control of the alkalinity, it has proved possible
to achieve excellent yields of aldehydes in the related oxidation of
vinylboranes.[34]

A detailed kinetic study of the reaction of alkaline hydrogen peroxide with trialkylboranes has yet to be made. However, there is available the related study by Kuivila and his co-workers of the reactions
of benzeneboronic acid with hydrogen peroxide.[54-57] He proposes the
following mechanism (16.51).

$$H_2O_2 + {}^-OH \rightleftharpoons HO_2^- + H_2O$$

$$
\begin{array}{c}
C_6H_5 \\
| \\
HO\!-\!B \\
| \\
OH
\end{array}
+ {}^-O_2H \longrightarrow
\left[
\begin{array}{c}
C_6H_5 \\
| \\
HO\!-\!B\!-\!OOH \\
| \\
OH
\end{array}
\right]^{-}
\qquad 16.51
$$

$$
\left[
\begin{array}{c}
C_6H_5 \\
| \\
HO\!-\!B\!-\!OOH \\
| \\
OH
\end{array}
\right]^{-}
\longrightarrow
\begin{array}{c}
HO\!-\!B\!-\!OC_6H_5 + OH^- \\
| \\
OH
\end{array}
$$

An identical mechanism, in three successive stages, is consistent
with all the available data for the oxidation with alkaline hydrogen
peroxide.[58] In this mechanism the organic group shifts with its pair
of electrons from boron to oxygen. This is consistent with both the

[54] H. G. Kuivila, *J. Amer. Chem. Soc.*, **76,** 870 (1954).

[55] H. G. Kuivila, *J. Amer. Chem. Soc.*, **77,** 4014 (1955).

[56] H. G. Kuivila and R. A. Wiles, *J. Amer. Chem. Soc.*, **77,** 4830 (1955).

[57] H. G. Kuivila and A. G. Armour, *J. Amer. Chem. Soc.*, **79,** 5659 (1957).

[58] W. J. Wechter, *Chem. and Ind.* (London), 294 (1959).

retention of configuration that is observed in the hydroboration-oxidation of cyclic olefins and the remarkable freedom from rearrangements that has been experienced to date.

9. Oxidation—Chromic Acid

The alkaline hydrogen peroxide oxidation of organoboranes provides an almost ideal means of converting the organometallic into the corresponding alcohol. These alcohols can be isolated and converted into other products, such as ketones and carboxylic acids, by standard methods. However, it was of interest to explore the possibility of a direct conversion of the organoborane into such oxidized products. Since aqueous chromic acid is the preferred reagent for the oxidation of secondary alcohols to ketones, we explored its action on organoboranes contained in the usual hydroboration solvents.[59]

In our study, the olefin was hydroborated in diglyme, tetrahydrofuran, or ethyl ether. To the reaction mixture was added a small quantity of water (to destroy residual hydride) followed by the slow addition of 10 percent excess of aqueous chromic acid (from sodium dichromate and sulfuric acid) at 25 to 35°. After 2 hours the product was isolated. Yields of ketone in the range 65–85 percent were realized (16.52).

$$3 \quad \text{(organoborane)} + 2Na_2Cr_2O_7 + 8H_2SO_4 \xrightarrow[25-35°]{2\ hr} \qquad\qquad 16.52$$

$$3 \quad \text{(2-methylcyclohexanone)} + 2Na_2SO_4 + 2Cr_2(SO_4)_3 + H_3BO_3$$

[59] H. C. Brown and C. P. Garg, *J. Amer. Chem. Soc.*, **83**, 2951 (1961).

Organoboranes

Among the transformations realized were (16.53–16.55):

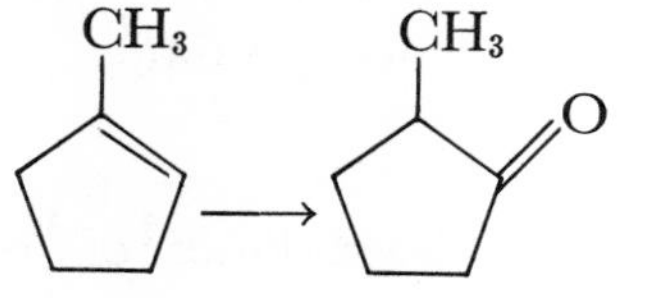

83%   16.53

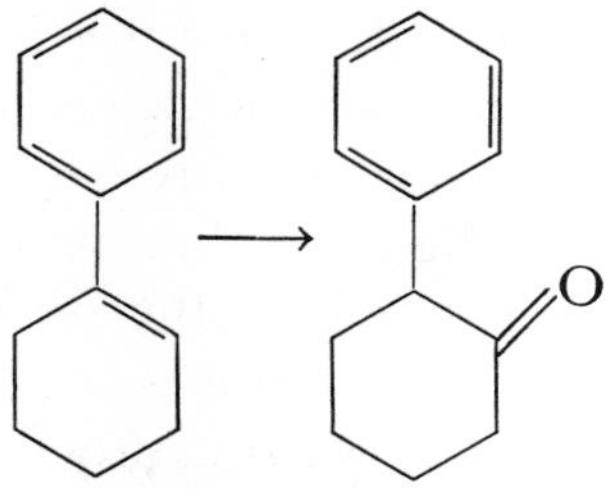

63%   16.54

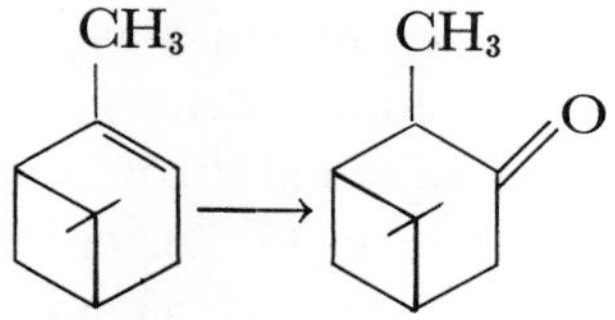

72%   16.55

An unexpected by-product of this research may be mentioned. The oxidation of secondary alcohols to ketones by aqueous chromic acid has long been a standard synthetic procedure. In seeking to improve this procedure, various workers have had recourse to solvents, such as acetic acid and acetone, that are miscible with water and resist oxidation by chromic acid. However, in the course of the above study[59] we observed that the use of the immiscible solvent, ethyl ether, offered major advantages for chromic acid oxidations, especially in cases where the ketone is subject to epimerization.[60]

For example, the use of ethyl ether made possible the oxidation of *l*-menthol to *l*-menthone in 97 percent yield, and isopinocampheol to isopinocampheone in 94 percent yield (yields from gas-chromatographic analysis). In each case only traces of the epimeric products were indicated, whereas the standard procedures resulted in the formation of several percent of the epimers.

[60] H. C. Brown and C. P. Garg, *J. Amer. Chem. Soc.*, **83**, 2952 (1961).

10. Amination

The remarkably facile conversion of organoboranes into alcohols by the action of alkaline hydrogen peroxide suggested the possibility of achieving a comparable conversion of the organoboranes into the corresponding amines by application of a suitable reagent (16.56).

$$R—B\diagdown^{\diagup} + HOOH + NaOH \longrightarrow ROH$$

$$\text{16.56}$$

$$R—B\diagdown^{\diagup} + \qquad ? \qquad \longrightarrow RNH_2$$

Hydrazine and hydroxylamine, which may be considered to be nitrogen analogs of hydrogen peroxide, were tested. They formed addition compounds with the organoboranes but showed no tendency to react further (16.57).

$$R_3B + NH_2NH_2 \longrightarrow R_3B:NH_2NH_2$$

$$\text{16.57}$$

$$R_3B + NH_2OH \longrightarrow R_3B:NH_2OH$$

It was reasoned that the presence of a better leaving group was required. Indeed, chloramine reacted to form the desired amine[61] (16.58).

$$\underset{|}{\overset{R}{\underset{|}{-B}}} + NH_2Cl \longrightarrow \underset{|}{\overset{R}{\underset{|}{-B}}}:NH_2Cl \longrightarrow \underset{|}{\overset{R}{\underset{|}{-B}}}:\overset{+}{N}H_2 + Cl^-$$

$$\text{16.58}$$

$$\underset{|}{-B-OH} + RNH_2 \overset{H_2O}{\longleftarrow} \underset{|}{-B}-\overset{+}{N}H_2R \longleftarrow$$

Unfortunately, the use of this reagent has disadvantages. It is unstable and must therefore be freshly prepared before use. The yields

[61] H. C. Brown, W. R. Heydkamp, E. Breuer, and W. S. Murphy, *J. Amer. Chem. Soc.*, **86,** 3565 (1964).

in the preparation of the reagent are only moderate (about 50 percent) and somewhat erratic. Consequently, such solutions should be
standardized before being used. There would be obvious advantages
to being able to carry out this synthesis with a relatively stable reagent, available commercially. Fortunately, we discovered that hydroxylamine-o-sulfonic acid was applicable to this synthesis, and a
highly satisfactory procedure was developed to permit the simple
synthesis of a wide variety of amines from the corresponding olefins
via hydroboration[62] (16.59–16.62).

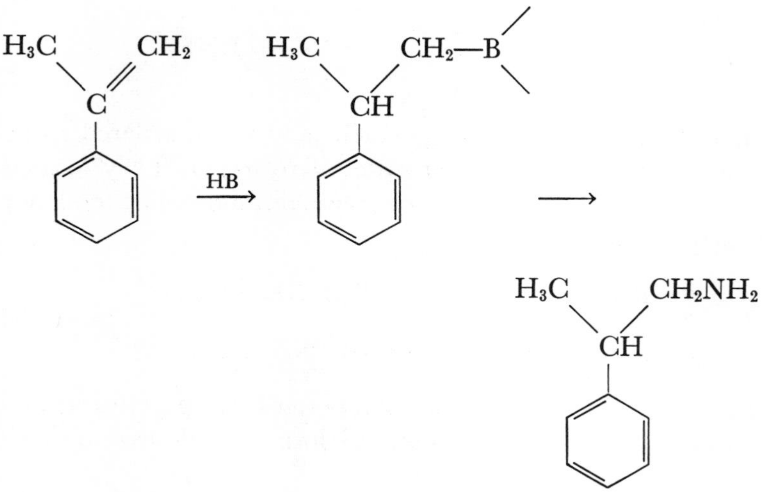

16.59

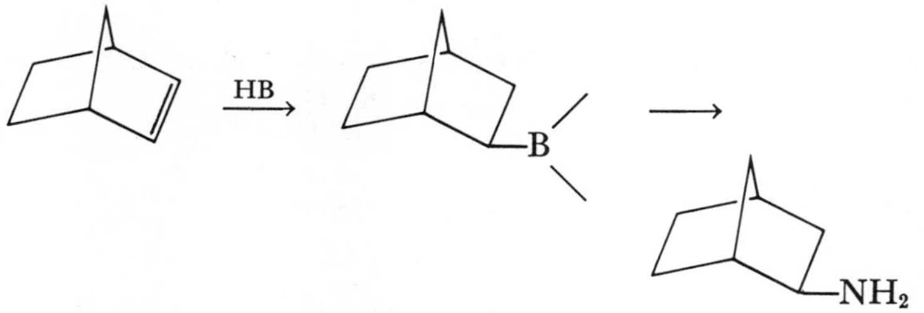

16.60

[62] M. W. Rathke, N. Inoue, K. R. Varma, and H. C. Brown, *J. Amer. Chem. Soc.*,
88, 2870 (1966).

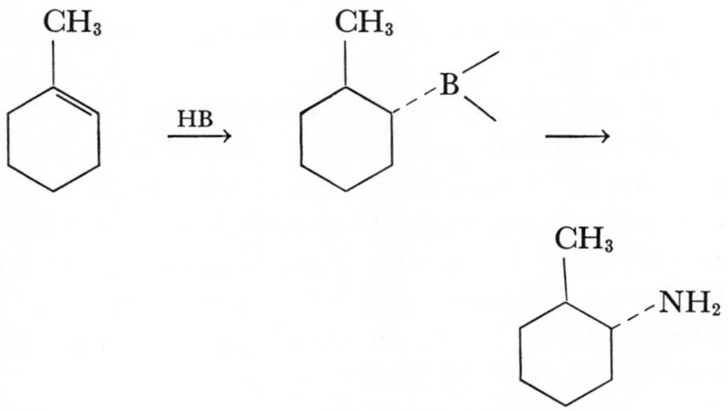

$$CH_2=CH(CH_2)_8CO_2C_2H_5 \longrightarrow \underset{\underset{NH_2}{|}}{CH_2}-CH_2(CH_2)_8CO_2C_2H_5 \qquad 16.62$$

It was previously reported that tri-*n*-butylborane failed to react with N-chlorodimethylamine to give *n*-butyldimethylamine,[63] in a manner analogous to the above reaction utilizing chloramine[61] (16.58). Instead the reaction proceeded to give *n*-butyl chloride in yields of 30 to 50 percent.

It is now evident that this latter reaction proceeds through a free-radical chain.[64] If this chain reaction is inhibited by galvinoxyl, the polar mechanism can proceed to the synthesis of *n*-butyldimethyl-amine (16.63).

$$n\text{-Bu}_3B + ClN(CH_3)_2 \xrightarrow{\text{galvinoxyl}} n\text{-BuN}(CH_3)_2 + n\text{-Bu}_2BCl \qquad 16.63$$

11. Metalation

The Grignard reagent is outstanding in the ease with which it transfers alkyl or aryl groups to the carbonyl or nitrile functions. However, it was early recognized that such transfers occur with difficulty, if at all, in the case of the organoboranes.

Another major application of the Grignard reagent is in the synthesis of other organometallics, such as diethylmercury and tetra-

[63] J. G. Sharefkin and H. D. Banks, *J. Org. Chem.*, **30**, 4313 (1965).

[64] A. G. Davies, S. C. W. Hook, and B. P. Roberts, *J. Organometal. Chem.*, **23**, C11 (1970).

ethyllead. For such applications, the organoboranes offer promise of being highly useful.

For example, Honeycutt and Riddle observed that the treatment of a mixture of triethylborane and mercuric chloride in aqueous suspension with sodium hydroxide led to a rapid reaction at 70 to 80°, with the formation of a 95 percent yield of diethylmercury. The reaction presumably involves the intermediate formation of mercuric oxide, and the reaction likewise proceeds when the oxide is utilized in place of the chloride.[65,66] In the same way it has been possible to achieve the synthesis of tetraethyllead.

The conversion of organoboranes to mercurials has been the subject of a detailed study.[67] Organoboranes derived from terminal olefins via hydroboration undergo a quantitative reaction in a matter of minutes with mercuric acetate at 0° or room temperature to give the corresponding alkyl mercuric acetate in essentially quantitative yields (16.64).

$$(RCH_2CH_2)_3B + 3Hg(OAc)_2 \xrightarrow{\text{THF}}$$
$$3RCH_2CH_2HgOAc + B(OAc)_3 \quad 16.64$$

Secondary alkyl groups fail to react under these conditions. Consequently, it is possible to use the high selectivity of dialkylboranes (Section XV-2) to prepare the corresponding organomercurial (16.65).

16.65

93% yield as the
mercuric chloride

[65] J. B. Honeycutt, Jr., and J. M. Riddle, *J. Amer. Chem. Soc.*, **81**, 2593 (1959).
[66] *Ibid.*, **82**, 3051 (1960).
[67] R. C. Larock and H. C. Brown, *J. Amer. Chem. Soc.*, **92**, 2467 (1970).

These alkylmercuric acetates are readily converted into the dialkyl-mercurials.[68] Consequently, this provides a convenient route from olefins (including functional groups) to organomercurials via hydroboration.

12. Formation of Carbon-Carbon Bonds

Up to this point we have been concerned with reactions that formed carbon-boron bonds, or replaced such bonds by bonds involving elements other than carbon. We may represent schematically as follows the transformations involved in the various reactions already discussed.

Isomerization	$-\overset{\mid}{\underset{\mid}{C_a}}-B\diagdown$	$\longrightarrow\ -\overset{\mid}{\underset{\mid}{C_b}}-B\diagdown$
Displacement	$R-\overset{\mid}{\underset{\mid}{C}}-\overset{\mid}{\underset{\mid}{C}}-B\diagdown$	$\longrightarrow\ R'-\overset{\mid}{\underset{\mid}{C}}-\overset{\mid}{\underset{\mid}{C}}-B\diagdown$
Cyclization	$-\overset{\mid}{C}-H\ \ H-B\diagdown$	$\longrightarrow\ -\overset{\mid}{C}\underset{\smile}{\quad}B\diagdown\ +\ H_2$
Protonolysis	$-\overset{\mid}{\underset{\mid}{C}}-B\diagdown$	$\longrightarrow\ -\overset{\mid}{\underset{\mid}{C}}-H\ +\ RCO_2B\diagdown$
Oxidation (O_2, NaOH), (H_2O_2 + NaOH)	$-\overset{\mid}{\underset{\mid}{C}}-B\diagdown$	$\longrightarrow\ -\overset{\mid}{\underset{\mid}{C}}-OH\ +\ HO-B\diagdown$
Oxidation (H_2CrO_4)	$-\overset{\mid}{\underset{\mid}{C}}-B\diagdown$ $\overset{\mid}{\underset{}{H}}$	$\longrightarrow\ --\overset{\mid}{C}=O\ +\ HO-B\diagdown$
Amination	$-\overset{\mid}{\underset{\mid}{C}}-B\diagdown$	$\longrightarrow\ -\overset{\mid}{\underset{\mid}{C}}-NH_2\ +\ HO-B\diagdown$
Metalation	$-\overset{\mid}{\underset{\mid}{C}}-B\diagdown\ +\ MAn$	$\longrightarrow\ -\overset{\mid}{\underset{\mid}{C}}-M\ +\ An-B\diagdown$

However, the most important use of organometallics in organic synthesis is the formation of new carbon-carbon bonds for the syn-

[68] Research in progress with Jerry D. Buhler.

thesis of the desired carbon structures. Accordingly, we next investigated the possibility of applying the organoboranes for this objective.

13. Coupling

The reaction of Grignard reagents with silver bromide provides a convenient means of bringing about the formation of carbon-carbon bonds, with yields of coupled products in the range of 40 to 60 percent.[69] Johnson and his co-workers reported that the treatment of *n*-butylboronic acid and *n*-hexylboronic acid with ammoniacal silver oxide brings about a similar coupling of the alkyl groups to form *n*-octane and *n*-dodecane, respectively.[70,71] It therefore appeared of interest to explore the possibility of achieving the coupling reaction in the trialkylboranes.

Neither silver bromide nor Tollens reagent yielded any significant quantity of product. Silver oxide was somewhat more favorable, bringing about the formation from triethylborane of 9 percent *n*-butane, 5 percent ethylene, and 15 percent ethane in 18 hours at room temperature. The addition of sodium hydroxide exerted a remarkable effect on the reaction, bringing about complete conversion to products (72 percent *n*-butane, 9 percent ethylene, and 9 percent ethane) in a matter of minutes at either 25 or 0°.[72] Gold compounds served similarly.[72]

We were able to achieve the coupling reaction directly in the hydroboration flask, permitting a combined hydroboration-coupling procedure as a convenient route to the products.[73] In a typical procedure 100 mmoles of 1-hexene was hydroborated in diglyme with sodium borohydride and boron trifluoride. A small quantity of water

[69] J. H. Gardner and P. Borgstrom, *J. Amer. Chem. Soc.*, **51,** 3375 (1929).

[70] H. R. Snyder, J. A. Kuck, and J. R. Johnson, *J. Amer. Chem. Soc.*, **60,** 105 (1938).

[71] J. R. Johnson, M. G. Van Campen, Jr., and O. Grummitt, *J. Amer. Chem. Soc.*, **60,** 111 (1938).

[72] H. C. Brown, N. C. Hebert, and C. H. Snyder, *J. Amer. Chem. Soc.*, **83,** 1001 (1961).

[73] H. C. Brown and C. H. Snyder, *J. Amer. Chem. Soc.*, **83,** 1001 (1961).

was added to destroy residual hydride, followed by 120 ml of 2.0 M aqueous potassium hydroxide. The reaction mixture was cooled to 0°, and 24.0 ml of a 5.0 M solution of silver nitrate was added over 10 minutes. After 1 hour at 0°, the product was isolated. A 70 percent yield of the coupled product was obtained.

The coupling reaction for more hindered olefins, such as 2-methyl-1-pentene, gives better yields in methanol solution, presumably because of the greater difficulty of reaction of potassium hydroxide with the more hindered organoborane in the two-phase aqueous system.

The reaction appears to be widely applicable. Terminal olefins undergo coupling with yields in the range of 60 to 80 percent (16.66).

$$
\begin{array}{c}
\text{C} \\
| \\
\text{C}-\text{C}-\text{C}{=}\text{C} \longrightarrow \\
| \\
\text{C}
\end{array}
$$

$$
\begin{array}{c}
\text{C} \qquad\qquad \text{C} \\
| \qquad\qquad | \\
\text{C}-\text{C}-\text{C}-\text{C}-\text{C}-\text{C}-\text{C}-\text{C} \\
| \qquad\qquad | \\
\text{C} \qquad\qquad \text{C}
\end{array}
$$

$$
\begin{array}{c}
\text{C} \\
| \\
\text{C}-\text{C}-\text{C}-\text{C}{=}\text{C} \longrightarrow \\
\end{array}
$$

$$
\begin{array}{c}
\text{C} \qquad\quad \text{C} \\
| \qquad\quad | \\
\text{C}-\text{C}-\text{C}-\text{C}-\text{C}-\text{C}-\text{C}-\text{C}-\text{C}-\text{C} \quad 16.66
\end{array}
$$

$$
\begin{array}{c}
\text{C} \quad\; \text{C} \\
| \quad\; | \\
\text{C}-\text{C}-\text{C}-\text{C}{=}\text{C} \longrightarrow \\
| \\
\text{C}
\end{array}
$$

$$
\begin{array}{c}
\text{C} \qquad \text{C} \qquad\quad \text{C} \qquad \text{C} \\
| \qquad | \qquad\quad | \qquad | \\
\text{C}-\text{C}-\text{C}-\text{C}-\text{C}-\text{C}-\text{C}-\text{C}-\text{C}-\text{C} \\
| \qquad\qquad\qquad\qquad | \\
\text{C} \qquad\qquad\qquad\qquad \text{C}
\end{array}
$$

Internal olefins form the coupled products in lower yield, in the range of 35 to 50 percent (16.67).

$$16.67$$

The reaction can be used without evident difficulties for a wide variety of structures (16.68).

$$16.68$$

The reaction has also been explored as a means of joining two different alkyl groups.[74] In a reaction involving a statistical coupling of two different groups, R and R$'$, the maximum yield of the desired product can only be 50 percent of the coupled material, R$_2$, RR$'$, R$'_2$. However, there is the possibility of improving the yield over this limit by utilizing a large excess of a relatively cheap olefin to achieve a more complete conversion of a second, more expensive olefin into the desired product. The practicality of this approach was demonstrated.[74]

The power of this new synthetic procedure is indicated by the following syntheses (16.69).

[74] H. C. Brown, C. Verbrugge, and C. H. Snyder, *J. Amer. Chem. Soc.*, **83,** 1002 (1961).

$$C-C-C-C-C=C + C-C-C-C=C \longrightarrow \textit{n}\text{-undecane}$$

$$C-C-C-C-C=C + C-C-\overset{\overset{\displaystyle C}{|}}{C}=C \longrightarrow 3\text{-methyldecane}$$

16.69

$$C-C-C-C-C=C + \text{(cyclopentene)} \longrightarrow \textit{n}\text{-hexylcyclopentane}$$

$$C-C-C-C-C=C + C-\overset{\overset{\displaystyle C}{|}}{C}=C-C \longrightarrow 2,3\text{-dimethylnonane}$$

The hydroboration reaction can tolerate many different functional groups not compatible with the Grignard reagent. Consequently this simple formation of carbon-carbon bonds should have very wide applicability in synthetic chemistry, as indicated by the following syntheses[75,76] (16.70).

$$2CH_2=CH(CH_2)_8CO_2C_2H_5 \longrightarrow \begin{matrix} CH_2CH_2(CH_2)_8CO_2C_2H_5 \\ | \\ CH_2CH_2(CH_2)_8CO_2C_2H_5 \end{matrix}$$ 16.70

$$\overset{\overset{\displaystyle CH_3}{|}}{CH_3}-C=CH_2 + CH_2=CH(CH_2)_8CO_2C_2H_5 \longrightarrow$$
$$(CH_3)_2CHCH_2CH_2CH_2(CH_2)_8CO_2C_2H_5$$

It is probable that the reaction proceeds through the formation of the silver alkyl. These are unstable under the reaction conditions, presumably breaking down into silver and free radical.[77] The free radicals present in high concentration unite to form the product.

It is apparent that this reaction offers promise as a new, simple route to free-radical reactions. For example, the presence of carbon tetrachloride in the reaction mixture diverts the reaction from dimerization to the formation of the corresponding alkyl chloride.[72,78]

[75] H. C. Brown and C. Verbrugge, manuscript in preparation.
[76] H. C. Brown and M. K. Unni, manuscript in preparation.
[77] C. E. H. Bawn and R. Johnson, *J. Chem. Soc.*, 3923 (1960).
[78] H. C. Brown and D. Burton, manuscript in preparation.

The scope of this new entry into free-radical chemistry is being explored.

14. Cyclopropane Synthesis

It was observed by Sommer and co-workers that γ-chloropropyl-trichlorosilane undergoes rapid conversion to cyclopropane under the influence of alkali[79] (16.71).

$$Cl_3SiCH_2CH_2CH_2Cl + 4NaOH \longrightarrow$$

$$+ 4NaCl + Si(OH)_4 \quad 16.71$$

The hydroboration of allyl chloride leads to the related derivatives, tri-(γ-chloropropyl)-borane and di-(γ-chloropropyl)-boron chloride[80] (16.72).

$$CH_2CH{=}CH_2 \xrightarrow{\ HB\ } (ClCH_2CH_2CH_2)_3B$$
$$|$$
$$Cl \qquad\qquad\qquad\qquad\qquad\qquad\qquad\qquad 16.72$$
$$+ (ClCH_2CH_2CH_2)_2BCl$$

Under the influence of aqueous alkali, these derivatives undergo an almost quantitative conversion to cyclopropane. Thus, 2.80 moles of cyclopropane were realized per mole of tri-(γ-chloropropyl)-borane and 1.90 moles of the cyclic hydrocarbon per mole of di-(γ-chloropropyl)-boron chloride.

The alkali is necessary to achieve a fast reaction at room temperature. For example, treatment of tri-(γ-chloropropyl)-borane at 100° with water alone produced only a 45 percent conversion into cyclopropane in 1 week. Presumably, the base coordinates with the organoborane, increasing the carbanionlike character of the boron-carbon linkage (16.73).

[79] L. H. Sommer, R. E. Van Strien, and F. C. Whitmore, *J. Amer. Chem. Soc.*, **71**, 3056 (1949).

[80] M. F. Hawthorne and J. A. DuPont, *J. Amer. Chem. Soc.*, **80**, 5830 (1958).

$$HO^- + \quad \diagdown B-CH_2CH_2CH_2Cl \rightleftharpoons$$

$$
\begin{array}{c}
HO-\overset{\diagdown\diagup}{\underset{\cdot}{B}}\overset{=}{}CH_2 \\[4pt]
H_2C\overset{\diagdown}{\underline{\hspace{1.2cm}}}CH_2 \\[4pt]
\mid \\
Cl
\end{array}
\qquad 16.73
$$

$$\downarrow$$

$$
HO-B\diagup \quad + \quad
\begin{array}{c}
CH_2 \\
\diagup \diagdown \\
H_2C\underline{\hspace{1.5cm}}CH_2
\end{array}
\quad + Cl^-
$$

The use of phenyllithium in nonaqueous systems likewise facilitated the reaction, providing an 84 percent yield.

This procedure has been explored as a general synthesis of cyclopropane derivatives.[81] Methallyl chloride was hydroborated and the reaction product treated *in situ* with aqueous sodium hydroxide. A 71 percent yield of methylcyclopropane was isolated. On a much smaller scale, phenylcyclopropane was obtained in a yield of 55 percent and benzylcyclopropane in a yield of 45 percent.

Since the cyclization reaction of the γ-chloropropylboranes appears to be essentially quantitative, the lower yields probably arise from one or more side reactions during the hydroboration stage. One such side reaction is the attachment of the boron atom in significant amounts at the secondary or tertiary position, instead of at the primary (16.74).

$$
\underset{\underset{Cl}{\overset{\mid}{\mid}}}{\overset{\overset{R}{\mid}}{CH_2C}}{=}CH_2 \xrightarrow{HB}
\underset{\underset{B}{\overset{\mid}{\diagup\diagdown}}}{\overset{\overset{R}{\mid}}{ClCH_2CH}}CH_2 +
\underset{\underset{B}{\overset{\mid}{\diagup\diagdown}}}{\overset{\overset{R}{\mid}}{ClCH_2C}}CH_3 \qquad 16.74
$$

[81] M. F. Hawthorne, *J. Amer. Chem. Soc.*, **82,** 1886 (1960).

This orientation is unimportant in simple olefins, but it can be markedly altered by aryl and halogen substituents (Section XIV-10). It has been established that allyl chloride undergoes hydroboration to place 40 percent of the boron at the secondary position.[82]

One means of controlling such undesired directive effects is the hydroboration with a dialkylborane, such as disiamylborane. Indeed the use of this reagent markedly improves the yield of cyclopropane realized in the combined hydroboration-cyclization of allyl chloride.[82]

Binger and Köster have proposed the combined use of dialkylboranes for the hydroboration stage and sodium hydride for the cyclization stage. This permits regeneration of the dialkylborane for subsequent utilization in a cyclic process[83] (16.75).

$$R_2BH + H_2C{=}CHCH_2Cl \longrightarrow R_2BCH_2CH_2CH_2Cl$$

$$NaH + R_2BCH_2CH_2CH_2Cl \longrightarrow Na^+[R_2\overset{..}{B}CH_2CH_2CH_2Cl]^- \quad 16.75$$

$$Na^+[R_2\overset{..}{B}CH_2CH_2CH_2Cl]^- \longrightarrow R_2BH + (CH_2)_3 + NaCl$$

Attempts to bring about a similar synthesis of cyclobutane from 4-chloro-1-butene were unsuccessful.[82,83] The hydroboration proceeded normally, but the product resisted cyclization.

9-BBN (Section XV-7) appears to possess major advantages in this cyclopropane synthesis.[84] Apparently the bridging boron atom in this bicyclic derivative is sterically far more available to the base. Consequently, coordination of the base occurs readily, followed by closure to the cyclopropane derivative (16.76).

$$CH_2{=}CHCH_2 \underset{Cl}{|} \xrightarrow{BH} CH_2CH_2CH_2 \underset{Cl}{|} \xrightarrow{OH^-} \triangle \quad 16.76$$

[82] H. C. Brown and K. A. Keblys, *J. Amer. Chem. Soc.*, **86,** 1791 (1964).
[83] P. Binger and R. Köster, *Tetrahedron Letters*, **156** (1961).
[84] H. C. Brown and S. P. Rhodes, *J. Amer. Chem. Soc.*, **91,** 2149 (1969).

This synthesis of cyclopropane derivatives appears capable of wide application. Thus it proceeds readily for the preparation of 1,1-dimethylcyclopropane (16.77), cyclopropyl chloride (16.78), and cyclopropylcarbinyl chloride (16.79).

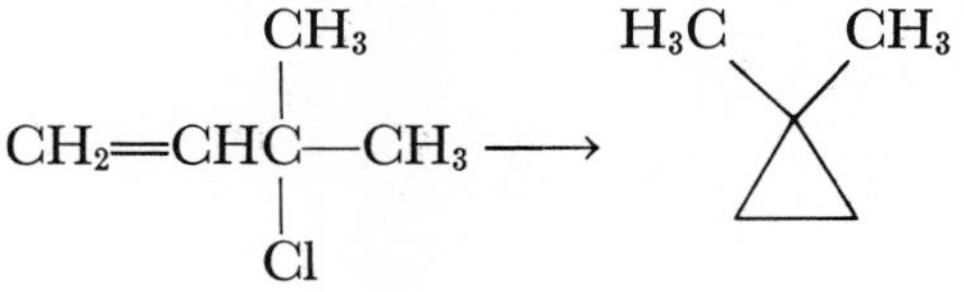

16.77

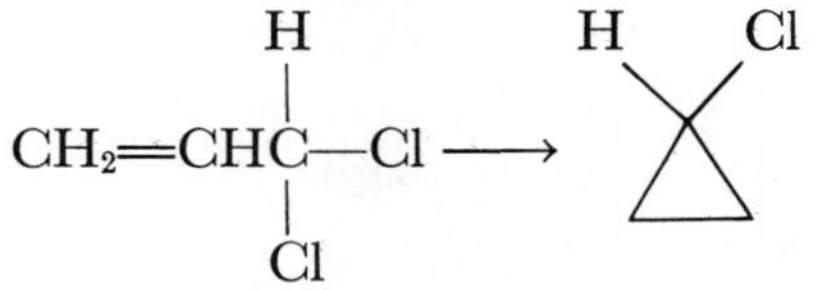

16.78

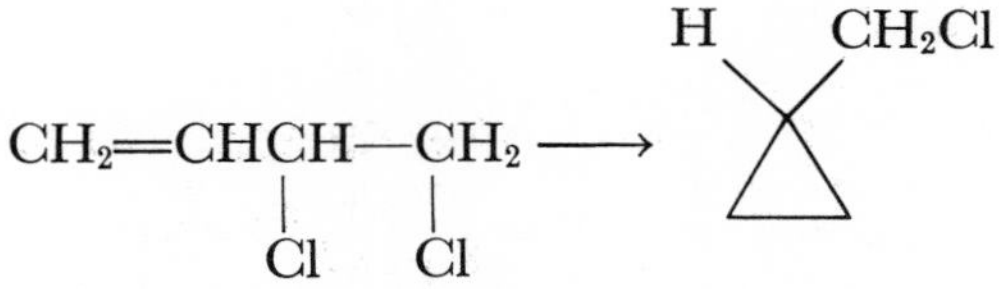

16.79

It is also possible to use this approach to synthesize B-cyclopropyl-9-BBN (16.80) and B-cyclobutyl-9-BBN (16.81) from open-chain intermediates.[85]

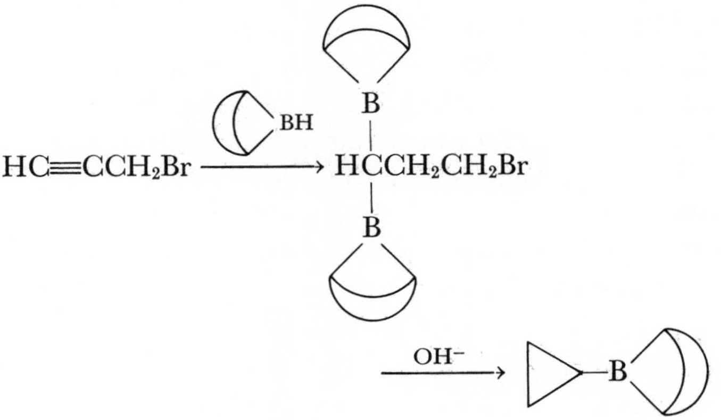

16.80

[85] H. C. Brown and S. P. Rhodes, *J. Amer. Chem. Soc.*, **91**, 4306 (1969).

$$HC\equiv CCH_2CH_2OTs \xrightarrow{\quad\bigcirc BH\quad} HCCH_2CH_2CH_2OTs$$

$$\xrightarrow{\text{LiMe}} \quad 16.81$$

These boron derivatives can be oxidized to the alcohols or utilized for alkylations (Chapter XVIII).

This ring synthesis is readily applicable to the preparation of 5- and 6-carbocyclic systems.[86]

15. Carbonylation

It was originally noted by Hillman that trialkylboranes react with carbon monoxide at high pressures to give products oxidizable to trialkylcarbinols[87] (16.82).

$$R_3B + CO \longrightarrow R_3CBO \xrightarrow[\text{NaOH}]{H_2O_2} R_3COH \qquad 16.82$$

A detailed study of the reaction revealed that in many cases it proceeds satisfactorily at atmospheric pressure,[88] especially conveniently utilizing the automatic hydrogenator,[48] modified for automatic carbonylation.[89] Moreover, it proved possible to control the reaction to provide a highly useful synthetic route not only to tertiary alcohols of a wide variety of structures, but also to ketones, secondary alcohols, carboxylic acids, methylol derivatives, and aldehydes. This promising reaction is discussed in detail in the next chapter.

[86] Research in progress with Charles G. Scouten.
[87] M. E. D. Hillman, *J. Amer. Chem. Soc.*, **84,** 4715 (1962).
[88] H. C. Brown and M. W. Rathke, *J. Amer. Chem. Soc.*, **89,** 2737 (1967).
[89] M. W. Rathke and H. C. Brown, *J. Amer. Chem. Soc.*, **88,** 2606 (1966).

16. Alkylation via Organoboranes

Under the influence of bases, trialkylboranes react rapidly at $0°$ with bromoacetic or chloroacetic esters to give the corresponding esters[90,91] (16.83, 16.84).

$$16.83$$

$$(n-C_{10}H_{21})_3B + CHBr_2CO_2C_2H_5 \longrightarrow n-C_{10}H_{21}CHCO_2C_2H_5 \quad 16.84$$

$$\underset{Br}{|}$$

92%

This reaction also reveals wide generality and has proven applicable to the alkylation in the *alpha* position of not only esters, but ketones nitriles, and other derivatives.

The reaction of organoboranes with various ylides[92] and with diazo esters and other diazo compounds[93] provides alternative routes for converting olefins into functional derivatives with lengthening of the chain.

These reactions will be reviewed in detail in Chapter XVIII.

17. Free-Radical Reactions of Organoboranes

Organoboranes do not add to the carbonyl group of simple aldehydes and ketones. However, an exceptionally fast reaction was ob-

[90] H. C. Brown, M. M. Rogić, M. W. Rathke, and G. W. Kabalka, *J. Amer. Chem. Soc.*, **90,** 818 (1968).

[91] H. C. Brown, M. M. Rogić, M. W. Rathke, and G. W. Kabalka, *J. Amer. Chem. Soc.*, **90,** 1911 (1968).

[92] J. J. Tufariello, L. T. C. Lee, and P. Wojtkowski, *J. Amer. Chem. Soc.*, **89,** 6804 (1967).

[93] J. Hooz and S. Linke, *J. Amer. Chem. Soc.*, **90,** 6891 (1968).

served between trialkylboranes and methyl vinyl ketone[94] (16.85) and with acrolein[95] (16.86).

$$(n-C_{10}H_{21})_3B \xrightarrow[\text{H}_2\text{O}]{\text{H}_2\text{C}=\text{CHCOCH}_3} n-C_{10}H_{21}CH_2CH_2\underset{\underset{\text{O}}{\|}}{C}CH_3 \qquad 16.85$$

16.86

This reaction is also widely applicable. Moreover, it has recently been shown to proceed through free-radical chains. This development reveals that trialkylboranes are an exceptionally fertile source of free-alkyl radicals; consequently, this reaction will be discussed in detail in the next chapter.

18. Conclusion

As pointed out earlier, the systematic study of the chemistry of the organoboranes is of very recent origin. Nevertheless, a number of highly interesting reactions of considerable value for synthetic chemistry have been uncovered. We may be confident that the continued study of this essentially virgin area will provide new chemical developments of major interest and utility. At that stage the organoboranes will take their place with the organomagnesium compounds and the complex hydrides as another major tool of the synthetic chemist.

[94] A. Suzuki, A. Arase, H. Matsumoto, M. Itoh, H. C. Brown, M. M. Rogić, and M. W. Rathke, *J. Amer. Chem. Soc.*, **89,** 5708 (1967).

[95] H. C. Brown, M. M. Rogić, M. W. Rathke, and G. W. Kabalka, *J. Amer. Chem. Soc.*, **89,** 5709 (1967).

XVII. The Reactions of Organoboranes with Carbon Monoxide[1]

1. Introduction

The discovery that diborane in ether solvents reacts practically instantaneously and quantitatively with alkenes and related unsaturated carbon compounds to convert them into the corresponding organoboranes (Chapters XIV and XV) made these intermediates readily available for the first time[2] (17.1, 17.2).

$$3RCH{=}CH_2 + BH_3 \xrightarrow{\text{THF}} (RCH_2CH_2)_3B \qquad 17.1$$

$$12RCH{=}CH_2 + 3NaBH_4 + 4BF_3 \cdot O(C_2H_5)_2$$
$$\xrightarrow[\text{or DG}]{\text{THF}} 4(RCH_2CH_2)_3B + 3NaBF_4 + 4(C_2H_5)_2O \qquad 17.2$$

The characteristics of this reaction, hydroboration, are highly favorable for synthetic work.[2] Not only is the reaction very fast and essentially quantitative, but practically all unsaturated compounds, with very rare exceptions,[3] undergo it. Rearrangements of the carbon structure are practically unknown. The reaction appears to involve an *anti*-Markovnikov *cis* addition from the less hindered side of the double bond (17.3, 17.4).

$$17.3$$

---

[1] H. C. Brown, *Accounts Chem. Research*, **2**, 65 (1969).

[2] H. C. Brown, *Hydroboration* (W. A. Benjamin, New York, 1962).

[3] M. Nussim, Y. Mazur, and F. Sondheimer, *J. Org. Chem.*, **29**, 1120, 1131 (1964).

$$99.5\% \ exo$$

Of equal significance is the fact that many functional groups are readily accommodated without difficulty[4] (17.5–17.7).

Thus the organic chemist has at his disposal for the first time a relatively reactive organometallic capable of bearing many different functional groups in the organic radical.

These developments have created intense interest in the discovery of new reactions of organoboranes of utility in organic synthesis. Among the more interesting possibilities to which attention has been called are (Chapter XVI): isomerization, displacement, contrathermodynamic isomerization of olefins, cyclization, protonolysis, oxidation to alcohols (alkaline hydrogen peroxide), oxidation to ketones

[4] H. C. Brown and M. K. Unni, *J. Amer. Chem. Soc.*, **90,** 2902 (1968); H. C. Brown and R. M. Gallivan, Jr., *J. Amer. Chem. Soc.*, **90,** 2906 (1968); H. C. Brown and R. L. Sharp, *J. Amer. Chem. Soc.*, **90,** 2915 (1968).

(chromic acid), amination, metalation, coupling with alkaline silver nitrate, 1,4-addition to α,β-unsaturated aldehydes, ketones, and Mannich bases, reaction with ylides, or with α-halo carbanions, and carbonylation to tertiary alcohols, secondary alcohols, ketones, methylol derivatives, aldehydes, ring ketones, and polycyclics.

The carbonylation reaction appears to be especially promising in its versatility, and this chapter will discuss its applications.

2. Early History

In 1937 Burg and Schlesinger reported that diborane reacts with carbon monoxide in a sealed tube at 100° under 20 atm to give a simple addition compound, borane carbonyl[5] (17.8).

$$\tfrac{1}{2}(BH_3)_2 + CO \rightleftharpoons H_3BCO \qquad\qquad 17.8$$

The product is a gas, bp $-64°$, which is largely dissociated into its components at atmospheric pressure[6] (17.9).

$$\tfrac{1}{2}(BH_3)_2 + CO \xrightarrow{\text{NaBH}_4} \tfrac{1}{3}(H_3CBO)_3 \qquad\qquad 17.9$$

At approximately the same time, Schlesinger and his co-workers studied the reaction of the methyl derivatives of diborane with carbon monoxide.[7] Dimethylborane (dimer) reacted with carbon monoxide at a pressure of 5 atm to produce a number of relatively complex materials. The major product was a liquid with the empirical formula $[(CH_3)_2BHCO]_2$. However, this compound did not liberate hydrogen when treated with water, indicating the absence of the original boron-hydrogen bond. The authors were unable to assign any structure to this intermediate, but in view of our own studies with dialkylboranes[8] it is probably **17.1**.

---

[5] A. B. Burg and H. I. Schlesinger, *J. Amer. Chem. Soc.*, **59**, 780 (1937).

[6] Under the catalytic influence of small amounts of sodium borohydride, the reaction between diborane (in tetrahydrofuran) and carbon monoxide proceeds very rapidly at room temperature and atmospheric pressure to yield trimethylboroxine: M. W. Rathke and H. C. Brown, *J. Amer. Chem. Soc.*, **88**, 2606 (1966).

[7] F. L. McKennon, Ph.D. Thesis, University of Chicago, 1937.

[8] Unpublished research with M. W. Rathke and M. M. Rogić.

$$\begin{array}{ccc}
CH_3 & O & H \\
\diagdown & \diagup\diagdown & \diagup \\
& B \quad\quad C & \\
H & | \quad\quad | & \\
\diagdown | & \quad | & CH_3 \\
& C \quad\quad B & \\
\diagup & \diagdown\diagup\diagdown & \\
CH_3 & O & CH_3
\end{array}$$

17.1

In 1961 a patent was issued to Reppe and Magin for the preparation of compounds of the type $(R_3BCO)_n$ by the carbonylation of trialkylboranes under pressure.[9] Typically, a trialkylborane reacted at a temperature of 10 to 20° with carbon monoxide at pressures of 100 to 200 atm to give, after distillation, a mixture of two products. The major component possessed a molecular weight and analysis in agreement with the empirical formula $(R_3BCO)_2$. The minor product corresponded to the empirical formula $(R_3BCO)_3$. No indication was presented of any understanding of the structures of these materials or of the nature of the chemical reaction taking place.

To Hillman belongs the credit for first establishing the nature of this reaction of carbon monoxide and trialkylboranes.[10] He treated the organoborane with carbon monoxide at very high pressures, usually about 10,000 psi, in the presence of water.[10] He reported that the products from carbonylation at relatively low temperature (50–75°) were 2,5-diboradioxanes (**17.2**), corresponding to the dimer of

$$\begin{array}{ccc}
& O & \\
& \diagup\diagdown & \\
RB & & CR_2 \\
| & & | \\
R_2C & & BR \\
& \diagdown\diagup & \\
& O &
\end{array}$$

17.2

[9] W. Reppe and A. Magin. U.S. Patent 3,006,961 (Oct. 31, 1961); *Chem. Abstr.*, **55**, 10386i (1961).

[10] M. E. D. Hillman, *J. Amer. Chem. Soc.*, **84**, 4715 (1962); *J. Amer. Chem. Soc.*, **85**, 982 (1963); *J. Amer. Chem. Soc.*, **85**, 1626 (1963).

Reppe and Magin.[9] Oxidation of these materials with alkaline hydrogen peroxide did not produce the expected dialkyl ketones, but rather the corresponding dialkylcarbinols.[11]

At higher temperatures, generally 150°, the 2,5-diboradioxanes (**17.2**) were converted into the corresponding boronic anhydrides (boroxines) (**17.3**).

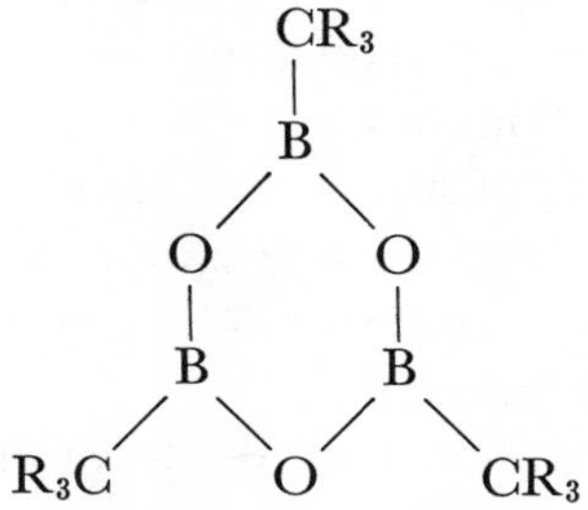

$$
\begin{array}{c}
CR_3 \\
| \\
B \\
/ \quad \backslash \\
O \qquad O \\
| \qquad | \\
B \qquad B \\
/ \ \backslash \ / \ \backslash \\
R_3C \quad O \quad CR_3
\end{array}
$$

17.3

Hillman examined the carbonylation reaction in the presence of various glycols.[10] He noted that whereas certain glycols led to the formation of polymeric products, the use of ethylene glycol yielded the corresponding cyclic esters (**17.4**) smoothly. Finally, carbonyla-

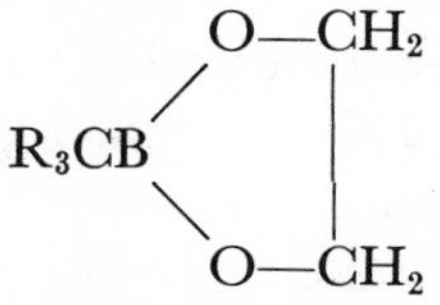

$$
\begin{array}{c}
O{-}CH_2 \\
/ \qquad | \\
R_3CB \qquad | \\
\backslash \qquad | \\
O{-}CH_2
\end{array}
$$

17.4

tion in the presence of aldehydes yielded 4-bora-1,3-dioxolanes[10] (**17.5**).

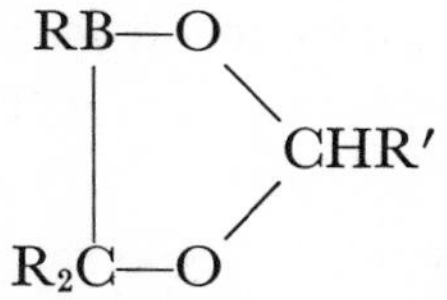

$$
\begin{array}{c}
RB{-}O \\
| \qquad \backslash \\
| \qquad \quad CHR' \\
| \qquad / \\
R_2C{-}O
\end{array}
$$

17.5

---

[11] As discussed later, we have found it possible to convert the carbonylation intermediate either to the ketone or to the carbinol, as desired, in excellent yields.

3. Carbonylation at Atmospheric Pressure

Practically all of Hillman's research had been done at very high pressures, generally about 10,000 psi. He had tried some carbonylations at atmospheric pressure but had noted that the reactions were relatively slow and incomplete.[10]

Carbonylation, combined with hydroboration, offered a highly promising synthetic route. However, very few laboratories are equipped to run carbonylations at 10,000 psi. Moreover, even if they were, the inconvenience of such a procedure would mitigate against its adoption unless no simple alternatives were available. Consequently, we undertook to find means to accomplish such carbonylations at atmospheric pressures.

Initially, our investigation took the form of a search for possible catalysts for the reaction. However, we soon discovered that if we merely raised the temperature to 100–125°, practically all of the organoboranes of interest reacted essentially quantitatively with carbon monoxide at atmospheric pressure, usually in 1–3 hours, although some of the more hindered derivatives required longer. This solved the problem of a convenient laboratory procedure, and we undertook a study of the new synthetic entry.

This study was greatly facilitated by our familiarity with the Brown□ hydrogenator.[12] This instrument, although originally developed for hydrogenations, is a very convenient automatic gas generator. It permits one to follow the rate of absorption of a gas by the system. Thus we have utilized it for hydrochlorinations.[13] In some cases the automatic feature has revealed that a given hydrochlorination is complete in 1 minute, with the initial product being very different from that present after 10 minutes.[14]

As adapted for carbonylation, the apparatus is assembled as shown in Figure XVII-1. In the buret A is placed formic acid. In the gen-

---

[12] C. A. Brown and H. C. Brown, *J. Amer. Chem. Soc.*, **84**, 2829 (1962); *J. Org. Chem.*, **31**, 3989 (1966). We used a commercial model from Delmar Scientific Laboratories, Maywood, Ill. 60154.

[13] H. C. Brown and M.-H. Rei, *J. Org. Chem.*, **31**, 1090 (1966).

[14] H. C. Brown and K.-T. Liu, *J. Amer. Chem. Soc.*, **89**, 3898 (1967).

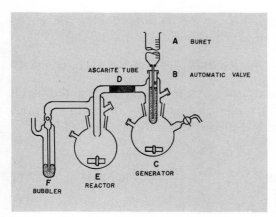

Figure XVII-1. Hydrogenation apparatus modified for generation of carbon monoxide.

erator flask C is placed concentrated sulfuric acid maintained at approximately 100° by a heating mantle. The hydroboration is carried out in the reactor E in the usual manner. The system is flushed with carbon monoxide by injecting formic acid through the serum cap into the generator C. Then the reactor temperature is raised to an appropriate level, depending on the reaction, and stirring is initiated. The mercury in the valve B is sufficient to support the column of formic acid in the buret as long as the pressure in the apparatus is atmospheric. However, as soon as carbon monoxide is absorbed by the organoborane solution in the reactor E, the pressure drops approximately 20 mm below atmospheric, and formic acid is drawn through the mercury seal into the generator. In this way the carbonylation proceeds automatically to completion. At any moment a mere reading of the buret will indicate the progress of the reaction. At the end of the reaction, the reactor vessel is cooled, an appropriate base is added, followed by 30 percent hydrogen peroxide, and the reaction product is then isolated.

4. Tertiary Alcohols[15]

It was observed that a wide variety of organoboranes, readily synthesized via hydroboration in diglyme (17.2), react readily with carbon monoxide at temperatures of 100–125°. Oxidation of the in-

---

[15] H. C. Brown and M. W. Rathke, *J. Amer. Chem. Soc.*, **89,** 2737 (1967).

termediates with alkaline hydrogen peroxide provides the corresponding trialkylcarbinols in excellent yield (17.10).

$$R_3B + CO \longrightarrow (R_3CBO)_x \xrightarrow[\text{NaOH}]{H_2O_2} R_3COH \qquad 17.10$$

In some cases the intermediate appeared to be polymeric and difficult to oxidize. However, by carrying out the carbonylation in the presence of ethylene glycol,[10] this difficulty could be avoided (17.11).

$$R_3B + CO \xrightarrow{(CH_2OH)_2} R_3CB \overset{\displaystyle O-CH_2}{\underset{\displaystyle O-CH_2}{\big|}} \xrightarrow[\text{NaOH}]{H_2O_2} R_3COH \qquad 17.11$$

The synthesis appears to be of wide applicability, as indicated by the following preparations (17.12–17.16).

$$CH_3CH{=}CHCH_3 \longrightarrow (CH_3CH_2\overset{\displaystyle CH_3}{\underset{\displaystyle |}{C}}H)_3COH \qquad 17.12$$

87%

$$(CH_3)_2C{=}CH_2 \longrightarrow [(CH_3)_2CHCH_2]_3COH \qquad 17.13$$

90%

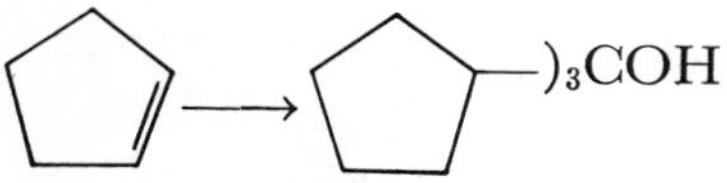

—)₃COH 17.14

90%

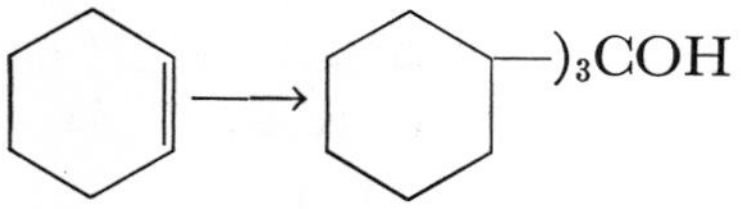

—)₃COH 17.15

80%

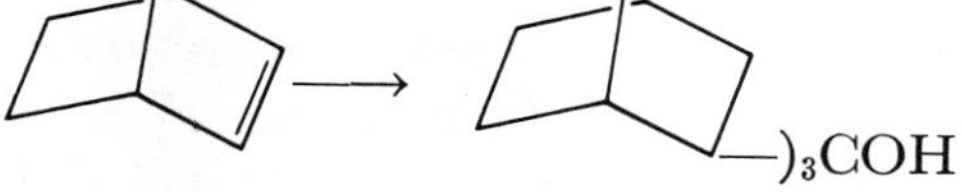

—)₃COH 17.16

80%

This makes readily available for the first time trialkylcarbinols containing bulky groups. For example, the yield of tricyclohexylcarbinol, through the Grignard reaction, is reported to be only 7 percent, and this was raised to 19 percent by use of a special procedure involving sodium.[16] No difficulty was experienced in achieving an isolated yield of 80 percent with the present procedure.

5. Mechanism of the Carbonylation Reaction

Carbonylation of an equimolar mixture of triethylboron and tri-*n*-butylboron gave an equimolar mixture of triethylcarbinol and tri-*n*-butylcarbinol (17.17).

$$(C_2H_5)_3B + (n\text{-}C_4H_9)_3B \xrightarrow{CO} \xrightarrow{[O]}$$
$$(C_2H_5)_3COH + (n\text{-}C_4H_9)_3COH \quad 17.17$$

There was no evidence for the presence of any mixed derivatives, such as diethyl-*n*-butylcarbinol or ethyl-di-*n*-butylcarbinol, which would have accompanied an intermolecular transfer of alkyl groups.[17]

Similarly, dicyclohexyl-*n*-octylborane, readily synthesized by the reaction of dicyclohexylborane and 1-octene, yielded dicyclohexyl-*n*-octylcarbinol in 82 percent yield, without evidence of any isomeric materials (17.18).

These results indicate that the groups are transferred intramolecularly from boron to carbon and are consistent with the mechanism (17.19–17.22) proposed by Hillman[10] and slightly modified.[18]

---

[16] P. D. Bartlett and A. Schneider, *J. Amer. Chem. Soc.*, **67**, 141 (1945).

[17] H. C. Brown and M. W. Rathke, *J. Amer. Chem. Soc.*, **89**, 4528 (1967).

[18] H. C. Brown and M. W. Rathke, *J. Amer. Chem. Soc.*, **89**, 2738 (1967).

$$R_3B + CO \rightleftharpoons R_3\overset{-}{B}\overset{+}{-}CO \qquad\qquad 17.19$$

$$R_3\overset{-}{B}\overset{+}{-}CO \rightleftharpoons R_2B-\underset{\underset{O}{\|}}{C}-R \qquad\qquad 17.20$$

$$R_2BCR \longrightarrow RB\overset{}{\underset{O}{\diagdown\diagup}}CR_2 \qquad\qquad 17.21$$

$$RB\overset{}{\underset{O}{\diagdown\diagup}}CR_2 \longrightarrow OBCR_3 \qquad\qquad 17.22$$

The boraepoxide (17.21) is presumably the precursor (by dimerization) of the 2,5-diboradioxane (**17.2**). Once the latter is formed, the transfer of the third alkyl group becomes very slow and requires heating at elevated temperatures in the presence of water or other materials to open up the stable diboradioxane system.[10] The boronic anhydride (17.22) trimerizes to the corresponding boroxine (**17.3**). Presumably it is the sluggish oxidation of some of these polymeric derivatives that makes it advantageous to carry out the reaction in the presence of ethylene glycol.[15]

6. Secondary Alcohols and Ketones[18]

It is evident from this mechanism that, were it possible to stop the reaction at the point where two alkyl groups had transferred from boron to carbon (17.21), hydrolysis would produce the secondary alcohol, R_2CHOH (17.23), and oxidation would produce the corresponding ketone, R_2CO (17.24).

$$R_2CHOH + RB(OH)_2 \qquad 17.23$$

$$RB\overset{}{\underset{O}{\diagdown\diagup}}CR_2 \quad\begin{array}{l} \overset{H_2O}{\nearrow} \\ \overset{NaOH}{} \\ \underset{NaOH}{\overset{H_2O_2}{\searrow}} \end{array}$$

$$R_2CO + ROH \qquad\qquad 17.24$$

Indeed, we discovered that the addition of equimolar quantities of water greatly inhibited the transfer of the third alkyl group, making possible a convenient synthesis of secondary alcohols and ketones (17.25–17.29).

$$CH_3CH_2CH{=}CH_3 \longrightarrow (CH_3CH_2CH_2CH_2)_2CO \qquad 17.25$$
$$85\%$$

$$CH_3CH{=}CHCH_3 \longrightarrow (CH_3CH_2\overset{\overset{\displaystyle CH_3}{|}}{C}H)_2CO \qquad 17.26$$
$$81\%$$

$$17.27$$
$$90\%$$

$$17.28$$
$$80\%$$

$$17.29$$
$$82\%$$

The water apparently converts the boraepoxide into the corresponding hydrate (17.30), and the latter is evidently less susceptible to the transfer of the third alkyl group.

$$RB{-}{-}{-}CR_2 + H_2O \xrightarrow{\text{fast}} RB{-}{-}{-}CR_2 \xrightarrow{\text{slow}} (HO)_2BCR_3 \qquad 17.30$$

Hydrolysis of the reaction mixtures with aqueous sodium hydroxide readily produces the corresponding secondary alcohols. Unless the oxidation is carried out carefully, in accordance with the recommended procedure,[18] hydrolysis can accompany the oxidation, so that the product can consist of a mixture of ketone and alcohol.[10]

7. Primary Alcohols [19] and Aldehydes [20]

It is evident that if it were possible to stop the reaction at the point where a single alkyl group had been transferred from boron to carbon (17.20), hydrolysis should yield the corresponding aldehyde (17.31) and oxidation the corresponding acid (17.32).

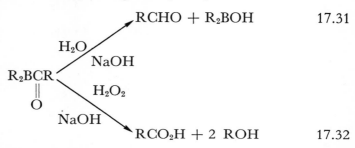

$$RCHO + R_2BOH \qquad 17.31$$

$$RCO_2H + 2\ ROH \qquad 17.32$$

Up to the present we have been unable to stop the reaction at this stage without simultaneously changing the oxidation stage of the intermediate. Thus in the presence of lithium borohydride or lithium trimethoxyaluminohydride an intermediate is obtained that hydrolyzes to the methylol derivative (17.33) and is oxidized to the corresponding aldehyde (17.34).

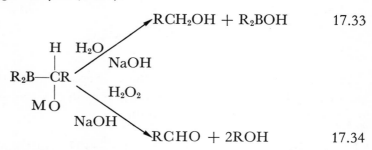

$$RCH_2OH + R_2BOH \qquad 17.33$$

$$RCHO + 2ROH \qquad 17.34$$

Not only do the complex hydrides reduce one of the intermediates (17.19 or 17.20), which permits the reaction to stop at the stage

---

[19] M. W. Rathke and H. C. Brown, *J. Amer. Chem. Soc.*, **89**, 2740 (1967).

[20] H. C. Brown, R. A. Coleman, and M. W. Rathke, *J. Amer. Chem. Soc.*, **90**, 499 (1968).

where only one of the three alkyl groups has been transferred from boron to carbon, but the reagents simultaneously facilitate greatly the absorption of carbon monoxide. Whereas the previous reactions are generally carried out at temperatures of 100–125°, the present reactions proceed relatively rapidly at temperatures of 25–45°.

The reaction with lithium trimethoxyaluminohydride[21] is especially interesting. Neither the reagent nor the trialkylborane individually reacts with carbon monoxide at atmospheric pressure at any appreciable rate at 25°. However, their presence together in the same reaction vessel (Figure XV-1) results in a rapid uptake of carbon monoxide that is complete in a matter of minutes. The reaction exhibits a simple $1:1:1$ stoichiometry for the three components (17.35).

$$
R_3B + CO + LiAlH(OCH_3)_3 \longrightarrow R_2B\!-\!\overset{\displaystyle H}{\underset{\displaystyle OAl(OCH_3)_3Li}{C}}\!-\!R \qquad 17.35
$$

$$
\overset{OH^-}{\swarrow} \qquad \overset{[O]}{\searrow}
$$

$$
RCH_2OH \qquad\qquad RCHO
$$

Based on this stoichiometry, the yields of aldehyde are almost quantitative (17.36–17.40).

$$
CH_3(CH_2)_3CH{=}CH_2 \longrightarrow CH_3(CH_2)_5CHO \qquad 17.36
$$
$$
98\%
$$

$$
CH_3CH{=}CHCH_3 \longrightarrow CH_3CH_2\overset{\displaystyle CH_3}{\underset{}{C}}HCHO \qquad 17.37
$$
$$
94\%
$$

[21] This reagent is readily synthesized by adding 3 mol of methanol to 1 mol of lithium aluminum hydride in tetrahydrofuran solution. The solution is used without isolation of the reagent: H. C. Brown and C. J. Shoaf, *J. Amer. Chem. Soc.*, **86**, 1079 (1964). For a survey of its reducing characteristics, see H. C. Brown and P. M. Weissman, *J. Amer. Chem. Soc.*, **87**, 5614 (1965); and Section XII–8.

$$(CH_3)_2C{=}CH_2 \longrightarrow (CH_3)_2CHCH_2CHO \qquad 17.38$$
$$91\%$$

17.39

93%

17.40

87%

Hydrolysis of the reaction mixture with alkali, without oxidation, produces essentially quantitative yields of the corresponding methylol derivatives. An obvious disadvantage of this procedure is the fact that only one of the original three alkyl groups on boron is utilized. Fortunately, it has proven possible to circumvent this difficulty, and the solution will be discussed later.

A word may be in order concerning the mechanism. The tremendous accelerating effect on the rate of uptake of carbon monoxide brought about by the reagent suggests that it must react rapidly with a small equilibrium concentration of an intermediate. If both stages 17.19 and 17.20 are reversible equilibria, as represented, then the reagent might operate by converting the carbonyl derivative, R_2BCOR, into the reduced product. On the other hand, if only the first stage, 17.19, is reversible, then the reagent must react with the small equilibrium concentration of the carbonyl intermediate, R_3BCO. Unfortunately, we have been too preoccupied with exploring the scope of these new reactions to investigate the mechanisms in detail.

8. Unsymmetrical Ketones [17, 22]

Certain dialkylboranes are readily synthesized by the controlled reaction of 2 moles of olefin with 1 mole of borane (Section XIV-7). Reaction of these dialkylboranes with 1 molar equivalent of another

[22] H. C. Brown, G. W. Kabalka, and M. W. Rathke, *J. Amer. Chem. Soc.*, **89**, 4530 (1967).

olefin gives a mixed trialkylborane. Carbonylation of such mixed organoboranes in the presence of water provides a synthetic route to ketones possessing two different organic moieties attached to the carbonyl group (17.41 and 17.42).

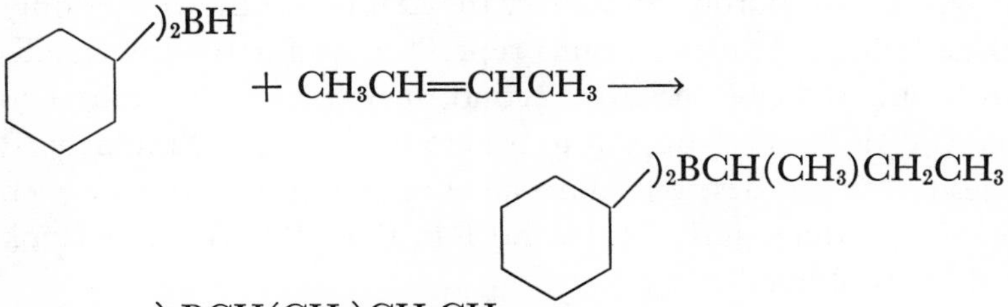

$)_2$BH

$+ \ CH_3CH\!=\!CHCH_3 \longrightarrow$

$)_2BCH(CH_3)CH_2CH_3$

17.41

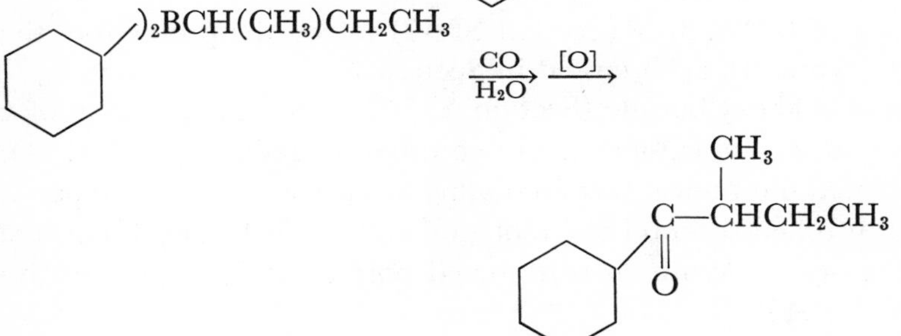

$)_2BCH(CH_3)CH_2CH_3$

$\xrightarrow[H_2O]{CO} \xrightarrow{[O]}$

$$\overset{CH_3}{\underset{}{|}}$$

$C\!-\!CHCH_2CH_3$ 17.42

$\overset{\|}{O}$

66%

Many functional groups can be tolerated, as indicated by the related synthesis of the derivatives **17.6–17.9**.

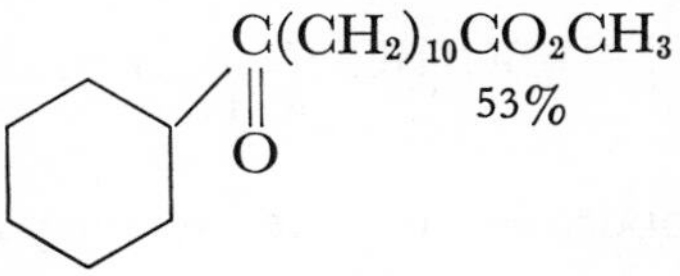

$C(CH_2)_{10}CO_2CH_3$

53%

$\overset{\|}{O}$

17.6

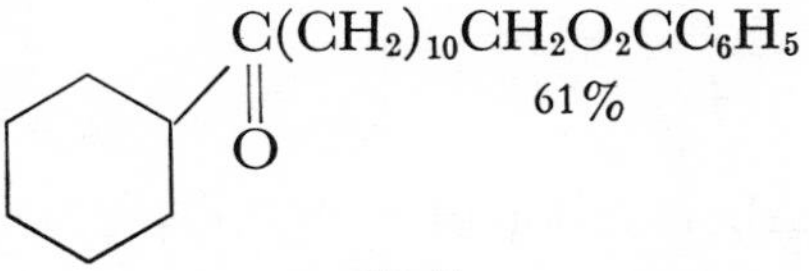

$C(CH_2)_{10}CH_2O_2CC_6H_5$

61%

$\overset{\|}{O}$

17.7

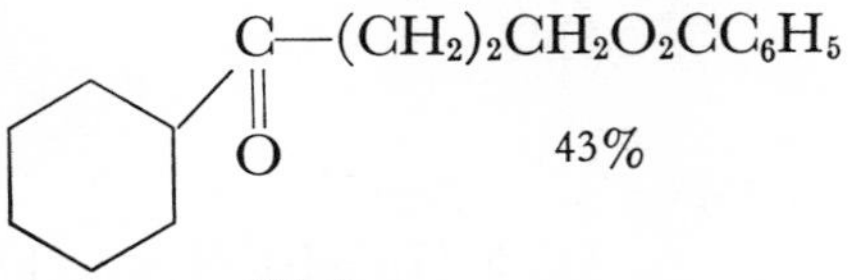

$C\!-\!(CH_2)_2CH_2O_2CC_6H_5$

$\overset{\|}{O}$ 43%

17.8

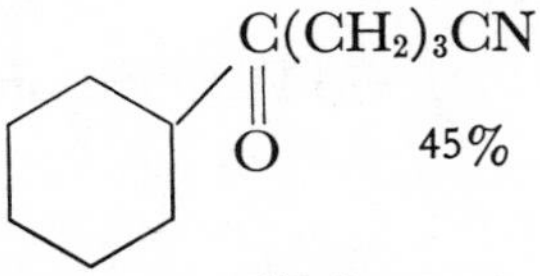

$C(CH_2)_3CN$

$\overset{\|}{O}$ 45%

17.9

However, derivatives containing substituents in the α- and β-positions relative to the boron atom possess unusual properties, and their use may pose difficulties.

9. Synthesis and Utilization of Mixed Organoboranes

It soon became apparent that this new synthetic approach bore two major handicaps. First, the aldehyde synthesis uses only one of the three groups on boron. Similarly, the ketone synthesis uses only two of these groups. If these groups represent a valuable intermediate, such losses could be serious. Second, it is frequently desired to introduce two different groups into a ketone, RCOR′. Similarly, it is often desired to synthesize tertiary alcohols containing two or even three different groups, $RR_2'COH$ and $RR'R''COH$. We undertook to solve these problems.

The use of B-R-9-BBN (Section XV-7) solved the problem of the aldehyde synthesis, as discussed in Section 10.

The use of thexylborane (Section XV-6) solved the problem of a general synthesis of ketones with two different groups (Section 11).

As pointed out above, certain dialkylboranes are readily prepared by the controlled reaction of 2 moles of olefin with 1 mole of borane. In such cases the synthesis of a mixed borane, R_2BR', presents no difficulty (17.43).

$$\langle hexyl \rangle -)_2BH + H_2C{=}\overset{\overset{\displaystyle CH_3}{|}}{C}{-}CH_3 \longrightarrow$$

$$\langle hexyl \rangle -)_2BCH_2CH(CH_3)_2 \quad 17.43$$

Carbonylation of suxh "mixed" organoboranes to mixed tertiary alcohols presents no difficulty (17.44).

$$\langle hexyl \rangle -)_2BCH_2CH(CH_3)_2 \longrightarrow$$

$$\langle hexyl \rangle -)_2\overset{\overset{\displaystyle }{|}}{C}CH_2CH(CH_3)_2 \quad 17.44$$
$$OH$$

Unfortunately, this simple formation of dialkylboranes is restricted to relatively hindered olefins, so it does not offer a general solution.

Organoboranes are readily exchanged with methyl borate in the presence of catalytic quantities of diborane to give the corresponding borinic acid ester[23] (17.45). Treatment of such borinic esters with

$$2R_3B + B(OCH_3)_3 \xrightarrow{\text{cat. BH}_3} 3R_2BOCH_3 \qquad 17.45$$

lithium aluminum hydride in the presence of the desired olefins provides mixed organoboranes that are readily converted to tertiary alcohols containing two different groups[24] (17.46, 17.47).

$$
3 \quad
\begin{array}{c}
\text{CH}_3 \\
| \\
\text{CH}_3\text{CH}_2\text{CH} \\
\diagdown \\
\qquad\qquad \text{BOCH}_3 + 3\text{CH}_3\text{CH}_2\text{CH}_2\text{CH}{=}\text{CH}_2 + \text{LiAlH}_4 \\
\diagup \\
\text{CH}_3\text{CH}_2\text{CH} \\
| \\
\text{CH}_3
\end{array}
$$

$$
\longrightarrow 3 \quad
\begin{array}{c}
\text{CH}_3 \\
| \\
\text{CH}_3\text{CH}_2\text{CH} \\
\diagdown \\
\qquad\qquad \text{B}{-}\text{CH}_2\text{CH}_2\text{CH}_2\text{CH}_2\text{CH}_3 + \text{LiAlH(OCH}_3)_3 \\
\diagup \\
\text{CH}_3\text{CH}_2\text{CH} \\
| \\
\text{CH}_3
\end{array}
$$

$$\qquad\qquad\qquad 17.46$$

90%

$$
\longrightarrow 3 \quad
\begin{array}{c}
\text{CH}_3 \qquad\qquad\qquad\qquad\\
| \qquad\qquad\qquad\qquad\qquad \\
\text{CH}_3\text{CH}_2\text{CH} \qquad \text{CH}_2\text{CH}_2\text{CH}_2\text{CH}_2\text{CH}_3 \\
\diagdown \quad\diagup \\
\text{C} \\
\diagup \quad\diagdown \\
\text{CH}_3\text{CH}_2\text{CH} \qquad \text{OH} \\
| \qquad\qquad\qquad\qquad \\
\text{CH}_3 \qquad\qquad\qquad\qquad
\end{array}
$$

88%

[23] B. M. Mikhailov and L. S. Vasil'ev, *Dokl. Akad. Nauk SSSR*, **139**, 385 (1961); *Zh. Obshch. Khim.*, **35**, 925 (1965).

[24] H. C. Brown, E. Negishi, and S. K. Gupta, *J. Amer. Chem. Soc.*, **92**, 6648 (1970).

$$3 \left\langle \overline{\bigcirc} \right. \text{B—OCH}_3 + 3 \left[\bigcirc\right] + \text{LiAlH}_4 \longrightarrow$$

$$3 \; \overset{\bigcirc}{\underset{\bigcirc}{\text{B}}} \; + \text{LiAlH(OCH}_3)_3 \longrightarrow 3 \; \overset{\bigcirc}{\underset{\bigcirc}{\text{C}}}\text{—OH} \qquad 17.47$$

90%   98%

Reduction of such borinic esters with aluminum hydride gives the dialkylborane as a complex with the aluminum methoxide (17.48). This intermediate can be used to hydroborate functional derivatives to make the corresponding mixed organoborane containing a functional group for conversion to the tertiary alcohol[25] (17.49).

$$3R_2BOCH_3 + AlH_3 \longrightarrow (R_2BH)_3 \cdot Al(OCH_3)_3 \qquad 17.48$$

$$(R_2BH)_3 \cdot Al(OCH_3)_3 + 3CH_2\text{=}CH(CH_2)_8CO_2C_2H_5$$
$$\longrightarrow 3R_2BCH_2(CH_2)_9CO_2C_2H_5 \qquad 17.49$$
$$\longrightarrow 3R_2\underset{\underset{OH}{|}}{C}CH_2(CH_2)_9CO_2C_2H_5$$

An alternate route to mixed organoboranes is provided by catechol borane (Section XV-7). This reagent readily hydroborates olefins to form the corresponding 2-alkyl-1,3,2-benzodioxaboroles[26] (17.50).

$$\overset{O}{\underset{O}{\bigcirc\hspace{-0.3em}\diagup}}\text{B—H} + \left[\bigcirc\right] \longrightarrow \overset{O}{\underset{O}{\bigcirc\hspace{-0.3em}\diagup}}\text{B—}\bigcirc \qquad 17.50$$

90%

[25] H. C. Brown and S. K. Gupta, *J. Amer. Chem. Soc.*, **93,** 1818 (1971).
[26] H. C. Brown and S. K. Gupta, *J. Amer. Chem. Soc.*, **93,** 1816 (1971).

Treatment of this intermediate with lithium aluminum hydride in the presence of the desired olefin readily provides the desired mixed organoborane (17.51) for conversion to the tertiary alcohol[27] (17.52).

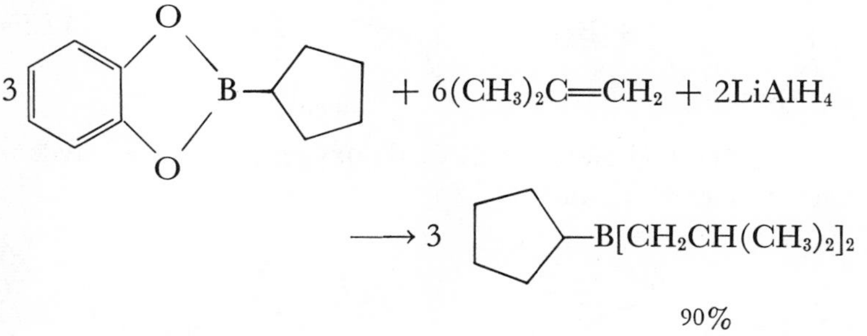

$$3 \quad \text{(catechol boronate–cyclopentyl)} + 6(CH_3)_2C=CH_2 + 2LiAlH_4$$

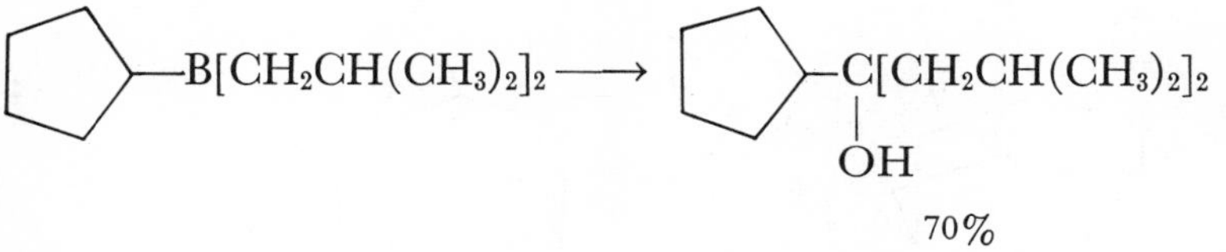

$$\longrightarrow 3 \quad \text{(cyclopentyl)}-B[CH_2CH(CH_3)_2]_2 \quad 17.51$$

90%

$$\text{(cyclopentyl)}-B[CH_2CH(CH_3)_2]_2 \longrightarrow \text{(cyclopentyl)}-\underset{\underset{OH}{|}}{C}[CH_2CH(CH_3)_2]_2 \quad 17.52$$

70%

Here also it is possible to use modifications of the procedure to permit preparation of mixed organoboranes containing one or two functional groups.[27]

The solution to the problem of providing an organoborane with three different alkyl groups, $RR'R''B$, convertible to the corresponding tertiary alcohol, $RR'R''COH$, is still ahead of us, but there is no apparent reason to doubt that a convenient solution will be developed.

10. Primary Alcohols and Aldehydes via B-R-9-BBN

As pointed out earlier (Section XV-7), treatment of 1,5-cyclooctadiene with borane under controlled conditions yields 9-borabicyclo[3.3.1]nonane (17.53), a material that is remarkably stable thermally and sufficiently insensitive to atmospheric oxygen and moisture that it can be handled with reasonable precautions in the open air.[28]

---

[27] H. C. Brown and S. K. Gupta, *J. Amer. Chem. Soc.*, **93,** 4062 (1971).

[28] E. F. Knights and H. C. Brown, *J. Amer. Chem. Soc.*, **90,** 5280 (1968).

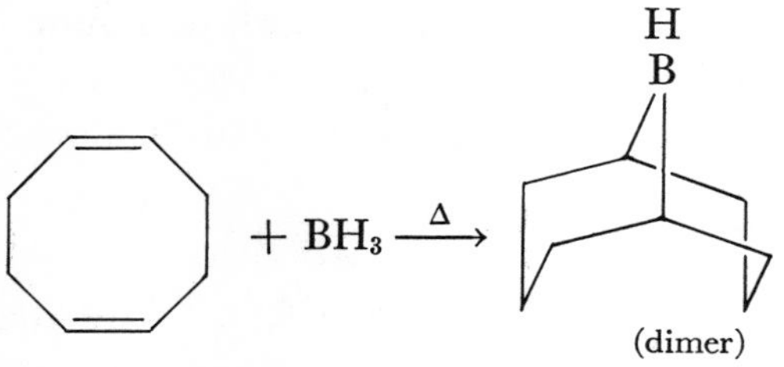

(dimer)

$$17.53$$

In spite of its chemical stability towards oxygen, it readily hydroborates even hindered olefins[29] (17.54).

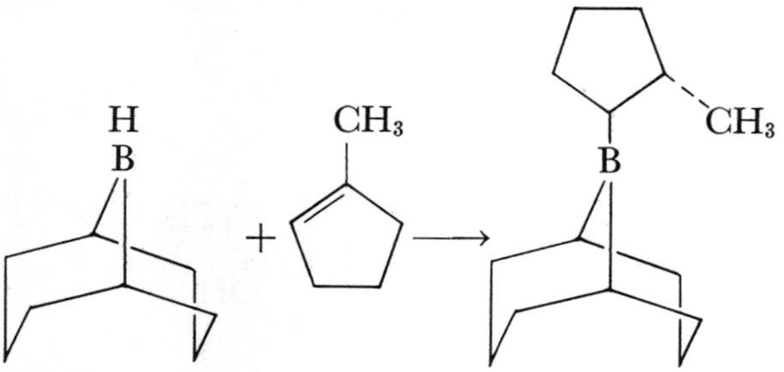

$$17.54$$

The 9-BBN moiety acts like an inert bifunctional substituent in the reaction with carbon monoxide and lithium trimethoxyaluminohydride. Only the 9-alkyl group reacts to produce the desired aldehyde[30] (17.55, 17.56).

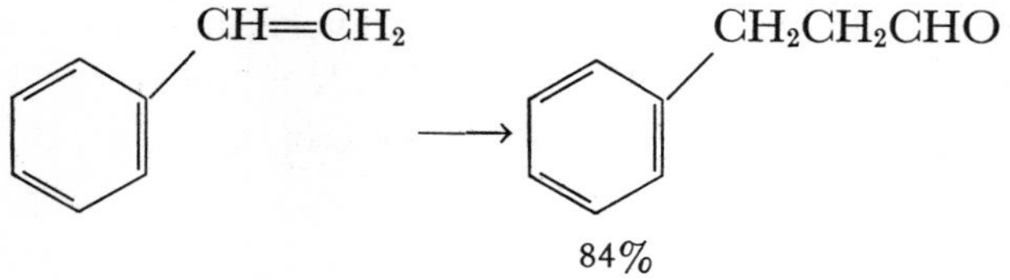

84%

$$17.55$$

79%

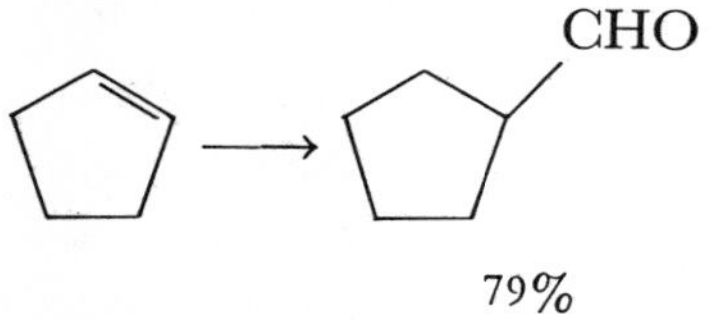

$$17.56$$

Hydrolysis of the reaction product with aqueous base, rather than oxidation with alkaline hydrogen peroxide, produces the correspond-

[29] E. F. Knights and H. C. Brown, *J. Amer. Chem. Soc.*, **90**, 1528 (1968).

[30] H. C. Brown, E. F. Knights, and R. A. Coleman, *J. Amer. Chem. Soc.*, **91**, 2144 (1969).

ing alcohol (17.35). Consequently, one can convert a given olefin into the corresponding aldehyde or methylol derivative.

In this reaction the aldehyde or methylol group is introduced with retention of configuration[31] (17.57).

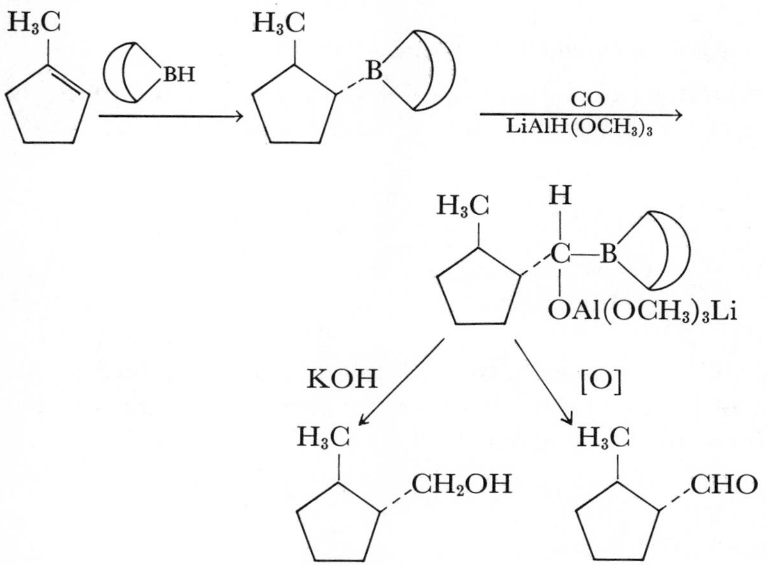

17.57

There are obvious advantages in such a simple, stereospecific homologation reaction.

Finally, this reaction can be extended to introduce the aldehyde group into olefins containing functional groups.[32] For such cases it is advantageous to utilize the gentler reducing agent, lithium tri-*t*-butoxyaluminohydride (Section XII-8). Some typical examples are shown in 17.58–17.61.

$$H_2C{=}CHCH_2CO_2C_2H_5 \longrightarrow OHCCH_2CH_2CH_2CO_2C_2H_5 \quad 17.58$$

83%

$$H_2C{=}CHCH_2CH_2O_2CCH_3 \longrightarrow$$
$$OHCCH_2CH_2CH_2CH_2O_2CCH_3 \quad 17.59$$

92%

[31] H. C. Brown, M. M. Rogič, M. W. Rathke, and G. W. Kabalka, *J. Amer. Chem. Soc.*, **91**, 2150 (1969).

[32] H. C. Brown and R. A. Coleman, *J. Amer. Chem. Soc.*, **91**, 4606 (1969).

$$H_2C\!=\!CHCH_2O_2CC_6H_5 \longrightarrow OHCCH_2CH_2CH_2O_2CC_6H_5 \quad 17.60$$
$$89\%$$

$$H_2C\!=\!CHCH_2CN \longrightarrow OHCCH_2CH_2CH_2CN \qquad 17.61$$
$$85\%$$

11. The Thexylborane Synthesis of Unsymmetrical Ketones [33]

Hydroboration of 2,3-dimethyl-2-butene can be controlled to yield the monoalkylborane, termed thexylborane[34] (Section XV-6) (17.62).

$$17.62$$

This valuable reagent is not isolated, but is formed and used *in situ*. Addition of 1 mole of a not too hindered olefin readily converts the thexylborane to a thexylmonoalkylborane (17.63).

$$17.63$$

Addition of a second mole of a suitable olefin results in the formation of an organoborane with three different groups attached to the boron atom (17.64).

$$17.64$$

[33] H. C. Brown and E. Negishi, *J. Amer. Chem. Soc.*, **89,** 5285 (1967).
[34] H. C. Brown and G. Zweifel, *J. Amer. Chem. Soc.*, **85,** 2066 (1963).

In the course of these studies we observed that the thexyl group greatly hindered the boron atom, so that atmospheric-pressure carbonylation was no longer practical. However, carbonylation at 1000 psi was satisfactory. Far more important was the observation that the thexyl group was very sluggish in migrating from boron to carbon. Consequently, we were now in position to achieve a specific ketone synthesis from almost any two olefins (17.65).

$$
\underset{CH_2CH_2CH_2CH_3}{\overset{CH_2CH(CH_3)_2}{H\!\!-\!\!B}} \quad \xrightarrow[\text{H}_2\text{O}]{\text{CO}} \xrightarrow{\text{[O]}}
$$

$$
(CH_3)_2CHCH_2\underset{\underset{O}{\|}}{C}(CH_2)_3CH_3 \quad 17.65
$$

Indeed, many functional groups can be tolerated, so that this provides a new, simple ketone synthesis of major promise (17.66–17.69).

$$
\underset{CH_3C=CH_2}{\overset{CH_3}{|}} + CH_2=CHCH_2CO_2C_2H_5 \longrightarrow
$$
$$
(CH_3)_2CHCH_2\underset{\underset{O}{\|}}{C}CH_2CH_2CH_2CO_2C_2H_5 \quad 17.66
$$

$$84\%$$

$$
\text{(phenyl)} \overset{CH=CH_2}{\diagup} + CH_2=CH(CH_2)_8CO_2CH_3 \longrightarrow
$$

$$
C_6H_5CH_2CH_2\underset{\underset{O}{\|}}{C}(CH_2)_{10}CO_2CH_3 \quad 17.67
$$

$$73\%$$

$$
\text{(cyclopentenyl)} + CH_2=CHCH_2CN \longrightarrow \text{(cyclopentyl)}\underset{\underset{O}{\|}}{-C}CH_2CH_2CH_2CN \quad 17.68
$$

$$45\%$$

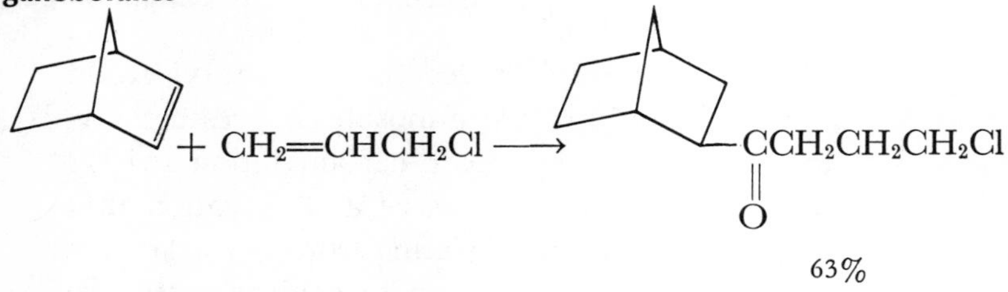
+ CH2=CHCH2Cl
CCH2CH2CH2Cl
O
63%
17.69

12. The Thexylborane Synthesis of Cyclic Ketones [35]

Thexylborane can be utilized for the cyclic hydroboration of appropriate dienes[36] (Section XV-6). Consequently, the carbonylation of such derivatives provides a convenient synthesis of cyclic ketones (17.70–17.72).

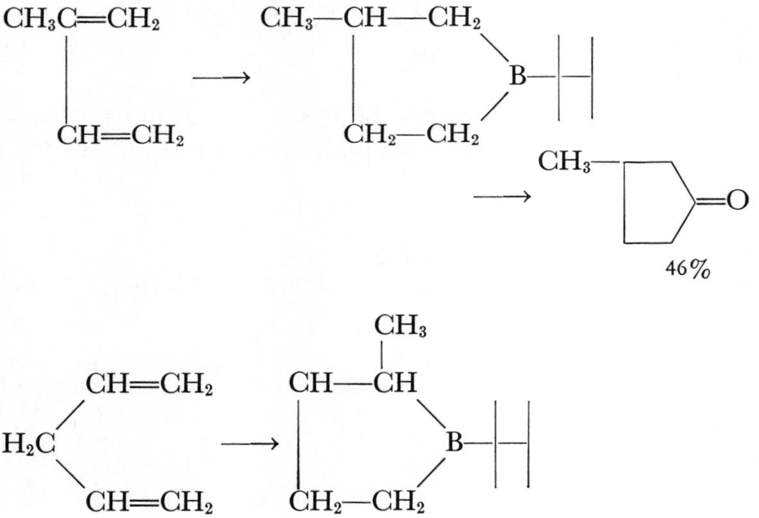
CH3C=CH2
CH=CH2
CH3—CH—CH2
B
CH2—CH2
CH3
=O
46%
17.70

CH=CH2
H2C
CH=CH2
CH3
CH—CH
B
CH2—CH2
CH3
=O
75%
17.71

[35] H. C. Brown and E. Negishi, *J. Amer. Chem. Soc.*, **89**, 5477 (1967).
[36] H. C. Brown and C. D. Pfaffenberger, *J. Amer. Chem. Soc.*, **89**, 5475 (1967).

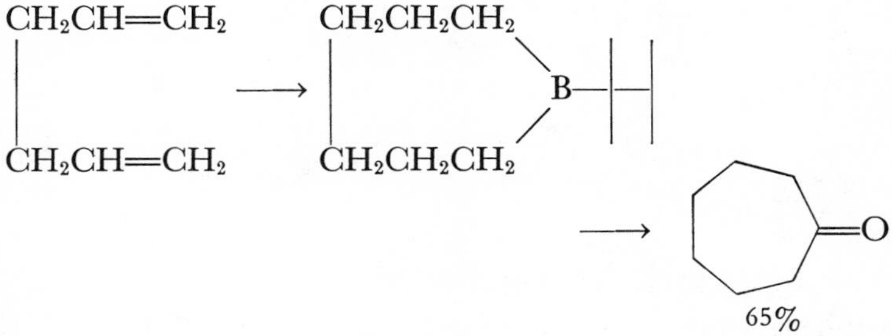
CH2CH=CH2
CH2CH=CH2
CH2CH2CH2
B
CH2CH2CH2
O
65%

17.72

13. A New Annelation Reaction [37]

It has proven possible to adapt this new synthesis of cyclic ketones to a new annelation reaction, as illustrated for the stereospecific conversion of cyclohexanone to the thermodynamically disfavored *trans*-perhydrol-1-indanone (17.73).

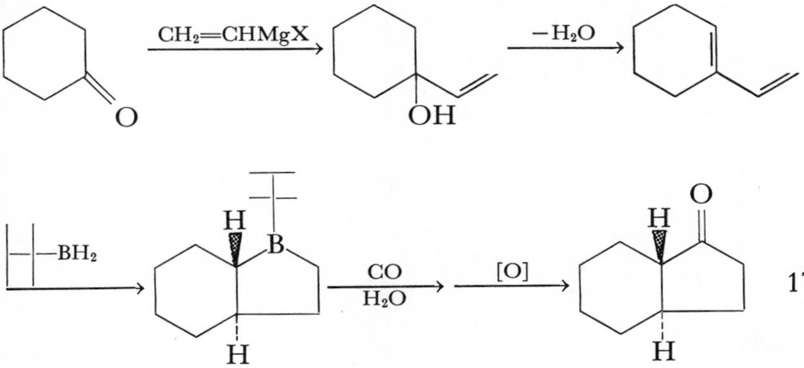
CH2=CHMgX
O
OH
−H2O
—BH2
H
B
H
CO
H2O
[O]
O
H
H

17.73

Similarly, by use of the allyl Grignard, one can convert cyclohexanone stereospecifically to the corresponding *trans*-1-decalone (17.74).

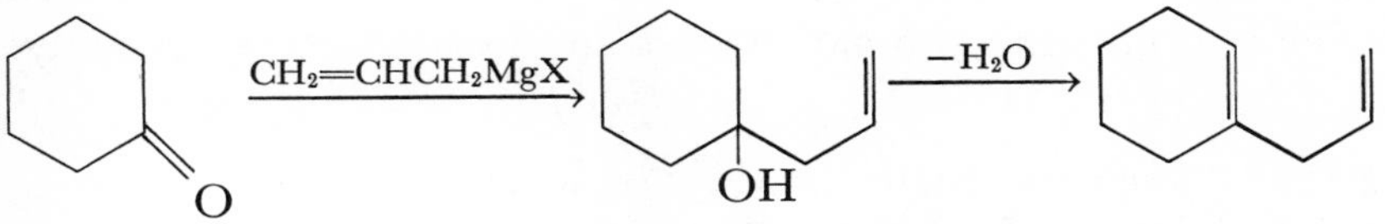
CH2=CHCH2MgX
O
OH
−H2O

[37] H. C. Brown and E. Negishi, *Chem. Commun.*, 594 (1968).

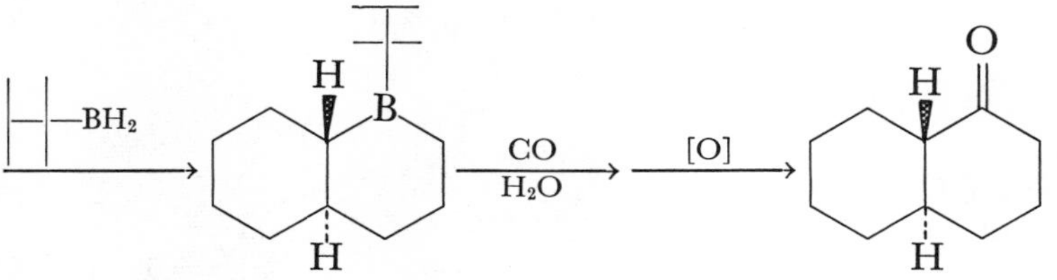

17.74

The reaction appears to be of considerable generality, as indicated by the following syntheses (the yields shown are based on diene) (**17.10–17.14**).

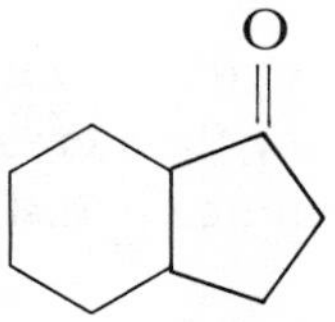

66%
17.10

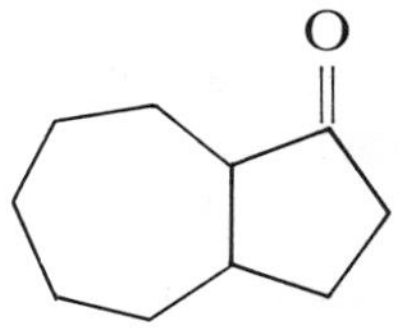

71%
17.11

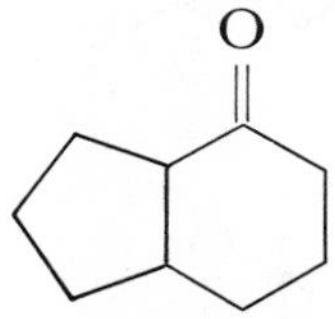

54%
17.12

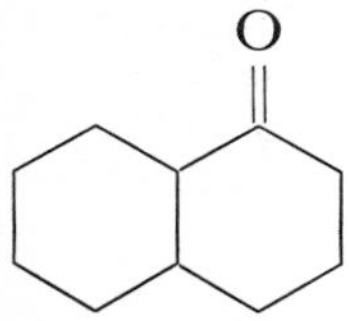

73%
17.13

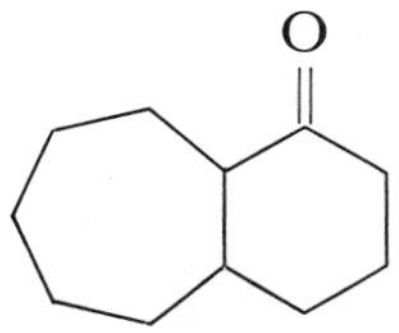

68%
17.14

14. Polycyclics [38, 39]

The remarkably easy addition of the boron-hydrogen bond to carbon-carbon double and triple bonds, the less easy but still facile substitution of carbon-hydrogen bonds by boron-hydrogen bonds,[40,41] and the ready isomerization of organoboranes (Chapter XV) all combine to give the organic chemist an unparalleled opportunity:

[38] H. C. Brown and E. Negishi, *J. Amer. Chem. Soc.*, **89,** 5478 (1967).

[39] E. F. Knights and H. C. Brown, *J. Amer. Chem. Soc.*, **90,** 5283 (1968).

[40] H. C. Brown, K. J. Murray, H. Müller, and G. Zweifel, *J. Amer. Chem. Soc.*, **88,** 1443 (1966).

[41] R. Köster, *Angew. Chem. Intern. Edit.*, **3,** 74 (1964).

he can use the unique characteristics of boron to bring together widely separated portions of a carbon structure into a more compact cyclic or polycyclic entity. Some representative examples follow[39-43] (17.75–17.79).

$$
\underset{\underset{CH_3}{|}}{\overset{\overset{CH_3}{|}}{H_3C-C}}-CH_2-\underset{}{\overset{\overset{CH_3}{|}}{C}}=CH_2 \xrightarrow{\quad \boxed{}-BH_3 \quad} \underset{\underset{CH_3}{|}}{\overset{\overset{CH_3}{|}}{H_3C-C}}-CH_2-\underset{\underset{CH_2}{|}}{\overset{\overset{CH_3}{|}}{CH}}
$$

$$\Delta \quad -H_2$$

$$17.75^{40}$$

$$CH_3(CH_2)_6CH=CH_2 \xrightarrow{BH_3} \xrightarrow[-2H_2]{\Delta}$$ $$\quad 17.76^{41}$$

$$\xrightarrow{Et_3N-BH_3}$$ $$\quad 17.77^{42}$$

[42] N. N. Greenwood, J. H. Morris, and J. C. Wright, *J. Chem. Soc.*, 4753 (1964).
[43] G. W. Rotermund and R. Köster, *Ann.*, **686,** 153 (1965).

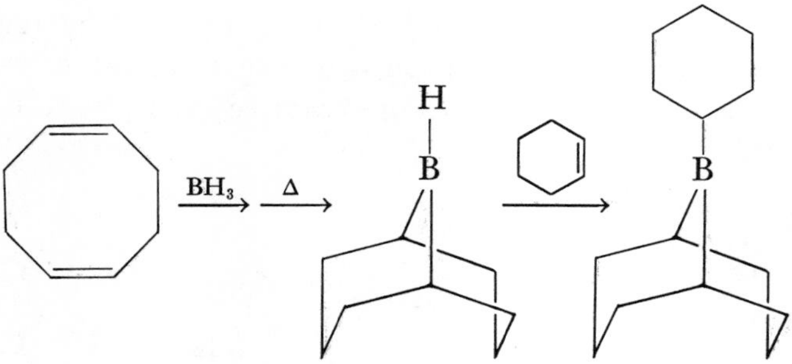

17.78[39]

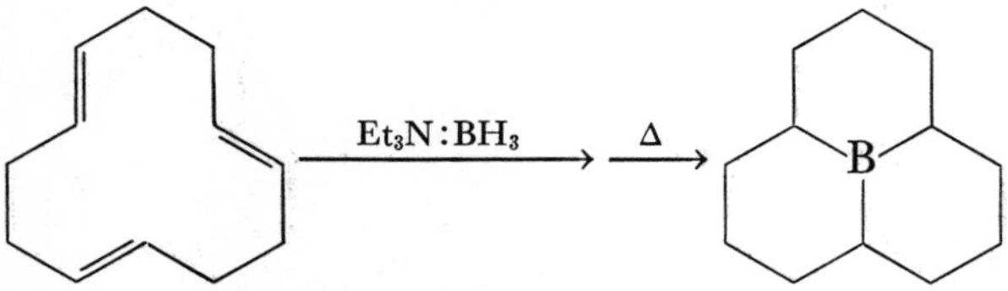

17.79[43]

It is evident that if the boron atom in these "stitched together" structures could be replaced by carbon, we could have a major new approach to the synthesis of complex carbon structures. We have tested this possibility on several representative systems and discovered that transformations proceed with remarkable ease[44] (17.80–17.82).

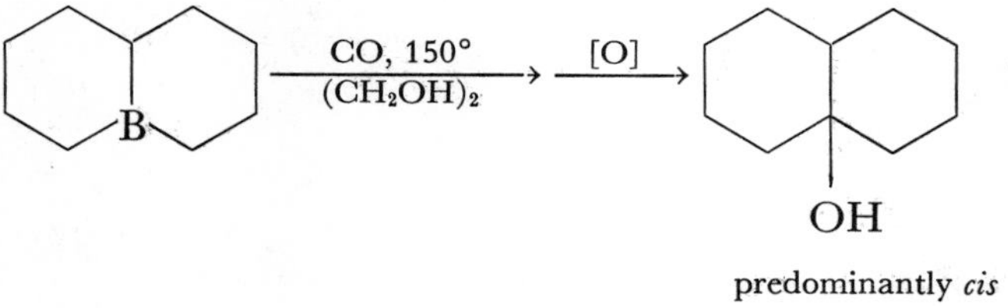

17.80[44]

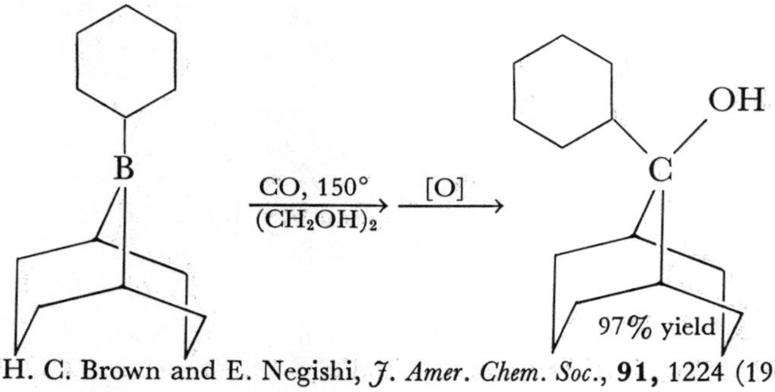

predominantly *cis*

17.81[39]

97% yield

[44] H. C. Brown and E. Negishi, *J. Amer. Chem. Soc.*, **91**, 1224 (1969).

70% yield

Thus it is now possible to "stitch" with boron and then "rivet" with carbon, providing a remarkably simple new route to complex carbon structures.

15. Conclusion

In 1967 we published the first of our studies describing the reaction of organoboranes with carbon monoxide at atmospheric pressure to form tertiary alcohols.[15] Consequently, the laboratory findings described in this chapter cover a period of three years. I wish to call this to the special attention of students. All too often textbooks give the impression that most of organic chemistry has been discovered and that little awaits even an industrious explorer. Clearly this chapter is a living refutation of that conclusion. Chapters XVIII and XIX should further persuade the student of the rich possibilities awaiting discovery by receptive investigators.

XVIII. Organoboranes as Alkylating and Arylating Agents

1. Introduction

The reaction of organoboranes with carbon monoxide in the presence of lithium borohydride[1] or trialkoxyaluminohydride[2] (Sections XVII-7 and XVII-10) provides a highly convenient means of achieving a one-carbon-atom homologation of olefins with the introduction of an aldehyde group (18.1).

$$R-B\diagup_{\diagdown} \xrightarrow{\text{CO, LiAlH(OR')}_3} \xrightarrow{\text{[O]}} RCHO \quad 18.1$$

Soon afterward we discovered that organoboranes undergo a remarkably facile 1,4-addition to α,β-unsaturated ketones[3] and aldehydes.[4] The reaction with acrolein (Chapter XIX) provides a new simple means of achieving a three-carbon-atom homologation to produce the corresponding aldehyde (18.2).

$$R-B\diagup_{\diagdown} \xrightarrow[\text{H}_2\text{O}]{\text{CH}_2=\text{CHCHO}} RCH_2CH_2CHO \quad 18.2$$

We thought it would be nice to have a two-carbon-atom homologation, leading to the corresponding aldehyde (18.3).

$$R-B\diagup_{\diagdown} \xrightarrow{?} RCH_2CHO \quad 18.3$$

---

[1] M. W. Rathke and H. C. Brown, *J. Amer. Chem. Soc.*, **89**, 2740 (1967).

[2] H. C. Brown, R. A. Coleman, and M. W. Rathke, *J. Amer. Chem. Soc.*, **90**, 499 (1968).

[3] A. Suzuki, A. Arase, H. Matsumoto, M. Itoh, H. C. Brown, M. M. Rogić, and M. W. Rathke, *J. Amer. Chem. Soc.*, **89**, 5708 (1967).

[4] H. C. Brown, M. M. Rogić, M. W. Rathke, and G. W. Kabalka, *J. Amer. Chem. Soc.*, **89**, 5709 (1967).

Unfortunately, we did not find a really satisfactory route to the aldehyde, in large part because of the difficulties in preparing and using the haloacetaldehydes.[5] However, we did find that the reaction of organoboranes with bromo- or chloroacetic acid esters under the influence of potassium t-butoxide provides a highly convenient two-carbon-atom homologation[6] (18.4).

$$R{-}B{\Big\langle} \quad \xrightarrow[\text{t-BuOK}]{\text{CH}_2\text{BrCO}_2\text{C}_2\text{H}_5} \quad RCH_2CO_2C_2H_5 \qquad 18.4$$

This development opened up a major new area for exploration—the reaction of α-halocarbanions with organoboranes. The results of this exploration are discussed in the present chapter.

Before we begin that discussion, it should be mentioned that we now have two four-carbon-atom homologations, the base-induced reaction of organoboranes with 4-bromocrotonic acid ester[7] (Section XVIII-6) (18.5) and the free-radical reaction of organoboranes with 1,3-butadiene monoxide[8] (Section XIX-5) (18.6).

$$R{-}B{\Big\langle} \quad \xrightarrow[\text{t-BuOK}]{\text{CH}_2\text{BrCH}{=}\text{CHCO}_2\text{C}_2\text{H}_5} \quad RCH{=}CHCH_2CO_2C_2H_5 \quad 18.5$$

$$R{-}B{\Big\langle} \quad \xrightarrow[\text{O}_2]{\underset{\text{O}}{\overset{\text{H}_2\text{C}=\text{CHCH}{-}{-}{-}\text{CH}_2}{\diagdown\diagup}}} \quad RCH_2CH{=}CHCH_2OH \quad 18.6$$

It should also be pointed out that the thexylborane synthesis of ketones (Section XVII-11) provides a general method of lengthening the chain by five, six, or more carbon atoms[9] (18.7).

[5] The reaction of organoboranes with diazoacetaldehyde with organoboranes does provide such a route. J. Hooz and G. F. Morrison, *Canadian J. Chem.*, **48**, 868 (1970). However, diazoacetaldehyde is not readily available as an intermediate. See Section XVIII–11.

[6] H. C. Brown, M. M. Rogić, M. W. Rathke, and G. W. Kabalka, *J. Amer. Chem. Soc.*, **90**, 818 (1968).

[7] H. C. Brown and H. Nambu, *J. Amer. Chem. Soc.*, **92**, 1761 (1970).

[8] A. Suzuki, N. Miyaura, M. Itoh, H. C. Brown, G. W. Holland, and E. Negishi, *J. Amer. Chem. Soc.*, **93**, 2792 (1971).

[9] E. Negishi and H. C. Brown, manuscript in preparation.

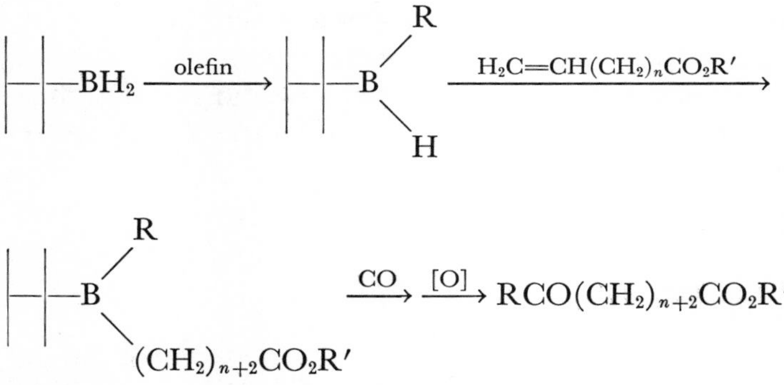

$$\text{\Large{|}}\!-\!BH_2 \xrightarrow{\text{olefin}} \text{\Large{|}}\!-\!B\!\!\begin{array}{c} R \\ \diagup \\ \diagdown \\ H \end{array} \xrightarrow{H_2C=CH(CH_2)_nCO_2R'}$$

18.7

$$\text{\Large{|}}\!-\!B\!\!\begin{array}{c} R \\ \diagup \\ \diagdown \\ (CH_2)_{n+2}CO_2R' \end{array} \xrightarrow{CO} \xrightarrow{[O]} RCO(CH_2)_{n+2}CO_2R'$$

where $n \geq 1$.

Consequently, we are now in position to achieve an exceedingly wide range of homologation through organoborane chemistry.

2. Alkylation of α-Haloalkanoic Esters with Organoboranes

The reaction of ethyl bromoacetate or chloroacetate with organoboranes in the presence of potassium t-butoxide proceeds rapidly and simply at 0°.[6] In the usual procedure the olefin is converted into the organoborane by treatment with the calculated quantity of borane in tetrahydrofuran. An equimolar amount of ethyl bromoacetate is added, followed by the addition at 0° of an equimolar quantity of potassium t-butoxide in t-butyl alcohol. As far as we could tell, the reaction was over immediately at completion of the addition.

The reaction appears to be of wide generality, as indicated by our examination of representative olefin types (18.8–18.11).

$$(CH_3)_2C=CH_2 \longrightarrow (CH_3)_2CHCH_2CH_2CO_2C_2H_5 \qquad 18.8$$
$$98\%$$

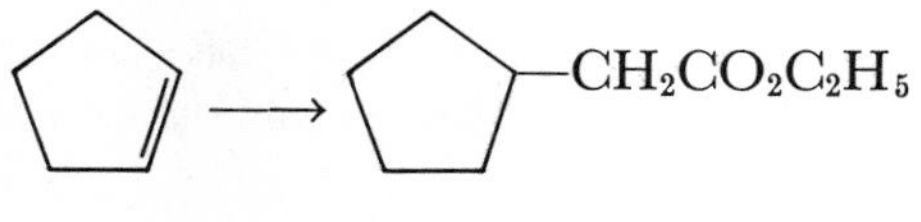

18.9

95%

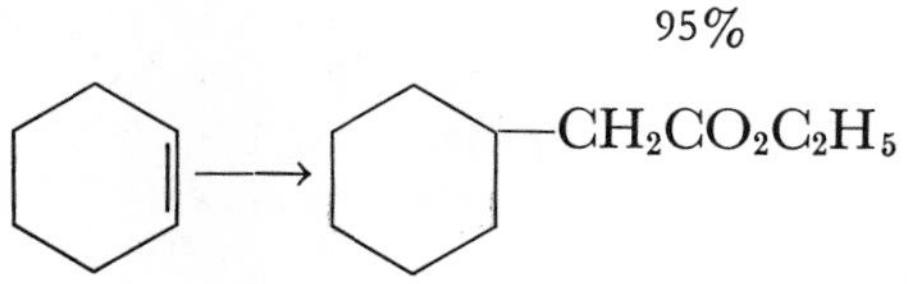

18.10

80%

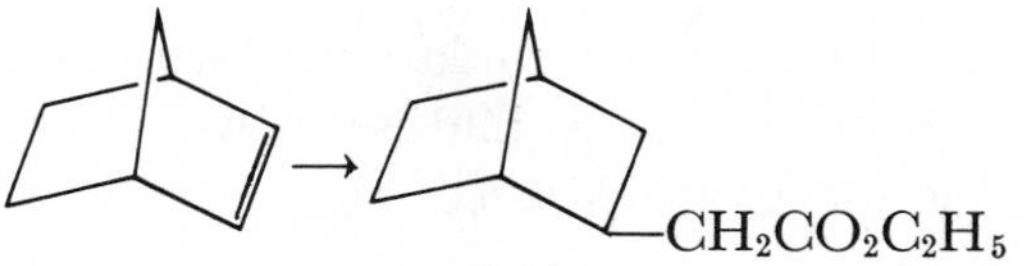

$$-CH_2CO_2C_2H_5 \qquad 18.11$$

85%

The reaction is also applicable to dichloro- and dibromoacetic acid esters, providing a simple synthesis of the corresponding α-halocarboxylic acid esters[10] (18.12, 18.13).

$$CH_3(CH_2)_3CH{=}CH_2 \xrightarrow{CHCl_2CO_2C_2H_5} CH_3(CH_2)_5CHCO_2C_2H_5$$
$$\underset{Cl}{|}$$
$$18.12$$

98%

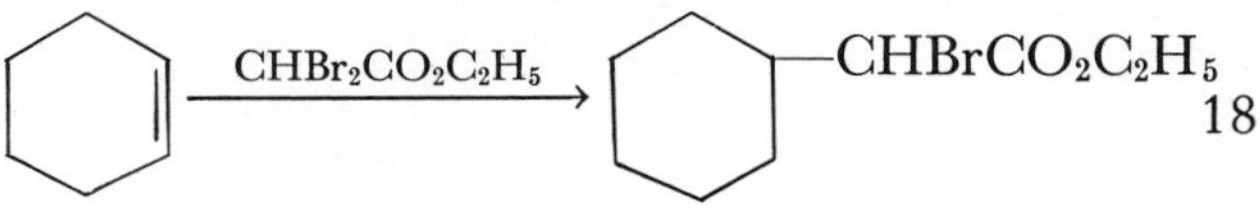

$$\xrightarrow{CHBr_2CO_2C_2H_5}$$

$$-CHBrCO_2C_2H_5$$
$$18.13$$

93%

Moreover, the use of two molar equivalents of potassium t-butoxide with two molar equivalents of the organoborane achieves the dialkylation of ethyl dibromoacetate (18.14).

$$2R_3B + CHBr_2CO_2C_2H_5 \xrightarrow[t\text{-BuOH}]{2t\text{-BuOK}} R_2CHCO_2C_2H_5 \quad 18.14$$

The dialkylation can be achieved in two successive stages, permitting the introduction of two different organic groups into the acetic acid moiety (18.15).

$$R_3B + CHBr_2CO_2C_2H_5 \xrightarrow[t\text{-BuOH}]{t\text{-BuOK}} RCHBrCO_2C_2H_5$$
$$18.15$$
$$R'_3B + RCHBrCO_2C_2H_5 \xrightarrow[t\text{-BuOH}]{t\text{-BuOK}} RR'CHCO_2C_2H_5$$

Consequently, this development provides an alternative to the commonly used malonic ester synthesis for the preparation of disubstituted acetic acids. Unfortunately, the dialkylation reaction proved to be quite sensitive to the structure of the organoborane and could not

[10] H. C. Brown, M. M. Rogić, M. W. Rathke, and G. W. Kabalka, *J. Amer. Chem. Soc.*, **90**, 1911 (1968).

be used for the introduction of secondary alkyl groups. However, this difficulty was later overcome by the use of B-R-9-BBN derivatives with dichloroacetonitrile, as discussed later (Section XVIII-6).

Although we have not yet undertaken a detailed study of the mechanism, it is probable that the reaction involves the following steps: (a) formation of the α-halocarbanion from the ester (18.16); (b) coordination of the α-halocarbanion with the trialkylborane (18.17); (c) rapid rearrangement of the quaternary boron intermediate (18.18); (d) protonolysis of the organoborane intermediate by the *t*-butyl alcohol present in the reaction mixture (18.19).

$$t\text{-BuO}^-\text{K}^+ + \text{CH}_2\text{BrCO}_2\text{Et} \longrightarrow$$
$$\text{K}^{+-}\text{CHBrCO}_2\text{Et} + t\text{-BuOH} \quad 18.16$$

$$\text{R}_3\text{B} + \text{K}^{+-}\text{CHBrCO}_2\text{Et} \longrightarrow$$

$$\text{K}^+\left[\begin{array}{c} \text{R} \\ | \\ \text{R}-\text{B}-\text{CHCO}_2\text{Et} \\ | \quad | \\ \text{R} \quad \text{Br} \end{array}\right]^- \quad 18.17$$

$$\text{K}^+\left[\begin{array}{c} \text{R} \\ | \\ \text{R}-\text{B}-\text{CHCO}_2\text{Et} \\ | \quad | \\ \text{R} \quad \text{Br} \end{array}\right]^- \longrightarrow$$

$$\begin{array}{c} \text{R} \\ | \\ \text{R}-\text{B}-\text{CHCO}_2\text{Et} + \text{KBr} \quad 18.18 \\ | \\ \text{R} \end{array}$$

$$\begin{array}{c} \text{R} \\ | \\ \text{R}-\text{B}-\text{CHCO}_2\text{Et} + t\text{-BuOH} \longrightarrow \\ | \\ \text{R} \end{array}$$

$$\text{RCH}_2\text{CO}_2\text{Et} + t\text{-BuOBR}_2 \quad 18.19$$

This mechanism suggests that the reaction might be extended to other α-halo compounds convertible to the α-halocarbanions by base. However, these are highly reactive species, subject to rapid condensations and internal rearrangements. Success could not be predicted with any assurance, so we undertook a systematic survey of the possible range of applicability.

3. Extension to Alkylation of Ketones [11]

Our attempt to extend the procedure developed for the bromo-acetate esters to the similar alkylation of α-bromo ketones proved unsatisfactory. Thus, treatment of a 1:1 molar mixture of phenacyl bromide and triethylborane with potassium t-butoxide in t-butyl alcohol at 0° resulted in only a 25 percent yield of n-butyro-phenone.

The difficulty was surmounted by shifting to potassium t-butoxide in tetrahydrofuran. At 0° a 1:1 molar mixture of phenacyl bromide and triethylborane reacted practically instantly upon the addition of an equivalent amount of the base to give an essentially quantita-tive yield of n-butyrophenone (18.20).

The reaction of α-bromocyclohexanone is much slower, but even this alkylation proceeds relatively smoothly to give 68 percent of α-ethylcyclohexanone (18.21) in approximately 12 hours at 0° (the reaction is essentially complete in 1 hour at 25°).

The reaction was readily extended to the introduction of other n-alkyl groups, such as n-butyl and n-decyl. However, the reaction failed with triisobutylborane and tri-sec-butylborane. The lifetime of

[11] H. C. Brown, M. M. Rogić, and M. W. Rathke, *J. Amer. Chem. Soc.*, **90,** 6218 (1968).

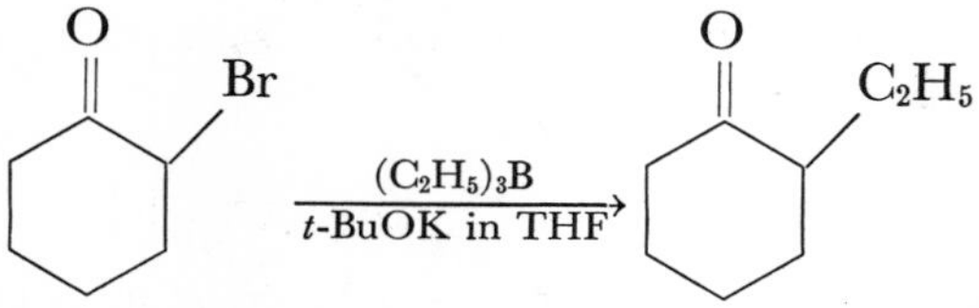

$$\text{18.21}$$

the α-halocarbanion must be exceedingly short in these keto derivatives and its rate of capture by the more hindered trialkylboranes must be too slow. We also attempted to extend the reaction to the alkylation of bromoacetone and chloroacetonitrile, but neither system participated satisfactorily. Fortunately, some of these problems were readily overcome by use of the less hindered B-R-9-BBN derivatives (Section XVIII-4) and a more gentle base, potassium 2,6-di-t-butylphenoxide (Section XVIII-5).

4. Application of B-R-9-BBN [12, 13]

The B-R-9-BBN derivatives readily participated in the reaction with the halo- (18.22) and dihaloacetic acid esters (18.23).

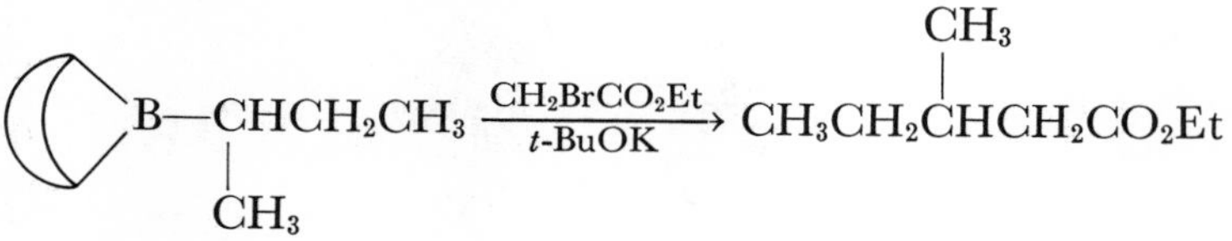

$$\text{18.22}$$

68%

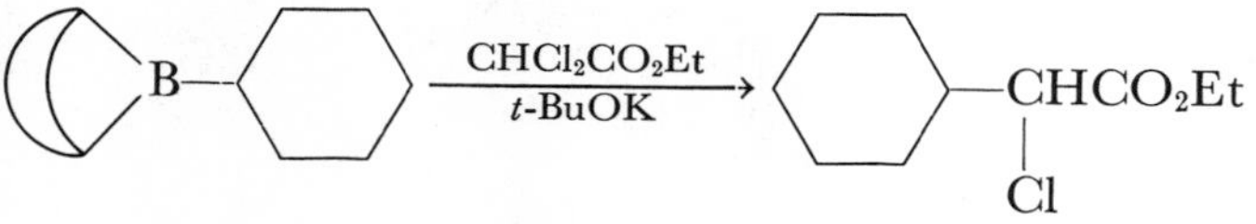

$$\text{18.23}$$

88%

Consequently, this overcomes the problem of utilizing only one of the three alkyl groups in the trialkylborane, as well as the problem previously encountered of steric difficulties in applying organoboranes as sterically demanding as tricyclohexylborane.

The B-R-9-BBN reagents also overcame many of the difficulties

[12] H. C. Brown and M. M. Rogić, *J. Amer. Chem. Soc.*, **91**, 2146 (1969).

[13] H. C. Brown, M. M. Rogić, H. Nambu, and M. W. Rathke, *J. Amer. Chem. Soc.*, **91**, 2147 (1969).

previously encountered in introducing the more sterically demanding alkyl groups into the α-position of ketones (18.24, 18.25).

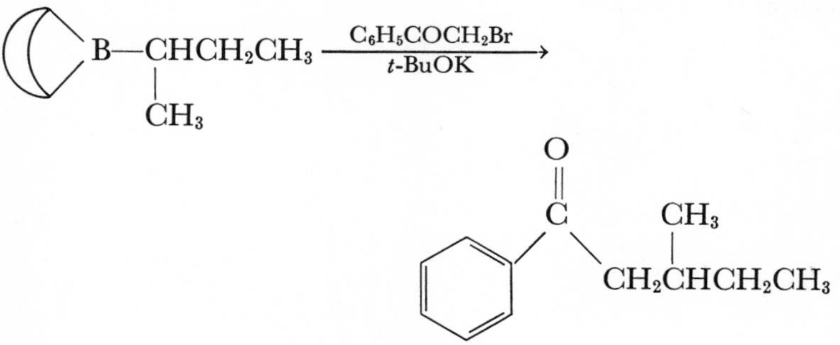
B—CHCH₂CH₃
CH₃
C₆H₅COCH₂Br
t-BuOK
O
C
CH₃
CH₂CHCH₂CH₃

18.24

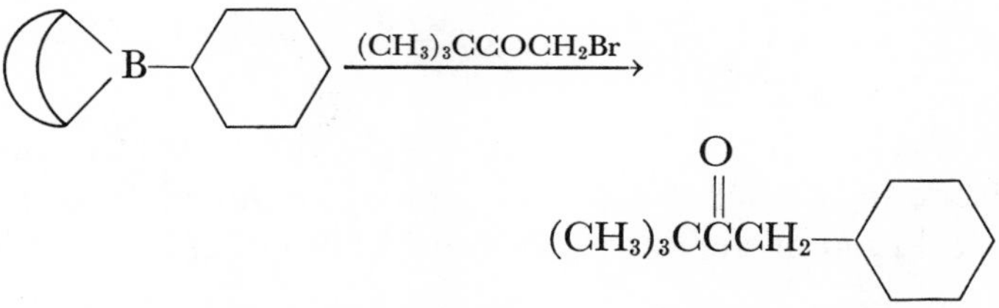
B
(CH₃)₃CCOCH₂Br
O
(CH₃)₃CCCH₂

18.25

Many of the selective features of the hydroboration reaction can be used in introducing alkyl groups into the *alpha* position of appropriate compounds through this procedure (18.26).

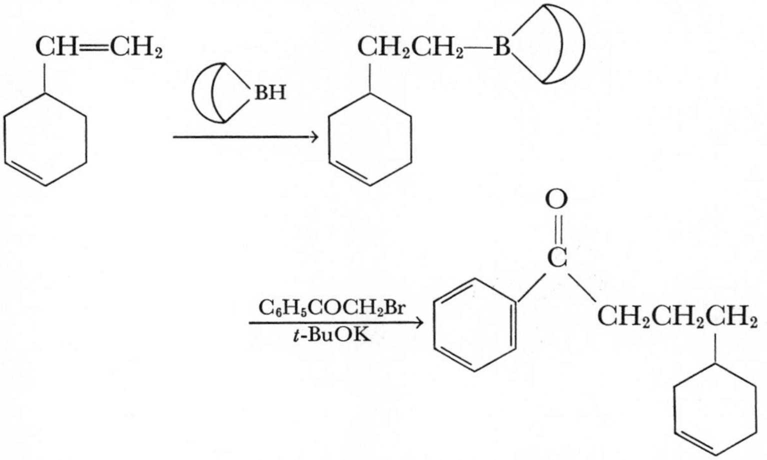
CH=CH₂
BH
CH₂CH₂—B
O
C
C₆H₅COCH₂Br
t-BuOK
CH₂CH₂CH₂

18.26

This procedure is by no means limited to the introduction of alkyl groups. B-Aryl-9-BBN derivatives are readily synthesized through the reaction of the aryllithium or arylmagnesium bromide with 9-BBN, followed by destruction of the substituted borohydride[14] (18.27).

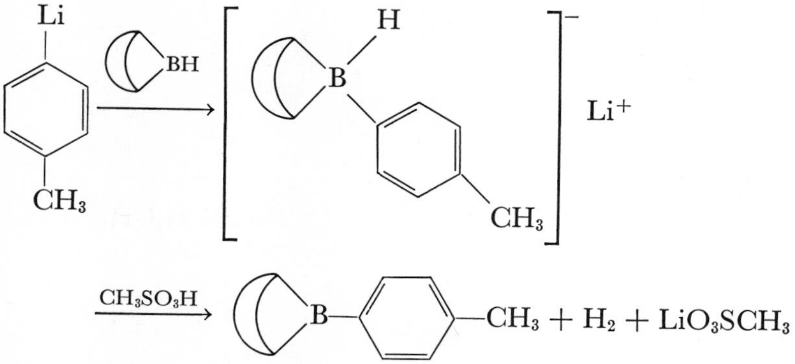

18.27

Alternatively, treatment of B-methoxy-9-BBN with the aryl- or alkyl-lithium in hexane precipitates lithium methoxide and provides a hexane solution of the desired B-R-9-BBN[15] (18.28).

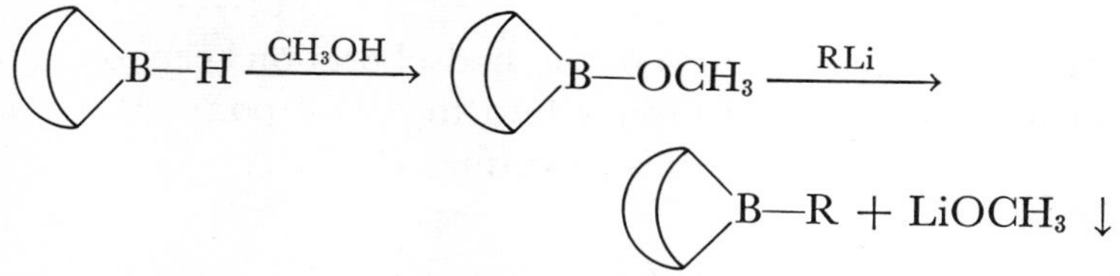

18.28

These B-aryl-9-BBN derivatives can be used for the *alpha* arylation of ketones and esters[14] (18.29).

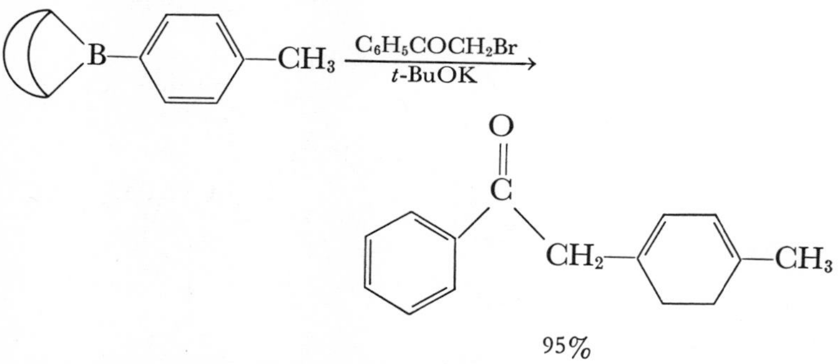

18.29

95%

[14] H. C. Brown and M. M. Rogić, *J. Amer. Chem. Soc.*, **91**, 4304 (1969).
[15] Research in progress with G. Kramer.

We were left with only one serious difficulty. We were unable to extend this alkylation and arylation reaction to the more labile structures, such as bromoacetone, chloroacetonitrile, and ethyl bromocyanoacetate. Perhaps the difficulty resided in the base. Potassium *t*-butoxide is an exceedingly strong base, perhaps too strong for the more labile structures. We decided to search for one more favorable.

5. The New Base [16]

We discovered that we could overcome practically all of the remaining difficulties by adopting potassium 2,6-di-*t*-butylphenoxide (**18.1**) as the base.[16]

18.1

In view of one of the special objectives I have set for this book, it may be desirable to indicate how we found this base—not all solutions in science come about through a quick "flash of genius."

Hirohiko Nambu was a relatively new member of our research group. He had been sent to work with us by the Mitsui Petrochemical Industries of Japan. For several weeks he had worked closely with Dr. Milorad M. Rogić, an experienced member of our group, learning the techniques and background for research in this area. He was now ready to tackle problems on his own.

As it happened, I was about to start out on a seven-week lecture trip to India and Israel. I discussed our problem with Mr. Nambu, who expressed interest in searching for a new base that might solve our remaining problems. Since I would not be in day-to-day contact with the research during my absence, I suggested that Mr. Nambu prepare a list of bases that should be tested. He did so, listing 29 possibilities, including practically all of the bases that have been

[16] H. C. Brown, H. Nambu, and M. M. Rogić, *J. Amer. Chem. Soc.*, **91**, 6855 (1969).

used for base-catalyzed condensations.[17] On the spur of the moment I suggested adding potassium phenoxide, potassium 2,6-dimethyl-phenoxide, and potassium 2,6-di-*t*-butylphenoxide, and I left him with this list of 32 bases. As a test system we selected the reaction of triethylborane with phenacyl bromide (18.20).

I returned seven weeks later to find that he had reached number 30 on the list without finding a base as favorable as potassium *t*-butoxide. Number 30, potassium phenoxide, provided a yield of 2% *n*-butyrophenone. Number 31, potassium 2,6-dimethylphenoxide, was much more favorable. It gave a yield of 75 percent. But we really hit the jackpot, 98 percent, with the last member, potassium 2,6-di-*t*-butylphenoxide.

This approach, the so-called Edisonian method, is sneered at in some circles. When one is young and full of theories, there is a tend-ency to feel that all such problems could and should be solved by theoretical deductions. With the years comes a better appreciation of the need for the Edisonian method. After one discovers a solution using such trial-and-error methods, it is always possible to apply theory to explain the solution. Unfortunately, this explanation often finds its way into the literature with no mention of the trial-and-error experiments, leaving the impression that the solution was actually arrived at through theoretical deductions—an impression that may mislead the less experienced reader.

The base not only worked well with phenacyl bromide, but it also proved to be highly satisfactory for α-bromo esters (18.30).

18.30

78%

---

[17] H. O. House, *Modern Synthetic Methods* (W. A. Benjamin, New York, 1965), Chap. 7.

Perhaps the most unexpected feature about this development is the great speed and ease with which the base can bring about these reactions. Certainly, the low pK_a value of the phenol, 11.7,[18] does not suggest that it would be so effective in removing a proton from the *alpha* position of an α-halo ester (pK_a 24[17]). Potassium *t*-butoxide is a far stronger base in tetrahydrofuran and dimethyl sulfoxide than in *t*-butyl alcohol. Possibly the reason is that the activity of the anion is not reduced by strong solvation. Perhaps the bulky substituents in 2,6-di-*t*-butylphenoxide serve a similar function in separating both cation and solvent from the phenoxide oxygen atom, so that the base is more effective in removing an active methylene hydrogen atom than one would predict from its pK_a value.[19] Moreover, the large steric requirements should prevent the base from coordinating with the trialkylboranes. Such coordination would be unfavorable for the rapid capture of the α-halocarbanions. These considerations suggest that potassium 2,6-di-*t*-butylphenoxide must be an ideal type of base for the reactions under consideration.

It is most regrettable that we could not have arrived at this conclusion purely from theoretical considerations and thereby saved Mr. Hirohiko Nambu seven weeks of effort at the bench. However, there is an intense pleasure in arriving at a desirable solution, regardless of the elegance of the experimental approach.

6. Extension to New Systems

One of the real advantages of the new base proved to be its relative mildness. In utilizing potassium *t*-butoxide it was essential that no more than one molar equivalent of the base be added to a 1:1 molar mixture of the organoborane and the α-halo compound. Excess base destroyed the product. Similarly it was not feasible to carry out the reaction by adding the α-halo compound to a 1:1 molar mixture of the organoborane and the base, for then the product would be produced in the presence of the base and be destroyed.

Practically all of the compounds produced in these reactions, esters, ketones, and nitriles, proved to be stable to potassium 2,6-di-*t*-butylphenoxide in tetrahydrofuran at 0°. Consequently, there

[18] L. A. Cohen and W. M. Jones, *J. Amer. Chem. Soc.*, **85,** 3397 (1963).

[19] The unusual behavior of 2,6-di-*t*-butylpyridine toward hydrogen chloride, boron trifluoride, and methyl iodide should be recalled (Section VIII–5).

was now no difficulty in using an excess of base or in varying the precise order of bringing the reactants together. We were now in position to explore the full utility of this improved alkylation procedure.

α-Bromoacetone, which previously had produced unsatisfactory results, now produced excellent yields of the methyl ketones[20] (18.31–18.33).

$$(C_2H_5)_3B + CH_2BrCOCH_3 \longrightarrow CH_3CH_2CH_2COCH_3 \qquad 18.31$$
$$88\%$$

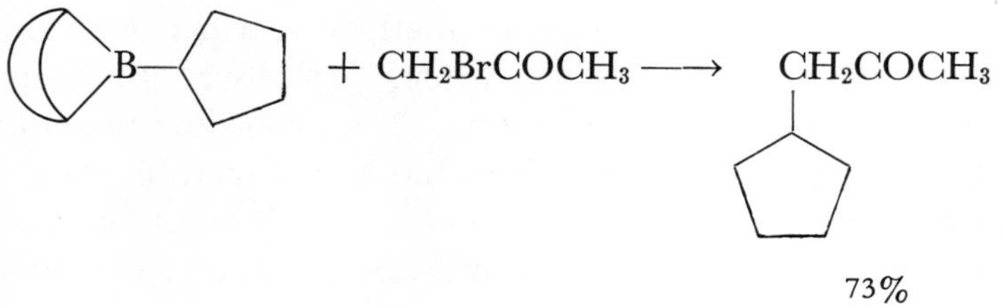

18.32

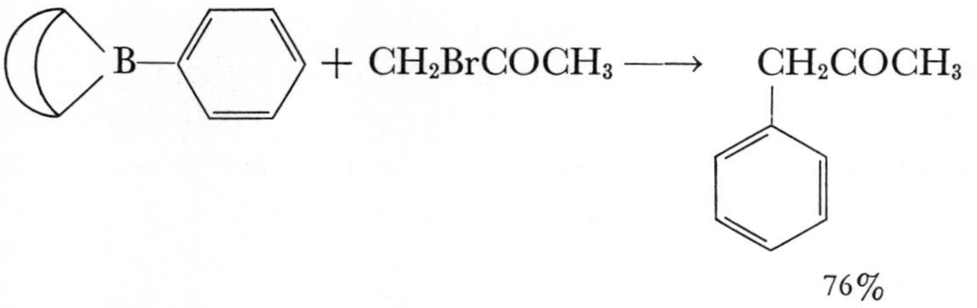

73%

18.33

76%

Chloroacetonitrile also proved useful in this procedure for the preparation of a wide variety of nitriles[21] (18.34–18.36).

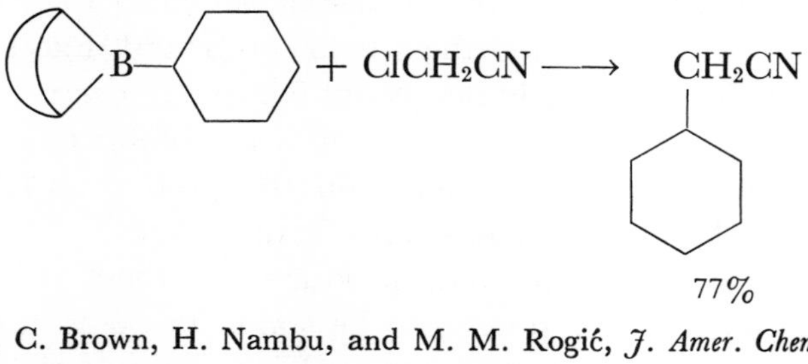

18.34

77%

[20] H. C. Brown, H. Nambu, and M. M. Rogić, *J. Amer. Chem. Soc.*, **91,** 6852 (1969).

[21] H. C. Brown, H. Nambu, and M. M. Rogić, *J. Amer. Chem. Soc.*, **91,** 6854 (1969).

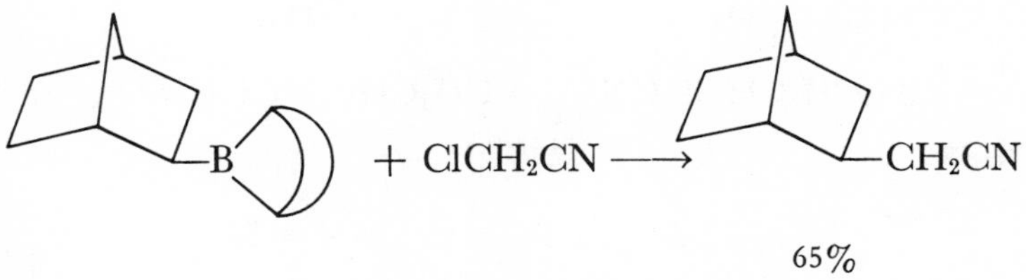

$$+ \text{ClCH}_2\text{CN} \longrightarrow \quad \text{—CH}_2\text{CN} \quad 18.35$$

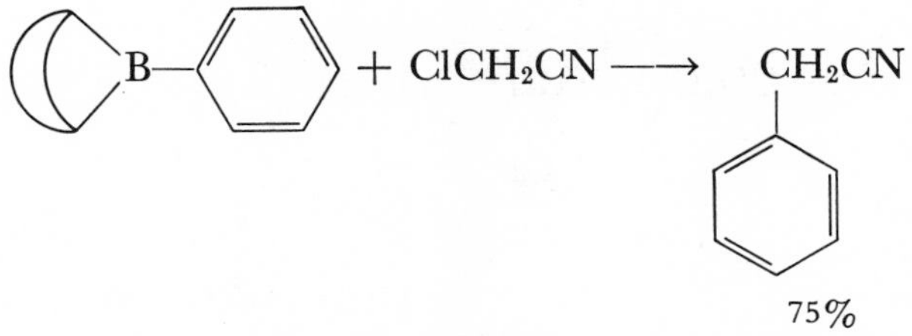

65%

$$\text{B—} \quad + \text{ClCH}_2\text{CN} \longrightarrow \quad \text{CH}_2\text{CN} \quad 18.36$$

75%

The alkylation of dichloroacetonitrile could be readily controlled to proceed to either the mono- (18.37, 18.38) or the dialkyl stage[22] (18.39).

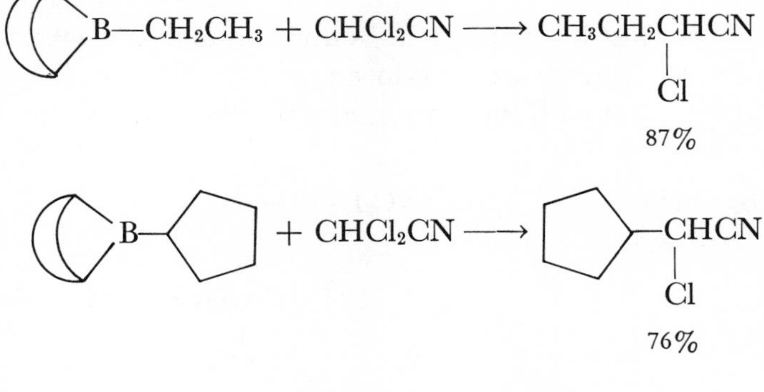

$$\text{B—CH}_2\text{CH}_3 + \text{CHCl}_2\text{CN} \longrightarrow \text{CH}_3\text{CH}_2\underset{\underset{\text{Cl}}{|}}{\text{CHCN}} \quad 18.37$$

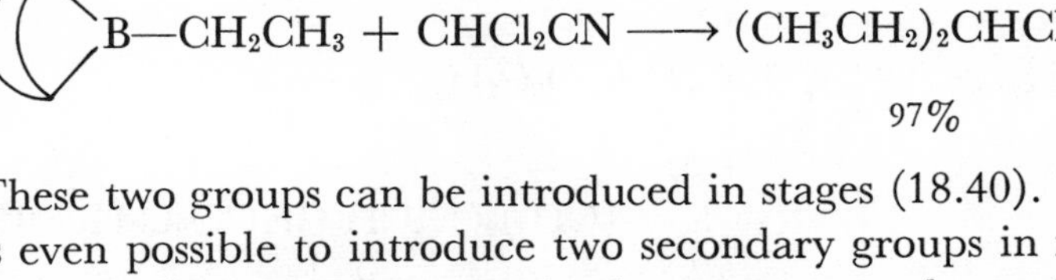

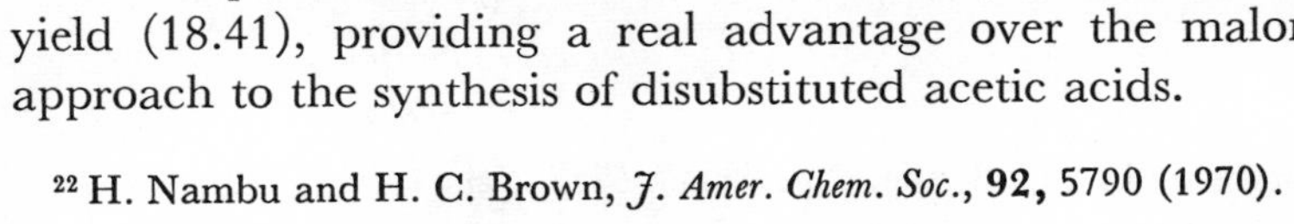

87%

$$\text{B—} \quad + \text{CHCl}_2\text{CN} \longrightarrow \quad \text{—}\underset{\underset{\text{Cl}}{|}}{\text{CHCN}} \quad 18.38$$

76%

$$2 \quad \text{B—CH}_2\text{CH}_3 + \text{CHCl}_2\text{CN} \longrightarrow (\text{CH}_3\text{CH}_2)_2\text{CHCN} \quad 18.39$$

97%

These two groups can be introduced in stages (18.40). Moreover, it is even possible to introduce two secondary groups in reasonable yield (18.41), providing a real advantage over the malonic ester approach to the synthesis of disubstituted acetic acids.

[22] H. Nambu and H. C. Brown, *J. Amer. Chem. Soc.*, **92,** 5790 (1970).

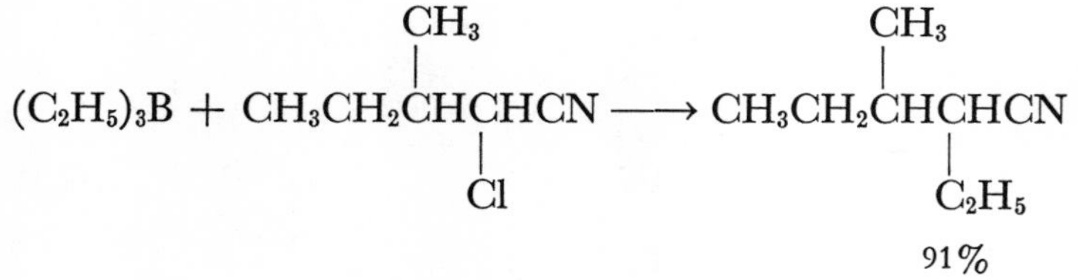

$$(C_2H_5)_3B + CH_3CH_2\underset{\underset{Cl}{|}}{CH}CHCN \longrightarrow CH_3CH_2\underset{\underset{C_2H_5}{|}}{\overset{\overset{CH_3}{|}}{CH}}CHCN \qquad 18.40$$

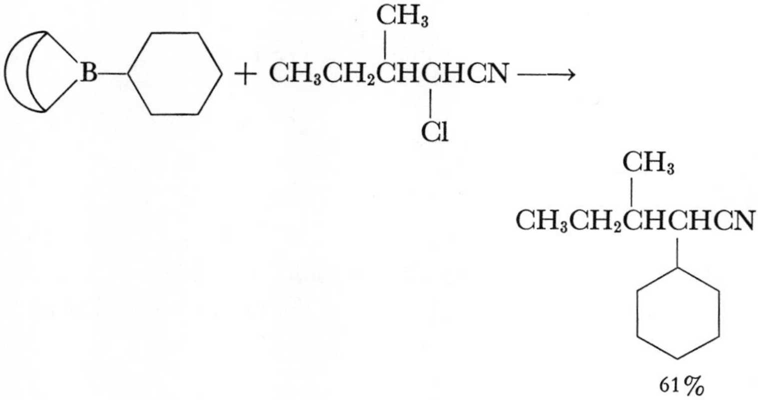

91%

$$\begin{array}{c}\text{B}-\end{array} + CH_3CH_2\underset{\underset{Cl}{|}}{\overset{\overset{CH_3}{|}}{CH}}CHCN \longrightarrow$$

$$CH_3CH_2\overset{\overset{CH_3}{|}}{CH}CHCN \qquad 18.41$$

61%

The reaction can be applied to ethyl bromocyanoacetate.[23] Straight-chain alkyl groups are introduced without difficulty (18.42), but the yield drops off with the more hindered alkyl groups (18.43).

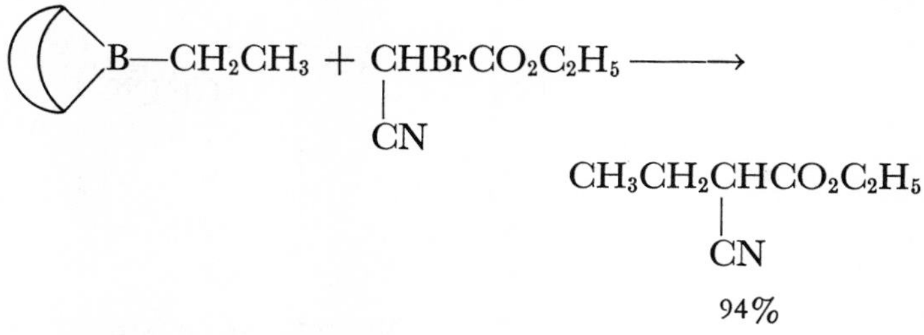

$$\begin{array}{c}\text{B}-\end{array}CH_2CH_3 + \underset{\underset{CN}{|}}{CH}BrCO_2C_2H_5 \longrightarrow$$

$$CH_3CH_2\underset{\underset{CN}{|}}{CH}CO_2C_2H_5 \qquad 18.42$$

94%

$$\begin{array}{c}\text{B}-\end{array}CH_2CH(CH_3)_2 + \underset{\underset{CN}{|}}{CH}BrCO_2C_2H_5 \longrightarrow$$

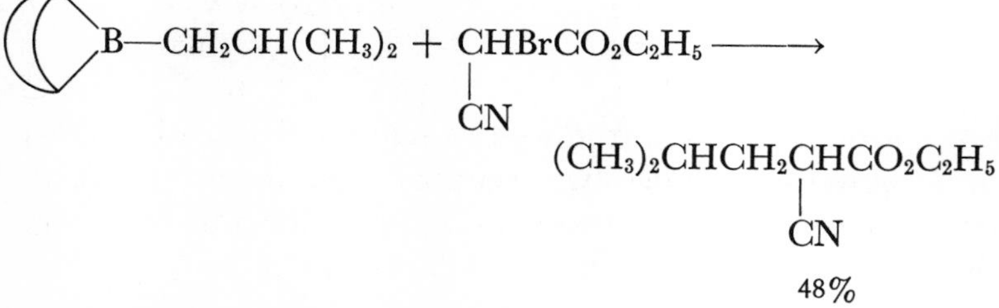

$$(CH_3)_2CHCH_2\underset{\underset{CN}{|}}{CH}CO_2C_2H_5 \qquad 18.43$$

48%

[23] H. Nambu and H. C. Brown, *Organometal. Chem. Syn.*, **1,** 95 (1970/1971).

The alkylation is much more favorable with bromomalononitrile,[23] accommodating both straight-chain and more sterically requiring alkyl groups without difficulty (18.44, 18.45).

$$\text{B-CH}_2\text{CH(CH}_3)_2 + \text{CHBr(CN)}_2 \longrightarrow$$
$$(\text{CH}_3)_2\text{CHCH}_2\text{CH(CN)}_2 \quad 18.44$$
$$91\%$$

$$\text{B-}\bigcirc + \text{CHBr(CN)}_2 \longrightarrow$$
$$\bigcirc\text{-CH(CN)}_2 \quad 18.45$$
$$85\%$$

Finally, the procedure proved to be readily applicable to ethyl 4-bromocrotonate.[24] This provides a simple four-carbon-atom homologation (18.46).

$$\bigcirc\text{-)}_3\text{B} + \text{CH}_2\text{BrCH=CHCO}_2\text{C}_2\text{H}_5 \longrightarrow$$
$$\bigcirc\text{-CH=CHCH}_2\text{CO}_2\text{C}_2\text{H}_5 \quad 18.46$$
$$81\%$$

In this reaction the double bond shifts from the 2,3- to the 3,4-position. Thus the product from triethylborane proved to be ethyl 3-hexenoate (79 percent *trans*). This is attributed to an allylic shift of the double bond in the original allylic boron derivative[25] (18.47).

[24] H. C. Brown and H. Nambu, *J. Amer. Chem. Soc.*, **92,** 1761 (1970).

[25] Such shifts in the hydrolysis of allylic boron derivatives have been previously demonstrated: B. M. Mikhailov and A. Y. Bezmenov, *Bull. Acad. Sci. USSR, Div. Chem. Sci.*, 904 (1965).

$$Et_3B + [\bar{C}HBrCH=CHCO_2C_2H_5]K^+ \longrightarrow$$

$$[Et_3B-CH-CH=CHCO_2C_2H_5]^-K^+$$
$$| \atop Br$$

$$[Et_2B\overset{Et}{\underset{Br}{-}}CH-CH=CHCO_2C_2H_5]^-K^+ \longrightarrow$$

$$Et_2B-\overset{Et}{\underset{|}{C}}H-CH=CHCO_2C_2H_5 + KBr \quad 18.47$$

Consequently, this reaction not only provides a highly convenient procedure for a four-carbon-atom homologation, but it also provides a simple route for the synthesis of $\Delta^{3,4}$-olefinic esters.

7. Stereochemistry [26]

Protonolysis of organoboranes with carboxylic acids (Section XVI-5), oxidation with alkaline hydrogen peroxide (Section XVI-8), and amination with hydroxylamine-O-sulfonic acid (Section XVI-10) all proceed with retention of configuration at the carbon atom undergoing reaction. That is to say, in these reactions the replacement of the boron atom by hydrogen (18.48), oxygen (18.49),

[26] H. C. Brown, M. M. Rogić, M. W. Rathke, and G. W. Kabalka, *J. Amer. Chem. Soc.*, **91**, 2150 (1969).

and nitrogen (18.50) proceeds with the new function taking the identical place on carbon previously occupied by boron.

$$\text{(structure)} \xrightarrow{CH_3CO_2D} \text{(structure)} \quad 18.48$$

$$\text{(structure)} \xrightarrow[NaOH]{H_2O_2} \text{(structure)} \quad 18.49$$

$$\text{(structure)} \xrightarrow{H_2NOSO_3H} \text{(structure)} \quad 18.50$$

Investigation revealed that the alkylation of ethyl bromoacetate with B-*trans*-2-methylcyclopentyl-9-BBN gives the essentially pure ethyl (*trans*-2-methylcyclopentyl)acetate (18.51).

$$\text{(structure)} \xrightarrow{HB} \text{(structure)} \xrightarrow{CH_2BrCO_2C_2H_5} \text{(structure)} \quad 18.51$$

The aldehyde synthesis (Section XVII-10) also proceeds with retention of configuration (18.52).

$$\text{(structure)} \xrightarrow[LiAlH(OCH_3)_3]{CO} \xrightarrow{[O]} \text{(structure)} \quad 18.52$$

Organoboranes

In these reactions the alkyl group is believed to migrate with its bonding pair from boron to another electron-deficient atom, as illustrated for oxidation (18.53) and the carboxymethylation reaction (18.54).

$$R_3B + {}^-O_2H \longrightarrow \left[\begin{array}{c} R \\ | \\ R-B-O-OH \\ | \\ R \end{array} \right]^-$$

$$\longrightarrow R_2BOR + OH^- \qquad 18.53$$

$$R_3B + {}^-\underset{\underset{Br}{|}}{C}HCO_2C_2H_5 \longrightarrow \left[\begin{array}{c} R \\ | \\ R-B-CHCO_2C_2H_5 \\ | \quad | \\ R \quad Br \end{array} \right]^- \longrightarrow$$

$$R_2B-\underset{\underset{}{}}{C}\overset{\overset{R}{|}}{H}CO_2C_2H_5 + Br^- \qquad 18.54$$

This mechanism (18.54) suggests that the migration of the group R from boron to carbon should involve an inversion at the receiving center. Indeed, Pasto and Hickman have demonstrated such an inversion in the migration of hydrogen from boron to carbon in the α-haloorganoboranes obtained in the hydroboration of appropriate vinyl halides[27] (18.55).

$$18.55$$

[27] D. J. Pasto and J. Hickman, *J. Amer. Chem. Soc.*, **89**, 5608 (1967).

It is now becoming possible to control substitution processes to achieve either substitution with inversion or substitution with retention merely by controlling the reagents and reactions utilized. These developments greatly add to the versatility of the procedures available for organic synthesis.

8. Photobromination of Organoboranes in the Presence of Water [28]

The facile migration of an organic grouping from boron to carbon in α-halo substituted organoboranes provides a promising new synthesis of carbon structures (18.56).

$$\underset{\underset{X}{|}}{\overset{\overset{R}{|}}{-C-B-}} \xrightarrow{\text{Base}} \underset{\underset{\text{Base}}{|}}{\overset{\overset{R}{|}}{-C-B-}} + X^- \qquad 18.56$$

There appears to be no serious limitation to the nature of R—it can be aliphatic, alicyclic, and aromatic, and can contain a wide variety of functional groups. Moreover, the migration frequently occurs under exceedingly mild conditions, such as $0°$, under the influence of tetrahydrofuran functioning as the base. Clearly the ability to form carbon-carbon bonds under such mild conditions deserves detailed exploration.

The reaction of α-halocarbanions with organoboranes (Section XVIII-2) provides one convenient route to the desired α-halo-organoboranes (18.57).

$$\underset{\underset{R}{|}}{\overset{\overset{R}{|}}{R-B}} + {}^-\underset{\underset{Br}{|}}{CHCO_2C_2H_5} \longrightarrow \left[\underset{\underset{R}{|}\ \underset{Br}{|}}{\overset{\overset{R}{|}}{R-B-CHCO_2C_2H_5}} \right]^-$$

$$\longrightarrow \underset{}{\overset{\overset{R}{|}}{R_2B-CHCO_2C_2H_5}} + Br^- \qquad 18.57$$

[28] C. F. Lane and H. C. Brown, *J. Amer. Chem. Soc.*, **93,** 1025 (1971).

391

Organoboranes

The hydroboration of vinyl halides[27,29] (Section XIV-10) provides another promising route to these derivatives (18.58).

$$
R-C=CHCl + H-B\diagup_{\diagdown}\quad \longrightarrow \quad R-\overset{\displaystyle R}{\underset{\displaystyle H}{\overset{|}{\underset{|}{C}}}}-\overset{|}{\underset{\displaystyle Cl}{\overset{|}{CH}}}-B-
$$

$$\Big\downarrow \text{Base}$$

$$
R-\overset{\displaystyle R}{\underset{\displaystyle H}{\overset{|}{\underset{|}{C}}}}-\overset{|}{CH}-\underset{\displaystyle Base}{\overset{|}{B}}-\ +\ Cl^-
$$

18.58

Still another promising route to these derivatives is the photochemical bromination of organoboranes[28] (18.59).

$$
-\overset{\displaystyle R}{\overset{|}{B}}-\overset{\displaystyle R}{\underset{\displaystyle H}{\overset{|}{\underset{|}{C}}}}-R' \xrightarrow{\ Br_2,\ h\nu\ } -\overset{\displaystyle R}{\overset{|}{B}}-\overset{|}{\underset{\displaystyle Br}{\overset{|}{C}}}-R' + HBr
$$

$$\Big\downarrow \text{Base}$$

$$
-\underset{\displaystyle Base}{\overset{|}{B}}\!\!-\!\!-\overset{\displaystyle R}{\overset{|}{\underset{|}{C}}}-R' + Br^-
$$

18.59

As described earlier (Section XVI-6), the dark reaction of organoboranes with bromine to produce alkyl bromide and dialkylboron bromide (18.60) is relatively slow. Investigation revealed that this

$$R_3B + Br_2 \longrightarrow R_2BBr + RBr \qquad\qquad 18.60$$

slow reaction proceeded through two distinct stages[30]—an initial free-radical bromination of the *alpha* position of one of the alkyl

[29] D. J. Pasto and R. Snyder, *J. Org. Chem.*, **31**, 2773 (1966); H. C. Brown and R. L. Sharp, *J. Amer. Chem. Soc.*, **90**, 2915 (1968).

[30] C. F. Lane and H. C. Brown, *J. Amer. Chem. Soc.*, **92**, 7212 (1970).

groups (18.61), followed by rupture of the carbon-boron bond by the hydrogen bromide produced in the bromination stage (18.62).

$$R_2B-\underset{\underset{H}{|}}{\overset{|}{C}}- \; + \; Br_2 \longrightarrow R_2B-\underset{\underset{Br}{|}}{\overset{|}{C}}- \; + \; HBr \qquad 18.61$$

$$R_2B-\underset{\underset{Br}{|}}{\overset{|}{C}}- \; + \; HBr \longrightarrow R_2BBr \; + \; H-\underset{\underset{Br}{|}}{\overset{|}{C}}- \qquad 18.62$$

This conclusion is supported by the results shown graphically in Figure XVIII-1.

It appeared reasonable that if the initial *alpha* bromination proceeds through a free-radical chain reaction, as originally proposed,[30] then the bromination stage should be greatly facilitated by light. Moreover, it appeared possible that if the photochemical bromi-

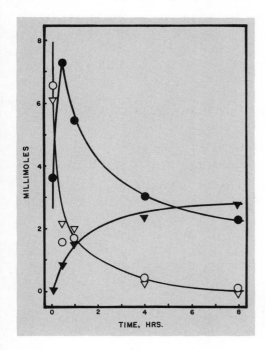

Figure XVIII-1. The dark reaction of tri-*sec*-butylborane (10 mmole) with bromine (10 mmole) in 20 ml of carbon tetrachloride solvent: $\bigcirc$, Br$_2$; $\bigtriangledown$, R$_3$B; $\bullet$, total HBr; $\blacktriangledown$, 2-bromobutane.

nation were carried out in the presence of a water phase, the latter should rapidly absorb the hydrogen bromide and circumvent the protonolysis stage. This procedure should produce the desired α-bromo derivative for the desired rearrangement.

Indeed, treatment of tri-*sec*-butylborane with a molar equivalent of bromine, followed by oxidation with alkaline hydrogen peroxide provided 3,4-dimethyl-3-hexanol (**18.3**) in a yield of 86 percent. Clearly the reaction is proceeding through the desired pathway (18.63).

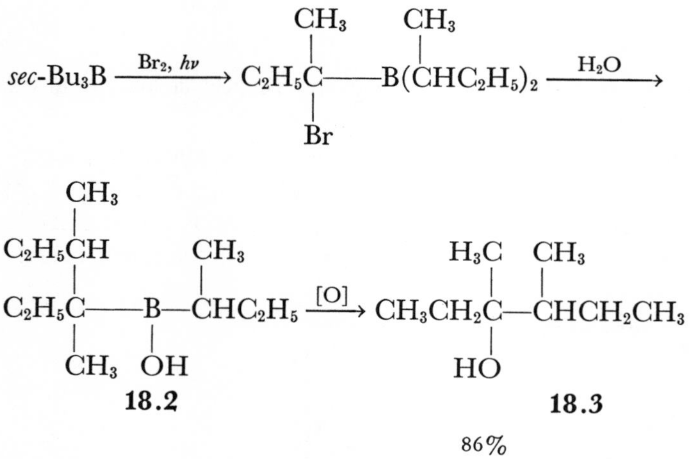

18.63

Similarly, tricyclohexylborane gives an 89 percent yield of 1-cyclohexylcyclohexanol (18.64).

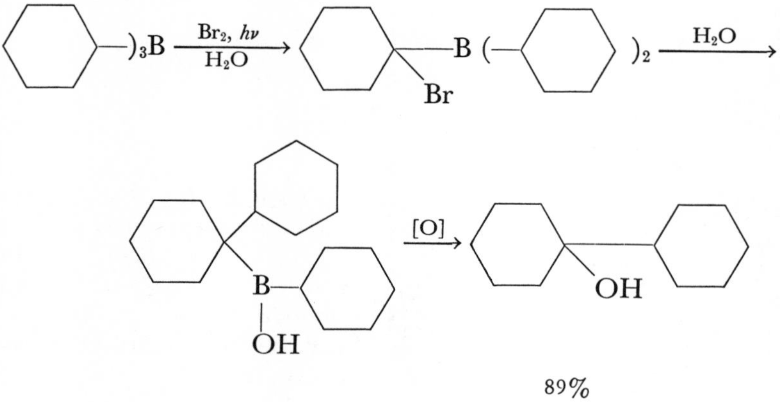

18.64

89%

Treatment with two molar equivalents of bromine can result in the migration of two alkyl groups, as shown by the synthesis of 4-ethyl-3,4,5-trimethyl-3-heptanol (**18.4**) from tri-*sec*-butylborane (18.65).

$$sec\text{-Bu}_3\text{B} \xrightarrow[\text{H}_2\text{O}]{\text{Br}_2,\ h\nu} \textbf{18.2} \xrightarrow[\text{H}_2\text{O}]{\text{Br}_2,\ h\nu}$$

$$
\begin{array}{c}
\text{CH}_3 \\
| \\
\text{C}_2\text{H}_5\text{CH} \text{CH}_3 \\
| | \\
\text{C}_2\text{H}_5\text{C}\!-\!\text{B}\!-\!\!-\!\!-\!\text{CC}_2\text{H}_5 \\
| | | \\
\text{H}_3\text{C} \text{OH} \text{Br}
\end{array}
$$

18.65

$$
\xrightarrow{\text{NaOH}}
\begin{array}{c}
\text{C}_2\text{H}_5\text{CHCH}_3 \\
| \\
\text{C}_2\text{H}_5\text{CCH}_3 \\
| \\
\text{C}_2\text{H}_5\text{CCH}_3 \\
| \\
\text{B(OH)}_2
\end{array}
\xrightarrow{[\text{O}]}
\begin{array}{c}
\text{C}_2\text{H}_5\text{CHCH}_3 \\
| \\
\text{C}_2\text{H}_5\text{CCH}_3 \\
| \\
\text{C}_2\text{H}_5\text{CCH}_3 \\
| \\
\text{OH}
\end{array}
$$

18.4, 46%

The lower yield realized in the case of **18.4** is evidently the result of a relatively sluggish migration of the bulky alkyl group in the last stage. Water apparently cannot induce the migration at any reasonable rate at 25°. However, addition of sodium hydroxide induces a rapid migration with a competitive attack by the base on α-bromo substituent.

Unexpectedly, the photobromination under these conditions of tri-*n*-alkylboranes, such as triethyl- and tri-*n*-butylborane, takes a somewhat different course. For example, treatment of triethylborane with one molar equivalent of bromine produces an 88 percent yield (based on the bromine) of 3-methyl-3-pentanol (**18.9**), rather than the 2-butanol anticipated for a reaction course similar to that shown in 18.63 and 18.64. Similarly, tri-*n*-butylborane provides 5-*n*-propyl-5-nonanol. These reactions are readily accounted for in terms of a rapid rearrangement of the initially formed α-bromotriethylborane (**18.5**) into *sec*-butylethylborinic acid (**18.6**). This then undergoes further bromination at the tertiary position (**18.7**) in preference to the remaining triethylborane. Migration (**18.8**), followed by oxidation (**18.9**), produces the product. Consequently, the major pathway for these tri-*n*-alkylboranes is a double alkyl migration even when only one molar equivalent of bromine is used (18.66).

$$(CH_3CH_2)_3B \xrightarrow{Br_2,\ h\nu} CH_3CH\underset{\overset{|}{Br}}{-}B(C_2H_5)_2 \xrightarrow{H_2O}$$

18.5

$$CH_3\overset{\overset{\displaystyle CH_3CH_2}{|}}{CH}\underset{\overset{|}{OH}}{-}B-C_2H_5 \xrightarrow{Br_2,\ h\nu} CH_3\overset{\overset{\displaystyle CH_3CH_2}{|}}{C}\underset{\overset{|}{Br}\ \ \overset{|}{OH}}{-}B-CH_2CH_3 \xrightarrow{H_2O} \quad 18.66$$

18.6   **18.7**

$$CH_3\overset{\overset{\displaystyle CH_3CH_2}{|}}{C}\underset{\overset{|}{CH_3CH_2}}{-}B(OH)_2 \xrightarrow{[O]} CH_3\overset{\overset{\displaystyle CH_3CH_2}{|}}{C}\underset{\overset{|}{CH_3CH_2}}{-}OH$$

18.8   **18.9**

These results are summarized in Table XVIII-1.

By carrying out the photochemical bromination in the absence of water, the rearrangement of the initially formed α-bromotriethylborane (**18.5**) can be avoided. Further bromination with a second molar equivalent of bromine produces the α,α′-dibromo derivative (**18.10**) predominantly, with a much smaller amount of the α,α-dibromo compound (**18.11**).

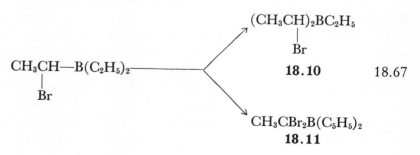

Table XVIII–1. Dimerization and Trimerization of Olefins into Alcohols via Hydroboration-Bromination-Oxidation

Organoborane[a]	Bromine mmole	Product	Yield[b] mmole	percent
Triethylborane	10	3-Methyl-3-pentanol[d]	4.4	88[c]
Triethylborane	20	3-Methyl-3-pentanol[d]	8.5	85
Tri-*n*-butylborane	20	5-Propyl-5-nonanol[d]	7.6	76
Tri-*sec*-butylborane	10	3,4-Dimethyl-3-hexanol[d]	8.6	86
Tri-*sec*-butylborane	20	3,4-Dimethyl-3-hexanol[d]	4.4	44
		4-Ethyl-3,4,5-trimethyl-3-heptanol[e]	4.6	46
Tricyclohexylborane	10	1-Cyclohexylcyclohexanol[d]	8.9	89

[a] All reactions involve bromination of 10 mmole of R_3B in methylene chloride in the presence of a water phase. [b] By glpc analysis based on the maximum product of 1 mole of alcohol from 1 mole of R_3B, except where otherwise indicated. [c] Based on bromine used. [d] Structure assigned by comparison with an authentic sample. [e] Exhibited analytical data and spectra in accordance with assigned structure.

Consequently, hydrolytic rearrangement followed by oxidation, after reaction with one molar equivalent of bromine, provides 2-butanol preferentially (via **18.6**). Similarly, treatment of the reaction product from two molar equivalents of bromine (**18.10** and **18.11**), gives 3-methyl-2-pentanol as the predominant product (via **18.10**).

It should be mentioned that the application of N-bromosuccinimide for the *alpha* bromination of organoboranes appears to offer advantages in convenience and yield over the use of elementary bromine.[31]

In these free-radical brominations there appears to be a remarkable activation of the α-position of the alkyl group to attack by bromine atoms. For example, the photochemical bromination of triethylborane in the gas phase appears to occur only in the α-position.[32] These authors have estimated that attack of the *alpha* carbon-hydro-

[31] Research in progress with Dr. Y. Yamamoto.

[32] J. Grotewold, E. A. Lissi, and J. C. Scaiano, *J. Organometal. Chem.*, **19**, 431 (1969).

gen bond of triethylborane is favored over attack of the secondary carbon-hydrogen bond of propane by some 14.5 kcal/mole. This enhanced reactivity presumably results from stabilization of the free radical produced through the overlap of the unpaired electron with the vacant p-orbital of the boron atom.

The remarkable activation of the α-position is also indicated by the fact that photochemical bromination of triethylborane can be carried out in cyclohexane solution. The reaction proceeds readily, with negligible formation of cyclohexyl bromide. Clearly, this major activation of the α-position of trialkylboranes should greatly facilitate application of this new synthesis of carbon structures.

The thexyl group (Section XV-6) has no hydrogen atoms in the α-position. Consequently, this group is not attacked in the α-position by bromine atoms in the bromination stage. This makes possible the application of the thexyl group as an inert, blocking group to achieve the union of two other, desired moieties.

Thus thexylborane can be used to react with two different alkenes, each of which can be substituted in appropriate positions, and these combined through the usual bromination-rearrangement-oxidation procedure[33] (18.68).

18.68

[33] Research in progress with C. F. Lane and Y. Yamamoto.

It is evident that this easy bromination of the α-position of organo-boranes and the remarkably facile rearrangement of the α-bromo derivatives combine to provide a highly promising new route for achieving the formation of carbon-carbon bonds under exceptionally mild conditions in the presence of a wide variety of functional groups. A detailed exploration of the possibilities is required.

9. Base-Induced Reaction of Organoboranes with Haloforms[34]

In Sections 2 through 6 of this chapter the base-induced reactions of α-halo esters, ketones, and nitriles were described. The base readily removes a proton from the activated α-position, providing the α-halocarbanion (18.16). This intermediate can readily be captured by the organoborane (18.17). There follows a rapid migration of an organic grouping from boron to the α-position of the intermediate, with loss of the α-halo substituent (18.18). Protonolysis of the re-arranged species gives the desired alkylated ester, ketone, or nitrile (18.19).

The question arose as to whether this reaction could be extended to organic halides not containing an activating substituent, such as a carbethoxy, carbonyl, or nitrile group, in the α-position. If this were feasible, the scope of this new alkylation and arylation reaction would be greatly broadened. Accordingly, we undertook to explore this possibility.

Indeed, considerable effort has been devoted to the synthesis of α-halomethyl organometallics for application in the conversion of olefins to cyclopropanes and for theoretical studies as to the impor-tance of carbene intermediates in the reactions of these deriva-tives.[35-37] Because of the special objectives of these earlier studies, the α-halomethyl organometallics were often prepared at relatively low temperatures by procedures that would not be convenient when carried out on the usual preparative scale.

[34] H. C. Brown, B. A. Carlson, and R. H. Prager, *J. Amer. Chem. Soc.*, **93**, 2070 (1971).

[35] G. Köbrich, K. Flory, and W. Drischel, *Angew. Chem. Internatl. Ed.*, **3**, 513 (1964); G. Köbrich, *Angew. Chem. Internatl. Ed.*, **6**, 41 (1967).

[36] W. Kirmse, *Carbene Chemistry* (Academic Press, New York, 1964).

[37] J. Hine, *Divalent Carbon* (Ronald Press, New York, 1964).

For example, Köbrich and Merkle[38] prepared dichloromethyllithium at $-74°$ by the reaction of n-butyllithium with dichloromethane. This material reacted with triarylboranes to provide the diarylcarbinols (18.69).

$$n\text{-BuLi} + CH_2Cl_2 \xrightarrow{-74°} LiCHCl_2 + n\text{-}C_4H_{10}$$

18.69

$$Ar_3B + LiCHCl_2 \xrightarrow{-74°} \xrightarrow{[O]} Ar_2CHOH$$

Accordingly, we explored this approach by preparing trichloromethyllithium from n-butyllithium and chloroform at $-110°$[35,39] and adding tri-n-butylborane. The reaction mixture was allowed to come to room temperature. Oxidation of the intermediate with alkaline hydrogen peroxide provided a 37 percent yield of tri-n-butylcarbinol. Clearly this approach was not competitive with the reaction of carbon monoxide with organoboranes (Section XVII-4).

For our purposes there appeared to be no real need to prepare the α-halomethyl organometallic under conditions where it is stable for considerable periods of time. It should be adequate to prepare the species as a fugitive intermediate that might be captured by the organoborane present in the reaction mixture almost immediately following generation. Accordingly, we decided to explore the possibility of inducing a reaction of chloroform and related trisubstituted methanes with a representative organoborane, such as tri-n-butylborane with typical alkoxide bases.

Various alkoxides were examined. The results indicated that the more hindered alkoxides offered advantages, presumably because coordination with the organoborane was less complete, leaving a higher concentration of the organoborane free to unite with the α-halocarbanion. Thus in the reaction of chloroform with tri-n-butylborane potassium t-butoxide provides a yield of 30 percent at both $0°$ and $65°$. The yield improves with potassium triethylcarboxide, 50 percent at $0°$ and 59 percent at $65°$. The reactions induced by the corresponding lithium bases are much slower, so that the higher temperature, $65°$ (refluxing THF) was adopted. Under these con-

[38] G. Köbrich and H. R. Merkle, *Chem. Ber.*, **100,** 3371 (1967).
[39] W. T. Miller and D. M. Whalen, *J. Amer. Chem. Soc.*, **86,** 2089 (1964).

ditions, lithium *t*-butoxide gives a yield of 55 percent in 1 hour, whereas lithium triethylcarboxide gives an 85 percent yield. (The yields quoted are realized with reaction mixtures containing one molar equivalent each of tri-*n*-butylborane and chloroform, and two molar equivalents of the alkoxide. Use of one molar equivalent of base results in lower yields.)

The 85 percent yield realized with lithium triethylcarboxide appeared promising. Accordingly, the latter conditions were adopted for a study of the relative effectiveness of related trisubstituted methanes. The results are summarized in Table XVIII-2.

Table XVIII–2. Reaction of Tri-*n*-butylborane with Representative Trisubstituted Methanes under the Influence of Lithium Triethylcarboxide[a]

	Oxidation products, percent[b]	
Reagent	n-Bu$_3$COH	n-Bu$_2$CO
Chloroform	85	5
Dichlorofluoromethane	79	20
Chlorodifluoromethane	98	trace
Fluoroform	0	0
1,1-Dichlorodimethyl ether[c]	80	0.5
Trimethyl orthoformate	0	0

[a] The reaction mixture containing 10 mmoles each of the organoborane and the reagent and 20 mmoles of the base was heated under reflux in THF—*n*-hexane for 1 hour. [b] Glpc analysis. [c] Only one molar equivalent of base was required and used.

The remarkably high yield realized with chlorodifluoromethane, 98 percent, is highly promising. However, the full range of applicability of this new synthesis remains to be explored, as well as the possibility of extending it to other types of unactivated alkyl halides.

The precise mechanism also awaits detailed exploration. Presumably, these reactions involve the formation of the trihalomethyllithium as highly unstable transient intermediates (18.70), followed by their rapid capture by the organoborane (18.71).

$$\text{Li}^{+-}\text{OCEt}_3 + \text{HCX}_3 \longrightarrow \text{LiCX}_3 + \text{HOCEt}_3 \qquad 18.70$$

$$R_3B + LiCX_3 \longrightarrow [R_3BCX_3]^-Li^+ \qquad 18.71$$

There would then ensue the usual migration of the alkyl groups from boron to carbon, producing an intermediate that is oxidized by alkaline hydrogen peroxide to the tertiary carbinol (18.72).

$$[R_3BCX_3]^-Li^+ \longrightarrow R_3CBX_2 + LiX \xrightarrow{[O]} R_3COH \qquad 18.72$$
$$\mathbf{18.12}$$

This mechanism does not account for the requirement of a second mole of base in the case of the haloforms, and the need for only one mole of base with 1,1-dichlorodimethyl ether. One could rationalize the need for a second mole of base by postulating a rapid reaction of **18.12** with one mole of alkoxide (18.73).

$$R_3CBX_2 + LiOCEt_3 \longrightarrow R_3CBX(OCEt_3) \qquad 18.73$$

This would correspond to the intermediate formed from 1,1-dichlorodimethyl ether, $R_3CBXOCH_3$. However, other interpretations are possible. Thus, one could postulate that the initial reaction involves the formation of dihalocarbene, followed by addition of this intermediate to the alkoxide[40] (18.74).

$$LiCX_3 \xrightarrow{\Delta} :CX_2 + LiX \xrightarrow{LiOCEt_3} Li[CX_2OCEt_3] \qquad 18.74$$

Clearly, considerable further research will be required to clarify the precise mechanism of these reactions.

Finally, these considerations raise a question as to the nature of the reaction of dihalocarbenes with organoboranes. It has been established that the thermal decomposition of bromodichloromethylphenylmercury proceeds through the formation of dichlorocarbene[41] (18.75), without the intermediate formation of the trihalomethyl anion postulated for the chloroform-base reaction (18.70).

$$PhHgCCl_2X \xrightarrow{\Delta} PhHgX + :CCl_2 \qquad 18.75$$

Indeed, the thermal reaction of bromodichloromethylphenylmercury with tri-n-butylborane produces not the tri-n-butylcarbinol realized

[40] J. Hine, A. D. Ketley, and K. Tanabe, *J. Amer. Chem. Soc.*, **82**, 1398 (1959).

[41] D. Seyferth, J. Y.-P. Mui, and J. M. Burlitch, *J. Amer. Chem. Soc.*, **89**, 4953 (1967).

in the base-induced reaction with chloroform (18.72), but 4-nonene as the major product[42] (18.76).

$$(n\text{-}C_4H_9)_3B \xrightarrow{\;:CCl_2\;} n\text{-}C_4H_9\!-\!\underset{\underset{Cl}{\big|}}{\overset{\overset{Cl}{\big|}}{C}}\!-\!B(n\text{-}C_4H_9)_2$$

$$\longrightarrow (n\text{-}C_4H_9)_2\underset{\underset{Cl}{\big|}}{C}\!-\!B(n\text{-}C_4H_9)Cl$$

$$\longrightarrow (n\text{-}C_4H_9)_2C\!: \; + \; n\text{-}C_4H_9BCl_2$$

$$\longrightarrow n\text{-}C_4H_9CH\!=\!CHCH_2CH_2CH_3 \qquad 18.76$$

These differences suggested that the reactions with organoboranes might provide a useful diagnostic test for the intermediacy of trihalomethyl anion in reactions proceeding through dihalocarbene intermediates.[43-45] We undertook to explore this possibility.

An equimolar mixture of tri-n-butylborane, tetramethylethylene, and chloroform was treated with potassium triethylcarboxide at 0°. The borane reacted practically exclusively. On the other hand, when the equimolar mixture of tri-n-butylborane and tetramethylethylene was treated in refluxing benzene with trichloromethylphenyl-mercury or bromodichloromethylphenylmercury, the tetramethyl-ethylene reacted exclusively.

These results support the conclusion that the chloroform–base reaction produces $MCCl_3$ (or the alkoxy derivative, 18.74) and that this anion is captured preferentially by the organoborane. On the other hand, the Seyferth reagents must dissociate directly into dichlorocarbene, and this uncharged intermediate must react preferentially with the tetramethylethylene.

To further test the utility of this approach, we prepared trichloromethyllithium from n-butyllithium and chloroform at $-110°$

[42] D. Seyferth and J. M. Burlitch, *J. Organometal. Chem.*, **4**, 127 (1965).

[43] D. F. Hoeg, D. I. Lusk, and A. L. Crumbliss, *J. Amer. Chem. Soc.*, **87**, 4147 (1965).

[44] P. S. Skell and M. S. Cholod, *J. Amer. Chem. Soc.*, **91**, 6035, 7131 (1969).

[45] G. Köbrich, H. Büttner, and E. Wagner, *Angew. Chem. Internatl. Ed.*, **9**, 169 (1970).

and added it to tri-*n*-butylborane and to tetramethylethylene individually, as well as to a mixture of the two. Reaction at this temperature occurred only with the organoborane.

Similarly, decomposition of sodium trichloroacetate in tetrahydrofuran at 70° in the presence of tri-*n*-butylborane and tetramethylethylene gave, after oxidation, 70 percent of tri-*n*-butylcarbinol and 25 percent of 1,1-dichloro-2,2,3,3-tetramethylcyclopropane. Consequently, this reaction must also proceed predominantly through the intermediate formation of trichloromethylsodium. Presumably, the minor amount of the cyclopropane species arises from a smaller amount of competitive dissociation of the intermediate at the elevated temperature, prior to the time the intermediate is captured by the organoborane.

To sum up, these results support the earlier conclusions that (1) the trihalomethyl anion is not an intermediate in the dissociation of the Seyferth reagents, and (2) the trihalomethyl (or dihaloalkoxymethyl) anion is an intermediate in the chloroform–base reaction. They further indicate that (3) the carbanions react preferentially with tri-*n*-butylborane, whereas the dihalocarbene reacts preferentially with tetramethylethylene. Consequently, the competitive reaction with a mixture of tri-*n*-butylborane and tetramethylethylene provides a useful tool to explore the possible intermediacy of carbanions in reactions proceeding through dihalocarbenes.

Although we initiated study in this area with relatively limited objectives, systematic examination is opening up major new areas of fascinating chemistry for further exploration.

10. Reaction of Organoboranes with Ylides

Consideration of the mechanism of the oxidation (Section XVI-8) and amination (Section XVI-10) of organoboranes led J. J. Tufariello and his co-workers to conclude that it should be possible to extend these migrations of an alkyl group from boron to oxygen and to nitrogen to the migration of the alkyl group to carbon by use of a suitable ylide[46] (18.77).

[46] J. J. Tufariello and L. T. C. Lee, *J. Amer. Chem. Soc.*, **88**, 4757 (1966).

$$R_3B + \bar{Y}\!-\!L^+ \qquad R_2\overset{\displaystyle R}{\overset{\displaystyle |}{B}}\!-\!\bar{Y}\!-\!L^+ \longrightarrow R_2BYR + L \quad 18.77$$

The first ylide they explored was dimethyloxosulfonium methylide[46] (**18.13**).

$$\bar{C}H_2\overset{+}{S}(CH_3)_2$$
$$\overset{\|}{O}$$

18.13

This ylide reacted readily with trialkylboranes in THF at 0° to afford, after oxidation, the desired homologated alcohol, accompanied unfortunately by an appreciable amount of the doubly homologated material (18.78).

$$Hex_3B + \bar{C}H_2\overset{+}{S}O(CH_3)_2 \xrightarrow[0°]{THF} \xrightarrow{[O]}$$
$$HexOH + HexCH_2OH + HexCH_2CH_2OH \quad 18.78$$
$$\quad\;\; 68\% \qquad\quad 26\% \qquad\qquad 6\%$$

Fortunately, the more reactive dimethylsulfonium methylide overcomes this difficulty[47] (18.79, 18.80).

$$Ph_3B + \bar{C}H_2\overset{+}{S}(CH_3)_2 \xrightarrow[-10°]{DMSO} \xrightarrow{[O]} PhOH + PhCH_2OH \qquad 18.79$$
$$\qquad\qquad\qquad\qquad\qquad\qquad\quad 69\% \qquad 31\%$$

$$18.80$$
$$\qquad\quad 76\% \qquad\qquad 24\%$$

It appears that the more basic the ylide, the more reactive it is toward organoboranes. Thus, the nitrogen ylide[48] (**18.14**) reacts well below 0°, whereas the phosphorus ylide[49] (**18.15**) reacts only at temperatures exceeding 100°.

[47] J. J. Tufariello, P. Wojtkowski, and L. T. C. Lee, *Chem. Commun.*, 505 (1967).
[48] W. K. Musker and R. R. Stevens, *Tetrahedron Letters*, 995 (1967).
[49] R. Köster and B. Rickborn, *J. Amer. Chem. Soc.*, **89**, 2782 (1967).

$$(CH_3)_3\overset{+}{N}-\overset{-}{CH_2} \qquad\qquad (C_6H_5)_3\overset{+}{P}-\overset{-}{CH_2}$$
$$\textbf{18.14} \qquad\qquad\qquad \textbf{18.15}$$

Many ylides containing functional groups have been synthesized, and this offers the possibility of functionalizing olefins via the organoboranes. Indeed, ethyl (dimethylsulfuranylidene)acetate (**18.16**) reacts readily with organoboranes to produce the corresponding esters[50] (18.81).

$$R_3B + (CH_3)_2\overset{+}{\underset{}{S}}\overset{-}{CH}CO_2C_2H_5 \longrightarrow R_2BCHRCO_2C_2H_5 + (CH_3)_2S$$
$$\textbf{18.16} \qquad\qquad\qquad\qquad\qquad\qquad\qquad 18.81$$

$$R_2BCHRCO_2C_2H_5 + H_2O \longrightarrow R_2BOH + RCH_2CO_2C_2H_5$$

Thus, reaction of an equimolar ratio of **18.16** and tri-*n*-heptylborane provides an 87 percent of ethyl nonanoate. Similarly, triphenylborane provides an 80 percent yield of ethyl phenylacetate.

Application of N,N-diethyl(dimethylsulfuranylidene)acetamide (**18.17**) proceeds similarly[50] (18.82).

$$R_3B + (CH_3)_2\overset{+}{\underset{}{S}}\overset{-}{CH}CON(C_2H_5)_2 \longrightarrow RCH_2CON(C_2H_5)_2 \qquad 18.82$$
$$\textbf{18.17}$$

This application of ylides offers great promise. An enormous amount of research has been done in developing ylides for use in organic synthesis.[51] It should be possible to draw upon the available information to use such ylides to introduce many functional groups into organic structures.

Unfortunately, there is a major difficulty. The reaction as it is now performed utilizes effectively only one of the three R groups of the borane. This problem, which has been successfully overcome in the aldehyde synthesis (Section XVII-10) and in the α-halocarbanion reaction (Section XVIII-4) through the use of 9-BBN, has not yet been solved for the ylide reaction.

[50] J. J. Tufariello, L. T. C. Lee, and P. Wojtkowski, *J. Amer. Chem. Soc.*, **89,** 6804 (1967).

[51] A. W. Johnson, *Ylide Chemistry* (Academic Press, New York, 1966).

11. Reaction of Organoboranes with Diazo Derivatives

It has long been known that diazomethane reacts with organo-boron compounds to produce polymethylene.[52] The reaction presumably involves an addition of the boron atom to the methylene group, followed by loss of nitrogen and a migration of the group from boron to carbon (18.83).

$$X_3B + CH_2N_2 \longrightarrow X_2BCH_2X + N_2 \qquad 18.83$$

The reaction can continue, producing polymethylene (18.84).

$$X_3B + yCH_2N_2 \longrightarrow X_2B(CH_2)_y X + yN_2 \qquad 18.84$$

J. Hooz and his co-workers have explored the reaction of organo-boranes with representative diazo derivatives, such as diazoacetaldehyde[53] (18.85), diazoacetone[54] (18.86), diazoacetonitrile[55] (18.87), and ethyl diazoacetate[55] (18.88).

$$\text{[(CH}_3)_2\text{CHCH}_2]_3\text{B} + \text{N}_2\text{CHCOCH}_3 \xrightarrow[\text{refl.}]{\text{THF}}$$

$$(CH_3)_2CHCH_2CH_2COCH_3 + N_2 \qquad 18.86$$

56%

[52] C. E. H. Bawn, A. Ledwith, and P. Matthies, *J. Polymer Sci.*, **34,** 93 (1959); A. G. Davies, D. G. Hare, O. R. Khan, and J. Sikora, *Proc. Chem. Soc.*, 172 (1961); C. E. H. Bawn and A. Ledwith, *Progress in Boron Chemistry*, **1,** 345 (Macmillan, New York, 1964).

[53] J. Hooz and G. F. Morrison, *Canadian J. Chem.*, **48,** 868 (1970).

[54] J. Hooz and S. Linke, *J. Amer. Chem. Soc.*, **90,** 5936 (1968).

[55] J. Hooz and D. M. Gunn, *J. Amer. Chem. Soc.*, **91,** 6195 (1969).

$$\underset{\text{CH}_3}{(\text{CH}_3\text{CH}_2\text{CH}_2\overset{|}{\text{C}}\text{HCH}_2)_3\text{B}} + \text{N}_2\text{CHCN} \xrightarrow[0°]{\text{THF}}$$

$$\underset{\text{CH}_3}{\text{CH}_3\text{CH}_2\text{CH}_2\overset{|}{\text{C}}\text{HCH}_2\text{CH}_2\text{CN}} + \text{N}_2 \quad 18.87$$

97%

$$\bigcirc\!\!-)_3\text{B} + \text{N}_2\text{CHCO}_2\text{C}_2\text{H}_5 \xrightarrow[25°]{\text{THF}}$$

$$\bigcirc\!\!-\text{CH}_2\text{CO}_2\text{C}_2\text{H}_5 + \text{N}_2 \quad 18.88$$

58%

These reactions are quite promising. They have a major advantage in that they frequently take place very readily at 0° or 25° in the absence of added bases or acids. Therefore they should be very useful to achieve the functionalization of labile groupings.

Unfortunately, they suffer from one major disadvantage and two minor ones. The major disadvantage is that the reaction uses but one of the three alkyl groups on boron. In these reactions the use of B-R-9-BBN fails to solve the problem, with the boron-cyclooctyl bond undergoing migration, rather than the desired boron-alkyl bond.[56,57]

A minor disadvantage is the fact that steric difficulties often result in large decreases in yield with the more bulky R groups. Thus tri-*sec*-butylborane reacts relatively sluggishly with diazoacetone, providing a yield of only 36%.[54] A second minor disadvantage is that often the diazo compounds are unavailable commercially and must be freshly prepared by delicate procedures.

Perhaps further research will overcome these handicaps.

12. Conclusion

The remarkably facile migration of both alkyl and aryl groups from boron to carbon provides a highly versatile new approach to the

[56] H. C. Brown and M. M. Rogić, *J. Amer. Chem. Soc.*, **91**, 2146 (1969).
[57] J. Hooz and D. M. Gunn, *Tetrahedron Letters*, 3455 (1969).

synthesis of carbon structures. The reaction takes place cleanly, with no rearrangement known as yet within the organic structure. The reaction can accommodate a wide range of functional substituents. Much has been done in a relatively short period—much more remains to be explored. This development promises to become a major route for the synthesis of organic structures.

XIX. Free-Radical Reactions of Organoboranes

1. Origins of the Discovery that Organoboranes Are Excellent Sources of Free Radicals

Recognition that the organoboranes are excellent sources of free radicals and can participate in free-radical chain reactions came about through detailed mechanistic studies of the reaction of organoboranes with oxygen[1-3] (Section XVI-7) and their reaction with α,β-unsaturated carbonyl compounds.[4]

Originally, it was proposed that the autoxidation of a trialkylborane proceeds through an intermediate "borine-peroxide" (**19.1**).[5]

$$R_3B\text{—}O\text{=}O$$
19.1

A similar intermediate was postulated for the peroxide formed in the autoxidation of n-butylboronic anhydride[6] (19.1).

$$(n\text{-BuBO})_3 + O_2 \longrightarrow n\text{-BuBO}(O_2) + 2n\text{-BuBO} \qquad 19.1$$

More recently it was recognized that these peroxides are true alkyl peroxide derivatives, containing the structure, $ROOB\big\langle$. Thus, the reaction of oxygen with tri-sec-butylborane in dilute solution proceeds to give the diperoxide (19.2) as the product.[7]

$$sec\text{-Bu}_3B + 2O_2 \longrightarrow sec\text{-BuB}(O_2Bu\text{-}sec)_2. \qquad 19.2$$

However, the formation of these peroxides was long interpreted to involve nonradical processes because many of the usual radical in-

---

[1] A. G. Davies and B. P. Roberts, *J. Chem. Soc.* (B), 17 (1967).

[2] *Ibid.*, 311 (1969).

[3] P. G. Allies and P. B. Brindley, *J. Chem. Soc.* (B), 1126 (1969).

[4] The reaction, a facile 1,4-addition, provides a major new route for the synthesis of carbon structures. It will be discussed in detail in the sections that follow.

[5] J. R. Johnson and M. G. Van Campen, Jr., *J. Amer. Chem. Soc.*, **60,** 121 (1938).

[6] O. Grummitt, *J. Amer. Chem. Soc.*, **64,** 1811 (1942).

[7] A. G. Davies and D. G. Hare, *J. Chem. Soc.*, 438 (1959).

hibitors, such as hydroquinone, had no apparent effect upon the reaction.[8]

Recognition that the reaction of oxygen with organoboranes produces free radicals could have been achieved through the studies of Furukawa on the catalysis of vinyl polymerization by organoboranes. Thus, it was noted that the polymerization of vinyl acetate in the presence of triethylborane is markedly accelerated by the introduction of small amounts of oxygen.[9] It was later pointed out that copolymers produced by such initiation have the same composition as those from typical radical copolymerization.[10] However, the authors failed to conclude that radicals were produced in the reaction of oxygen with organoboranes.

The oxidation of optically active 1-phenylethylboronic acid gave racemic product, and this suggested a process involving radicals.[1] Indeed, it was observed that the autoxidation of this boronic acid exhibits a remarkable induction period in the presence of added inhibitors, such as copper(II) N,N-dibutyldithiocarbamate, and galvinoxyl.[1] It was then discovered that galvinoxyl also effectively inhibits the autoxidation of triisobutylborane,[1] the epimeric trinorbornylboranes,[2] and other organoboranes.[3] These developments led to the proposal that the autoxidation of trialkylboranes proceeds through a free-radical chain mechanism[1-3] (19.3–19.5).

$$R_3B + O_2 \longrightarrow R\cdot \qquad\qquad 19.3$$

$$R\cdot + O_2 \longrightarrow RO_2\cdot \qquad\qquad 19.4$$

$$RO_2\cdot + BR_3 \longrightarrow RO_2BR_2 + R\cdot \qquad\qquad 19.5$$

As discussed later (Section XIX-6), iodine is a much more effective inhibitor than galvinoxyl and in high concentrations can effectively halt the absorption of oxygen by solutions of organoboranes.[11]

Perhaps the most unusual feature about this mechanism (19.3–

[8] M. H. Abraham and A. G. Davies, *J. Chem. Soc.*, 429 (1959).

[9] J. Furukawa and T. Tsuruta, *J. Polymer Sci.*, **28**, 227 (1958).

[10] J. Furukawa, T. Tsuruta, T. Imada, and H. Fukutani, *Makromol. Chem.*, **31**, 122 (1959).

[11] M. M. Midland and H. C. Brown, *J. Amer. Chem. Soc.*, **93**, 1506 (1971).

19.5) is the proposal that an alkylperoxy free radical is capable of a rapid attack on boron with the displacement of an alkyl radical (19.5) continuing the chain. Indeed, evidence for this reaction has accumulated rapidly. Thus, the photolysis of di-t-butylperoxide in the presence of trimethylborane in the cavity of an esr spectrometer provides an esr spectrum of methyl radicals[12] (19.6, 19.7).

$$(t\text{-BuO})_2 \xrightarrow{\ h\nu\ } 2t\text{-BuO}\cdot \qquad\qquad 19.6$$

$$t\text{-BuO}\cdot \ + \ \text{B(CH}_3)_3 \longrightarrow t\text{-BuOB(CH}_3)_2 + \text{CH}_3\cdot \qquad 19.7$$

Similar results have been realized with triethylborane[12] and tri-n-butylborane.[12,13]

This appears to be a reaction of wide generality for trialkylboranes, with displacement of an alkyl group readily occurring through reaction of a free radical containing the odd electron on oxygen[12,13] (RO·), nitrogen[14] (R$_2$N·), and sulfur[15] (RS·).

The marked tendency of such free radicals to react with organoboranes with displacement of a free radical has led to the suggestion that a similar reaction may be involved in the initiation step of the autoxidation (19.3). Oxygen is a diradical. It has been proposed that it also reacts with organoboranes with displacement of an alkyl radical[11] (19.8).

$$\overset{\cdot}{\text{O}}\text{—}\overset{\cdot}{\text{O}} + \text{BR}_3 \longrightarrow \cdot\text{O}_2\text{BR}_2 + \text{R}\cdot \qquad\qquad 19.8$$

The ease with which these displacement reactions proceed may be a result of the high strengths of the bonds of boron with oxygen, nitrogen, and sulfur, providing a potent driving force for the facile displacement of the alkyl free radical from the organoborane. However, irrespective of the precise reason for this new, facile reaction, the reaction itself opens up a major new area of application for the organoboranes in free-radical chemistry, examined in the present chapter.

[12] P. J. Krusic and J. K. Kochi, *J. Amer. Chem. Soc.*, **91**, 3942 (1969).

[13] A. G. Davies and B. P. Roberts, *Chem. Commun.*, 699 (1969).

[14] A. G. Davies, S. C. W. Hook, and B. P. Roberts, *J. Organometal. Chem.*, **22**, C37 (1970).

[15] Private communication from A. G. Davies, May 1, 1970.

2. The 1,4-Addition of Trialkylboranes to Methyl Vinyl Ketone [16] and Acrolein [17]

Organoboranes are much less reactive organometallic compounds than the organolithium and organomagnesium derivatives, and they do not add to the carbonyl groups of aldehydes and ketones in the manner of the more reactive organometallics.[18] However, we did observe that certain α,β-unsaturated carbonyl derivatives, such as methyl vinyl ketone and acrolein, undergo an exceptionally fast 1,4-addition reaction with representative organoboranes produced via hydroboration (19.9, 19.10).

$$R_3B + CH_2\!\!=\!\!CHCHO \xrightarrow[25°]{THF} RCH_2CH\!\!=\!\!CHOBR_2 \qquad 19.9$$
$$\mathbf{19.2}$$

$$R_3B + CH_2\!\!=\!\!CHCOCH_3 \xrightarrow[25°]{THF} RCH_2CH\!\!=\!\!\overset{\overset{\displaystyle CH_3}{|}}{C}OBR_2 \qquad 19.10$$
$$\mathbf{19.3}$$

Addition of water to the reaction mixture results in the rapid hydrolysis of the enol borinate intermediates ($\mathbf{19.2}$ $\mathbf{19.3}$), producing the corresponding aldehyde (19.11) or methyl ketone (19.12).

$$\mathbf{19.2} + H_2O \longrightarrow RCH_2CH_2CHO + R_2BOH \qquad 19.11$$

$$\mathbf{19.3} + H_2O \longrightarrow RCH_2CH_2COCH_3 + R_2BOH \qquad 19.12$$

The reaction provides an exceedingly simple means of homologation, lengthening the chain by three or more carbon atoms. Moreover, it appears to be broadly applicable to a wide range of olefin structures (19.13–19.16).

---

[16] A. Suzuki, A. Arase, H. Matsumoto, M. Itoh, H. C. Brown, M. M. Rogić, and M. W. Rathke, *J. Amer. Chem. Soc.*, **89**, 5708 (1967).

[17] H. C. Brown, M. M. Rogić, M. W. Rathke, and G. W. Kabalka, *J. Amer. Chem. Soc.*, **89**, 5709 (1967).

[18] The possibility of achieving such additions is currently under exploration with J. H. Buhler.

$$CH_2{=}CH_2 \longrightarrow CH_3CH_2CH_2CH_2COCH_3 \qquad \text{19.13}$$
$$99\%$$

$$\underset{\overset{|}{CH_3}}{CH_3{-}C}{=}CH_2 \longrightarrow \underset{\overset{|}{CH_3}}{CH_3CH}CH_2CH_2CH_2CHO \qquad \text{19.14}$$
$$87\%$$

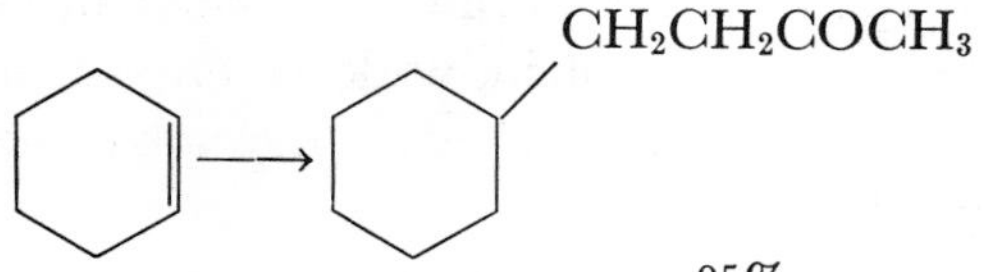

19.15

95%

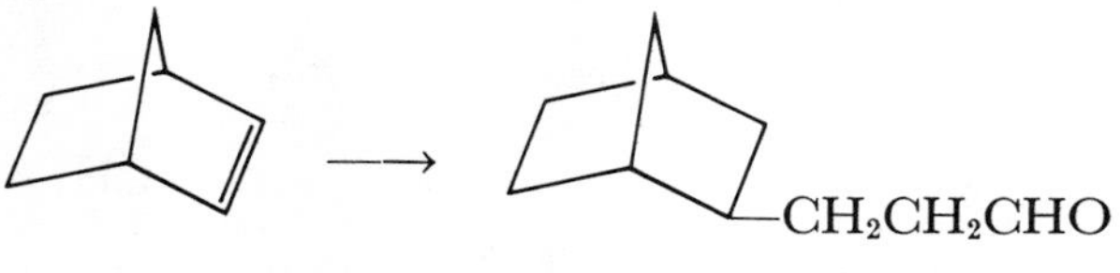

19.16

80%

Recycling of the R_2BOH intermediate on an industrial scale should offer no difficulty. However, it is inconvenient for laboratory syntheses to attempt such a recycle.

9-BBN solved this difficulty for the carbon monoxide-aldehyde synthesis (Section XVII-10) and the base-induced alkylation reaction (Section XVIII-4). However, in the present reaction it is the boron-cyclooctyl bond that becomes involved in the reaction rather than the desired boron-alkyl bond.[19]

Tri-*n*-butylborane from the hydroboration of 1-butene contains 94 percent *n*-butyl and 6 percent *sec*-butyl groups (Section XIV-5). However, the product from this tri-*n*-butylborane and acrolein contains 85 percent of *n*-heptanal and 15 percent of 4-methylhexanal.[17] Clearly the *sec*-butyl groups in the trialkylborane had transferred preferentially.

This observation suggested that B-alkylborinanes (**19.4**), or the more readily available methyl derivatives (**19.5, 19.6**), containing secondary or tertiary groups on the boron atom, might solve the difficulty.

[19] H. C. Brown, M. M. Rogić, M. W. Rathke, and G. W. Kabalka, *J. Amer. Chem. Soc.*, **91,** 2150 (1969).

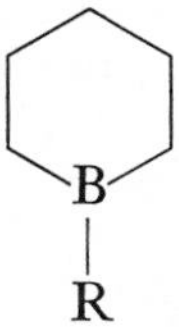

19.4	**19.5**	**19.6**

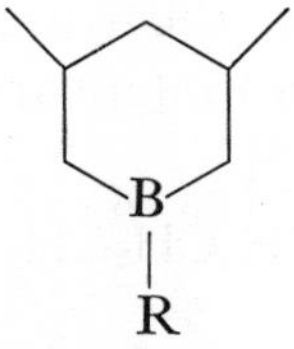

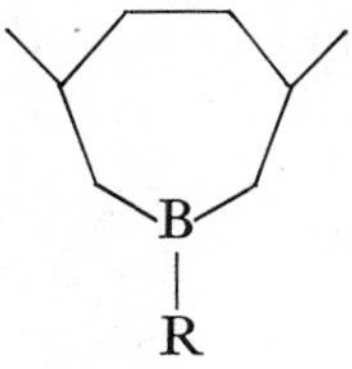

Convenient methods have been developed for the synthesis of the parent reagents (Section XV-7).

Indeed, these B-alkylboracyclanes react easily with methyl vinyl ketone and similar α,β-unsaturated carbonyl derivatives to give the desired products in excellent yields[20] (19.17, 19.18).

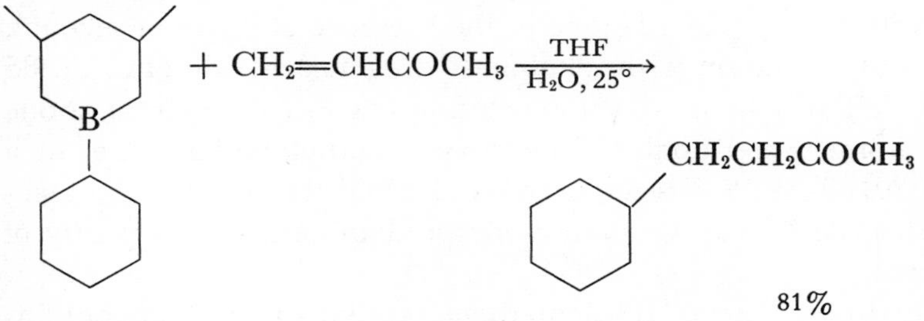

$$+ \ CH_2{=}CHCOCH_3 \xrightarrow[H_2O,\,25°]{THF}$$

$$CH_2CH_2COCH_3$$

19.17

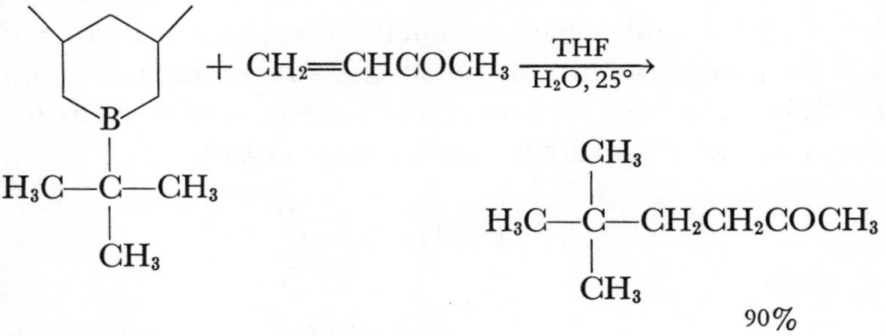

81%

$$+ \ CH_2{=}CHCOCH_3 \xrightarrow[H_2O,\,25°]{THF}$$

$$H_3C-\overset{\overset{\displaystyle CH_3}{|}}{\underset{\underset{\displaystyle CH_3}{|}}{C}}-CH_2CH_2COCH_3 \quad 19.18$$

90%

3. Scope of the 1,4-Addition Reaction

The 1,4-addition reaction of organoboranes to methyl vinyl ketone and acrolein is capable of accommodating a wide range of structures in the alkyl group, from acyclic, to cyclic, to bicyclic. Indeed, cyclic organoboranes, available from the hydroboration of dienes, readily participated in the reaction, providing a simple route to the synthesis of ω-hydroxy aldehydes and ketones.[21]

[20] H. C. Brown and E. Negishi, *J. Amer. Chem. Soc.*, **93**, 3777 (1971).

[21] A. Suzuki, S. Nozawa, M. Itoh, H. C. Brown, E. Negishi, and S. K. Gupta, *Chem. Commun.*, 1009 (1969).

For example, the product from the hydroboration of 1,3-butadiene contains both 1,4- and 1,3-diborabutane moieties[22] (**19.7, 19.8**). Re-

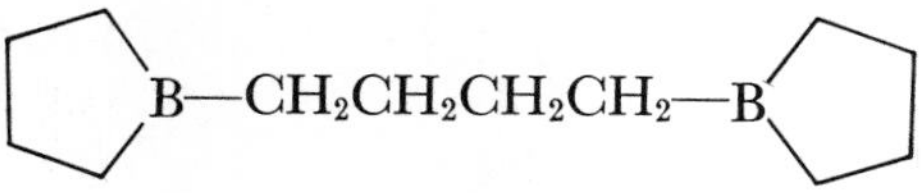

$$B—CH_2CH_2CH_2CH_2—B$$

19.7

$$B—CH_2CH_2\overset{\overset{\displaystyle CH_3}{|}}{C}H—B$$

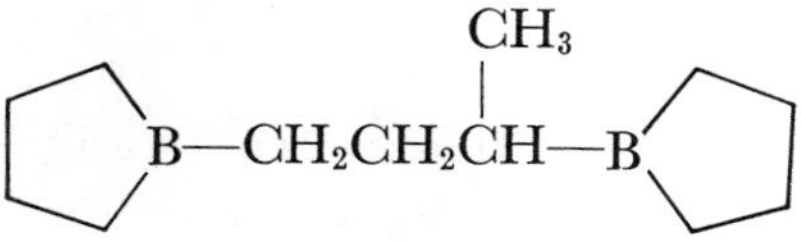

19.8

action with methyl vinyl ketone in the presence of water or alcohol, followed by oxidation with alkaline hydrogen peroxide gave an 85 percent yield of ω-hydroxy ketones containing 8-hydroxyoctan-2-one and its isomer, presumably 7-hydroxy-5-methylheptan-2-one, in a ratio of 70:30. If **19.8** is removed by thermal treatment,[22] the reaction yields the 8-hydroxyoctan-2-one predominantly, in a purity of 95 percent.

The hydroboration of 1,4-pentadiene produces a product containing both 1,5- and 1,4-diborapentane moieties.[23] Thermal treatment converts these predominantly to the species containing 1,5-dibora moieties (**19.9**). Treatment with methyl vinyl ketone, followed by the usual oxidation, yields 9-hydroxynonan-2-one[21] (19.19).

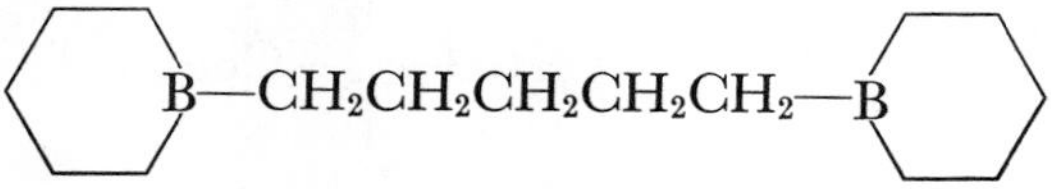

$$B—CH_2CH_2CH_2CH_2CH_2—B$$

19.9

$$B— + CH_2\!\!=\!\!CHCOCH_3 \xrightarrow[i\text{-PrOH}]{THF}$$

$$\xrightarrow{[O]} \underset{\overset{|}{OH}}{CH_2(CH_2)_6COCH_3}$$

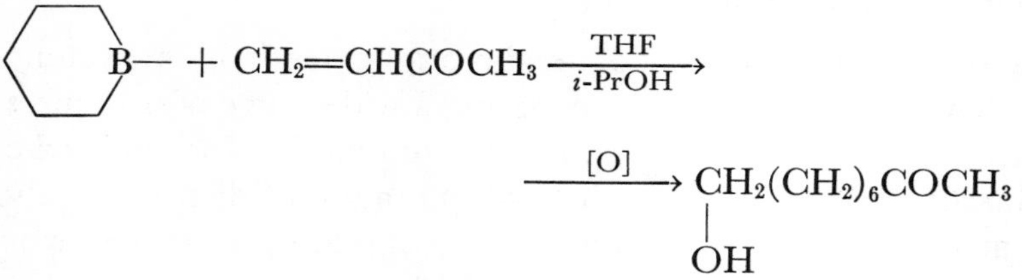

19.19

---

[22] H. C. Brown, E. Negishi, and S. K. Gupta, *J. Amer. Chem. Soc.*, **92**, 2460 (1970).
[23] Research in progress with E. Negishi and P. L. Burke.

We also undertook to explore the scope of this new synthesis with respect to the α,β-unsaturated carbonyl component. Derivatives such as acrolein and methyl vinyl ketone are often highly reactive and difficult to isolate in pure form. Organic chemists have frequently simplified the problems of working with these species by utilizing them in the form of the corresponding Mannich bases.[24,25] Accordingly, we examined the question of whether Mannich bases could be used in this synthesis.

We found that Mannich bases, derived from ketones such as cyclopentanone, cyclohexanone, and norbornanone, quaternized *in situ*, react smoothly in alkaline solution with organoboranes to produce the corresponding alkylated ketones[26] (19.20–19.22).

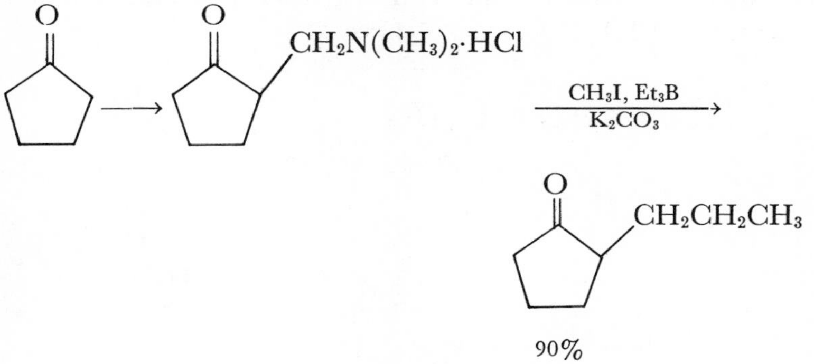

19.20

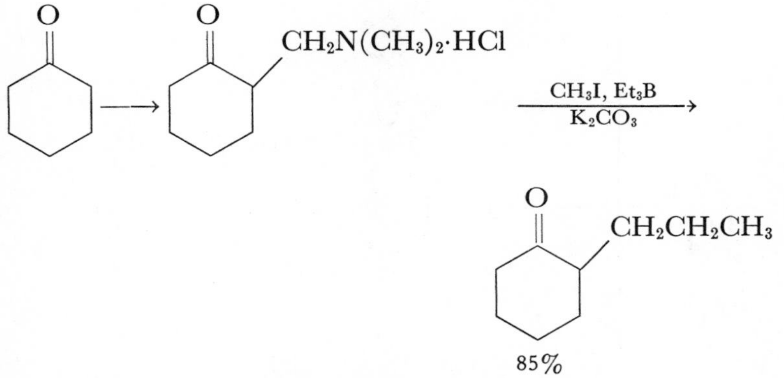

90%

19.21

85%

[24] F. F. Blicke, *Org. Reactions*, **1**, 303 (1942).

[25] J. H. Brewster and E. L. Eliel, *Org. Reactions*, **7**, 99 (1953).

[26] H. C. Brown, M. W. Rathke, G. W. Kabalka, and M. M. Rogić, *J. Amer. Chem. Soc.*, **90**, 4166 (1968).

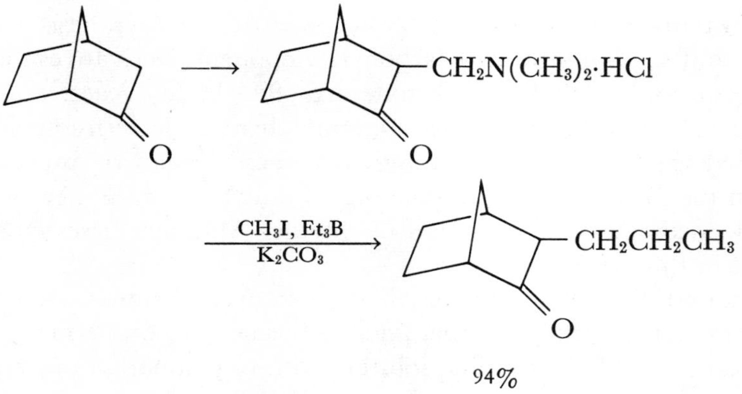

$$19.22$$

94%

A wide range of trialkylboranes can evidently be accommodated in this reaction (19.23).

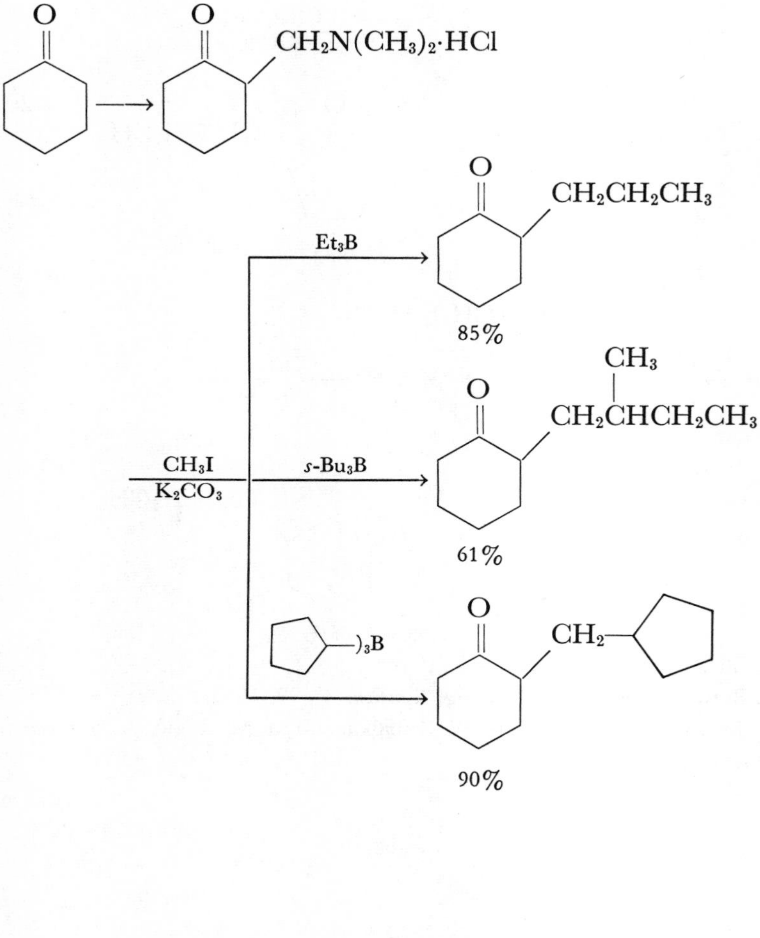

$$19.23$$

85%

61%

90%

These reactions presumably proceed through the *in situ* formation of the labile α,β-unsaturated carbonyl derivatives (**19.10–19.12**).

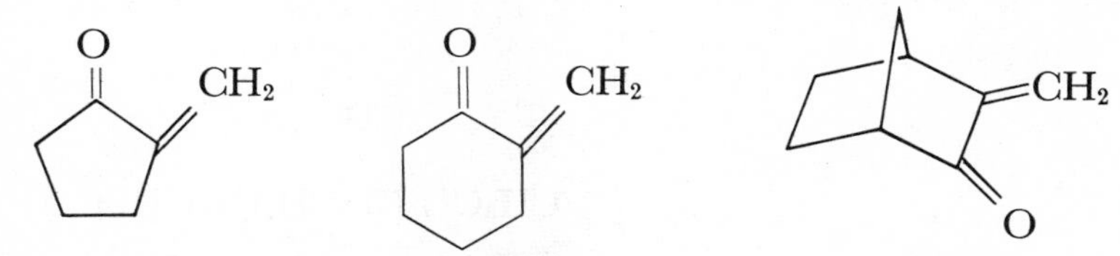

| **19.10** | **19.11** | **19.12** |

Substituents in the 2-position of acrolein do not affect the reaction adversely. Thus, both 2-methylacrolein and 2-bromoacrolein react smoothly[27] (19.24–19.27; figures are isolated yields).

$$CH_3\overset{\overset{\displaystyle CH_3}{|}}{C}HCH_2)_3B + CH_2\!\!=\!\!\overset{\overset{\displaystyle CH_3}{|}}{C}\!\!-\!\!CHO \xrightarrow[\text{H}_2\text{O, 25°}]{\text{THF}}$$

19.24

$$CH_3\overset{\overset{\displaystyle CH_3}{|}}{C}HCH_2CH_2\overset{\overset{\displaystyle CH_3}{|}}{C}HCHO$$

95%

$$+ CH_2\!\!=\!\!\overset{\overset{\displaystyle CH_3}{|}}{C}\!\!-\!\!CHO \longrightarrow$$

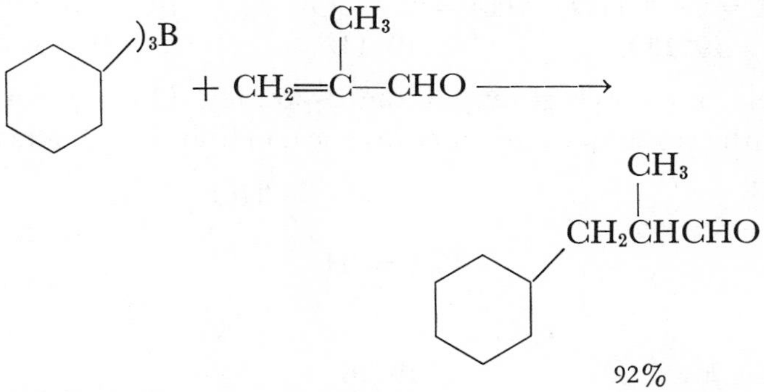

19.25

$$\overset{\overset{\displaystyle CH_3}{|}}{CH_2CHCHO}$$

92%

$$CH_3CH_2CH_2CH_2)_3B + CH_2\!\!=\!\!\overset{\overset{\displaystyle C}{|}}{\underset{\underset{\displaystyle Br}{|}}{C}}\!\!-\!\!CHO \longrightarrow$$

19.26

$$CH_3(CH_2)_4\overset{\overset{\displaystyle CHCHO}{}}{\underset{\underset{\displaystyle Br}{|}}{}}$$

85%

[27] H. C. Brown, G. W. Kabalka, M. W. Rathke, and M. M. Rogić, *J. Amer. Chem. Soc.*, **90**, 4165 (1968).

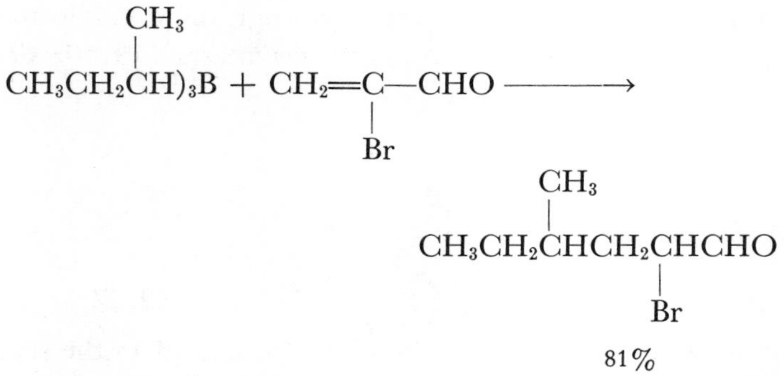

$$\underset{\substack{|\\CH_3}}{CH_3CH_2CH)_3B} + \underset{\substack{|\\Br}}{CH_2{=}C{-}CHO} \longrightarrow$$

19.27

$$\underset{\substack{|\\CH_3}}{CH_3CH_2CHCH_2CHCHO} \atop \underset{\substack{|\\Br}}{}$$

81%

This appears to provide the most favorable route now available to the α-bromoaldehydes.[27]

Neither crotonaldehyde nor 3-penten-2-one reacted with the organoborane under these conditions. This was surprising. Evidently, the introduction of a methyl group in the acrolein molecule (**19.13**) in the 1-position (**19.14**) or the 2-position (**19.15**) had no adverse effect on the ease of reaction.

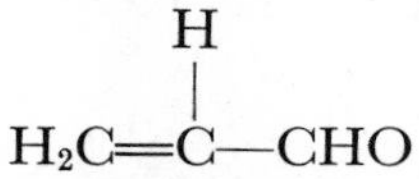

$$\underset{\substack{|\\H}}{H_2C{=}C{-}CHO} \qquad \underset{\substack{|\\H}}{H_2C{=}C{-}COCH_3} \qquad \underset{\substack{|\\CH_3}}{H_2C{=}C{-}CHO}$$
$$\textbf{19.13} \qquad\qquad\qquad \textbf{19.14} \qquad\qquad\qquad \textbf{19.15}$$

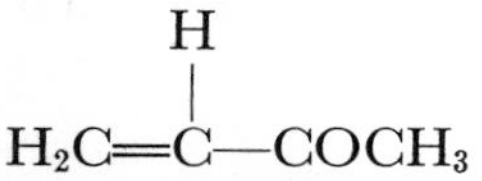

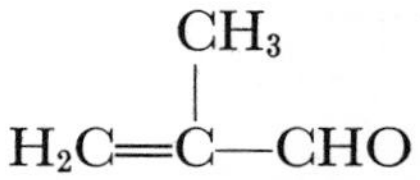

However, a methyl group in the 3-position (**19.16**) completely blocked the reaction even under drastic conditions, 24 hours at 125°.

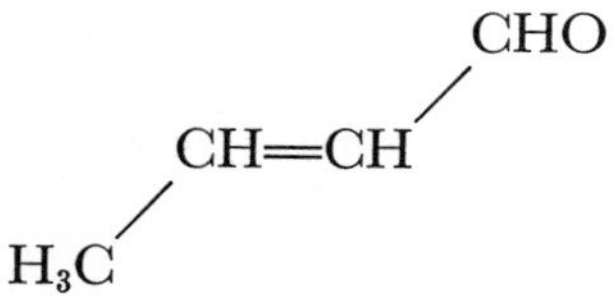

$$H_3C{-}CH{=}CH{-}CHO$$

19.16

This puzzling outcome led us to undertake an exploration of the reaction mechanism, discussed in the next section.

4. The 1,4-Addition as a Free-Radical Chain Reaction

The 1,4-addition of the Grignard reagent to α,β-unsaturated ketones is believed to proceed through a cyclic transition state.[28] As a

[28] R. E. Lutz and W. G. Reveley, *J. Amer. Chem. Soc.*, **63,** 3180 (1941).

working hypothesis, we originally assumed that the 1,4-addition of the organoboranes likewise involved such a cyclic mechanism (19.28).

$$R_3B + CH_2=CHCHO \longrightarrow \left[\begin{array}{c} R_2 \\ B \\ R \end{array} \begin{array}{c} O \\ CH \\ CH_2=CH \end{array} \right]^{\ddagger} \longrightarrow$$

19.28

$$\begin{array}{c} R_2 \\ B \longrightarrow O \\ R \qquad CH \\ CH_2-CH \end{array}$$
19.17

Indeed, in the absence of water, or other protonolyzing species, the reaction product is the enol borinate[16,17] (**19.17**). This is rapidly hydrolyzed to the product aldehyde (or ketone) by addition of water to the reaction mixture or by the carrying out of the reaction in the presence of water. The confirmed formation of the enol borinate was originally considered to be consistent with the hypothetical cyclic addition mechanism.

The failure of 2-cyclohexen-1-one (**19.18**) to react was not surprising and appeared to be consistent with this cyclic mechanism.[29] However, 1-cyclohexenyl methyl ketone (**19.19**) also failed to react, and this was difficult to rationalize in terms of the proposed cyclic mechanism.

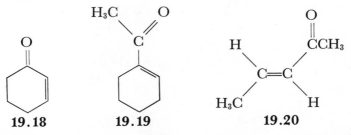

19.18   **19.19**   **19.20**

[29] G. W. Kabalka, H. C. Brown, A. Suzuki, S. Honma, A. Arase, and M. Itoh, *J. Amer. Chem. Soc.*, **92**, 710 (1970).

The failure of *trans*-crotonaldehyde (**19.16**) and the related *trans*-3-penten-2-one (**19.20**) to react was much more difficult to explain in terms of this mechanism. The difference in reactivity was not a small factor. Whereas acrolein and methyl vinyl ketone reacted with typical trialkylboranes in a matter of minutes at 25°, no significant reaction was observed for *trans*-crotonaldehyde (**19.16**) or *trans*-3-penten-2-one (**19.20**) at temperatures as high as 125° (diglyme solutions) for periods as long as 24 hours.

It did not appear possible to account for this enormous effect of the terminal (*trans*) methyl group in terms of the cyclic mechanism we were considering. Such enormous effects of terminal methyl groups are often observed in free-radical polymerizations. For example, esters of acrylic acid and methacrylic acid are readily polymerized by free-radical initiators, whereas esters of crotonic acid are relatively inert to such polymerization.[30] Consequently, we began to consider the possibility that the reaction mechanism involved free-radical intermediates.

Galvinoxyl is an efficient scavenger for free radicals.[31] As pointed out earlier (Section XIX-1), it has been successfully employed to inhibit the autoxidation of certain organoboranes.[1-3] We explored its effect on the 1,4-addition reaction. The presence of 5 mole percent of galvinoxyl effectively stopped the otherwise very fast reactions of trialkylboranes with acrolein and methyl vinyl ketone. Consequently, these reactions must be free-radical chain reactions.

The mechanism evidently involves the addition of a radical to the carbon-carbon double bond of the carbonyl compound (19.29), producing a radical intermediate (**19.21**).

$$\text{R}\cdot + \text{CH}_2{=}\text{CHCHO} \longrightarrow \text{RCH}_2\overset{\cdot}{\text{C}}\text{HCHO}$$

$$\updownarrow \qquad\qquad 19.29$$

$$\text{RCH}_2\text{CH}{=}\text{CHO}\cdot$$
$$\textbf{19.21}$$

[30] P. J. Flory, *Principles of Polymer Chemistry* (Cornell University Press, Ithaca, N.Y., 1953), p. 55.

[31] P. D. Bartlett and T. Funahashi, *J. Amer. Chem. Soc.*, **84,** 2596 (1962).

In the absence of a reactive organoborane this intermediate would add to the monomer through carbon to initiate a typical vinyl polymerization.[32] However, in the presence of free trialkylborane this intermediate (**19.21**) evidently reacts through the oxygen atom to form the enol borinate with its very strong boron-oxygen bond (19.30).

$$\mathbf{19.21} + R_3B \longrightarrow RCH_2CH{=}CHOBR_2 + R\cdot \qquad 19.30$$

This step thus resembles the reaction of alkoxy free radicals with organoboranes[12,13] discussed earlier (Section XIX-1).

The kinetic chain length for the addition of a free radical to a terminal unsubstituted grouping (**19.22**) must be much longer than for the corresponding addition to the methyl substituted grouping (**19.23**). It would appear that the reactions of organoboranes with

$$CH_2{=}\overset{|}{C}{-}\overset{|}{C}O$$
19.22

$$H_3C \diagup \overset{\displaystyle CH{=}\overset{|}{C}{-}\overset{|}{C}O}{}$$
19.23

derivatives that contain such an unsubstituted grouping (acrolein and methyl vinyl ketone) proceed readily, with a long chain length. Thus they appear to proceed spontaneously, without added catalysts. On the other hand, derivatives carrying a methyl substituent in the terminal position, such as *trans*-crotonaldehyde and *trans*-3-penten-2-one, fail to react spontaneously. Indeed, the difference in reactivity of the unsubstituted and substituted systems is illustrated by an experiment in which triethylborane was added to a typical reaction mixture containing both methyl vinyl ketone and *trans*-3-penten-2-one. The methyl vinyl ketone reacted, while *trans*-3-penten-2-one remained essentially unchanged.

If the difficulty with these terminally substituted α,β-unsaturated carbonyl compounds was the result of a relatively short chain length, it should be possible to overcome this difficulty by introducing an

---

[32] F. J. Welch, *J. Polymer Sci.*, **61**, 243 (1962).

efficient free-radical initiator. We observed that both photochemical activation and the addition of diacetyl peroxide were effective in extending the 1,4-addition reaction to these terminally substituted derivatives[33] (19.31–19.34).

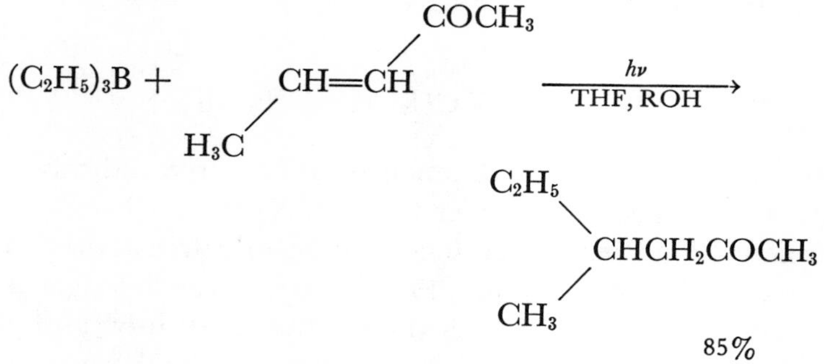

$$(C_2H_5)_3B + \underset{H_3C}{}CH{=}CH\underset{}{\overset{COCH_3}{}} \xrightarrow[\text{THF, ROH}]{h\nu}$$

19.31

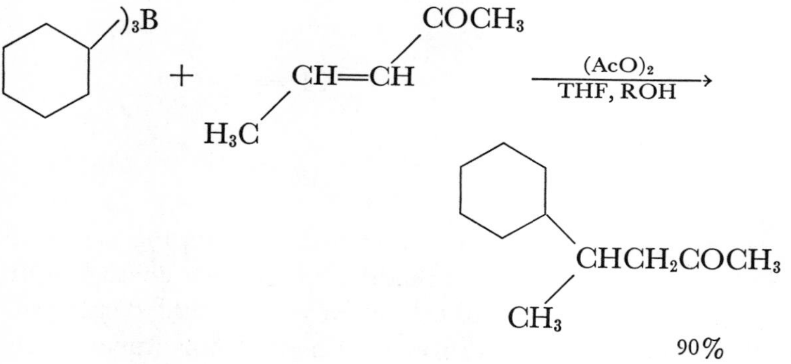

$$\underset{CH_3}{\overset{C_2H_5}{}}CHCH_2COCH_3$$

85%

19.32

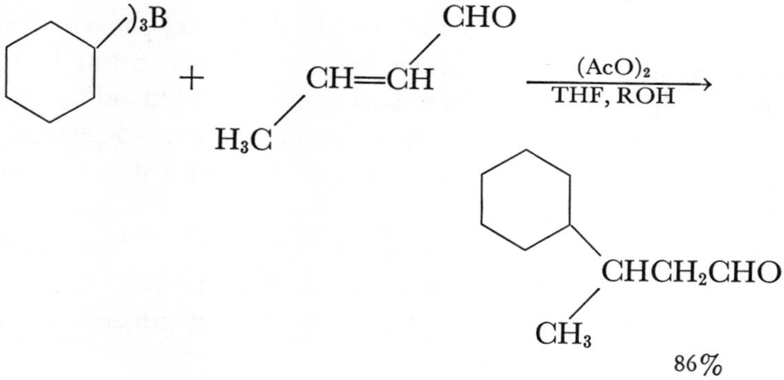

90%

19.33

86%

[33] H. C. Brown and G. W. Kabalka, *J. Amer. Chem. Soc.*, **92,** 712 (1970).

$$(C_2H_5)_3B + \quad \xrightarrow[\text{THF, ROH}]{h\nu} \quad \qquad 19.34$$

As pointed out earlier, the addition of oxygen to organoboranes produces free radicals. Consequently, it appeared that the addition of air to a mixture of the organoborane and the "inert" α,β-unsaturated carbonyl derivative might induce the desired 1,4-addition. When a slow stream of air was passed over a solution of triethylborane and *trans*-3-penten-2-one in aqueous tetrahydrofuran, a 70 percent yield of 4-methyl-2-hexanone was realized.[34]

It is essential to add the air slowly and in controlled amounts. In experiments in which oxygen was added rapidly, oxidation of the organoborane became the dominant reaction, and only a minor amount of the 1,4-addition product was realized. Even when the oxygen is added slowly, some of the organoborane is lost through oxidation. However, usually this loss is small enough to be insignificant, or a small excess of the organoborane can be added to compensate for that lost through oxidation.

The reaction appears to be one of very wide generality, accommodating a wide variety of organoboranes and "inert" α,β-unsaturated carbonyl compounds to yield the corresponding β-substituted aldehydes and ketones in excellent yields (19.35–19.38).

$$19.35$$

[34] H. C. Brown and G. W. Kabalka, *J. Amer. Chem. Soc.*, **92,** 714 (1970).

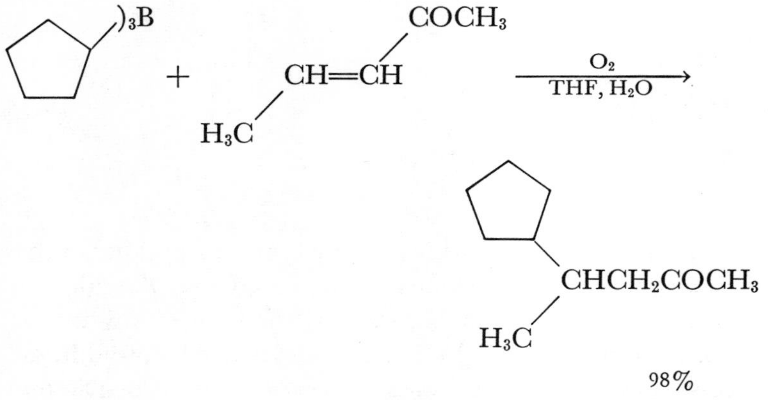

Organoboranes

19.36

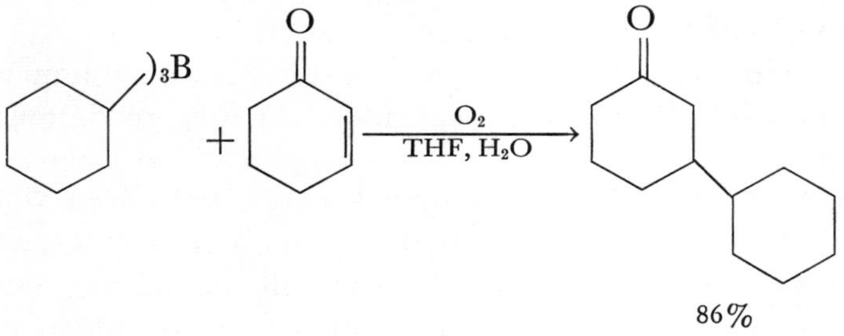

98%

19.37

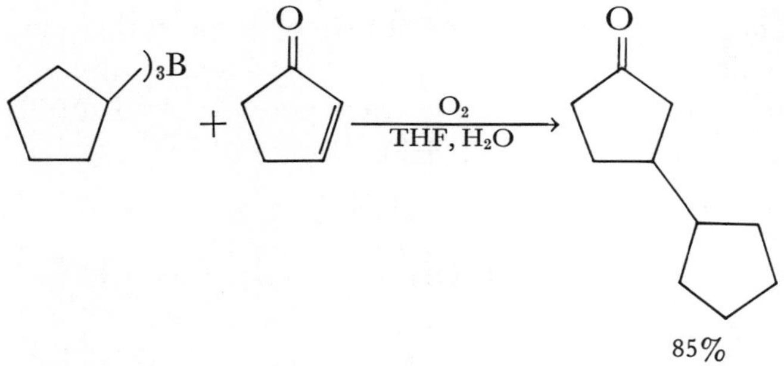

86%

19.38

85%

This development suggested that the slow introduction of oxygen or air into an organoborane produced free radicals to initiate free-radical chain reactions. In suitable systems the organoborane was also active for the chain propagation. Consequently, the organoboranes offered major promise for such free-radical chain reactions, and we undertook to explore them.

5. Extension of the 1,4-Addition Reaction to New Systems

Following our original discovery of the 1,4-addition reaction of trialkylboranes to methyl vinyl ketone[16] and acrolein[17] (Section XIX-2), we undertook to explore the possible extension of this new 1,4-addition reaction to other related systems, such as acetylacetylene, acrylonitrile, and 1,3-butadiene monoxide.[35] The results were disappointing. The fast, spontaneous 1,4-addition observed in the reactions of trialkylboranes with methyl vinyl ketone and acrolein did not occur in these related systems.

Then came the realization that these 1,4-addition reactions are free-radical chain processes (Section XIX-4). By utilizing free-radical initiators, such as small amounts of oxygen, we had been successful in extending the 1,4-addition reaction to α,β-unsaturated carbonyl derivatives, such as crotonaldehyde, 3-penten-2-one, and 2-cyclohexen-1-one, compounds which had previously failed to undergo this reaction. Consequently, we returned to explore the behavior of acetylacetylene, acrylonitrile, 1,3-butadiene monoxide, and other related systems to participate in the 1,4-addition reaction under the influence of suitable free-radical initiators. This program, in its initial stages at the time of this writing, has already achieved considerable success.

Thus, no reaction takes place between triethylborane and acetylacetylene when air and light are carefully excluded from the reaction mixture, even with an extended reaction time of 24 hours. However, a 79 percent yield of 3-hexen-2-one is obtained in 2 hours under identical reaction conditions provided the reaction mixture is exposed to a slow stream of air[36] (19.39, 19.40).

$$Et_3B + HC\equiv CCOCH_3 \xrightarrow[25°, THF]{cat. O_2} RCH=C=C\begin{smallmatrix} OBR_2 \\ \\ CH_3 \end{smallmatrix} \qquad 19.39$$

19.24

[35] Unpublished research with G. W. Kabalka.

[36] A. Suzuki, S. Nozawa, M. Itoh, H. C. Brown, G. W. Kabalka, and G. W. Holland, *J. Amer. Chem. Soc.*, **92,** 3503 (1970).

$$RCH=C=C \begin{smallmatrix} OBR_2 \\ \\ CH_3 \end{smallmatrix} + H_2O \longrightarrow \begin{smallmatrix} H \\ \\ R \end{smallmatrix} C=C \begin{smallmatrix} H \\ \\ COCH_3 \end{smallmatrix}$$

$$+ \begin{smallmatrix} R \\ \\ H \end{smallmatrix} C=C \begin{smallmatrix} H \\ \\ COCH_3 \end{smallmatrix} + R_2BOH \quad 19.40$$

In contrast to the behavior of methyl vinyl ketone,[17] the presence of water in the reaction mixture is required in order to achieve a good yield of the desired product. Presumably the allene intermediate (**19.24**) is highly susceptible to further attack by the free-radical intermediates and this side reaction is minimized by the *in situ* hydrolysis of the borinate. Solvolysis of the intermediate may also be achieved through the use of alcohols, such as methanol and ethylene glycol, especially convenient in cases where anhydrous conditions should be maintained.

The conclusion that the reaction is proceeding via a free-radical chain reaction is supported by the fact that other typical free-radical initiators may be used to induce the 1,4-addition reaction. As an example, the reaction of triethylborane can be induced photolytically or through the thermal decomposition of azobisisobutyronitrile at 40°. The yields obtained are comparable to those realized when air is used for the initiation.

The reaction appears to be one of wide generality, providing unsaturated methyl ketones from a wide variety of structural types of alkenes (19.41–19.43).

$$CH_3CH_2CH=CH_2 \longrightarrow CH_3(CH_2)_3CH=CHCOCH_3 \quad 19.41$$
$$72\%$$

$$19.42$$
$$65\%$$

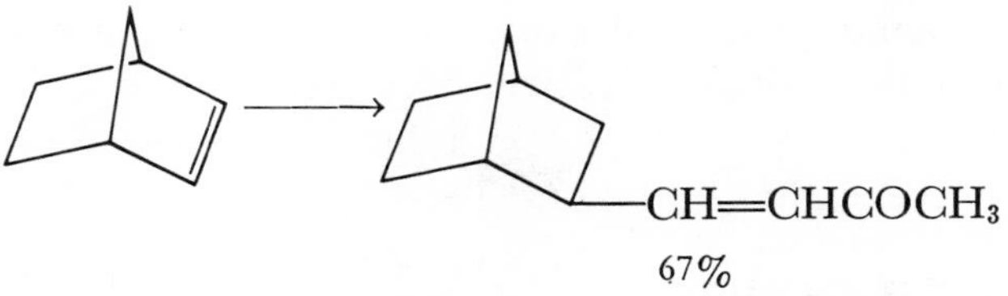

$$19.43$$

CH=CHCOCH₃ → $-CH{=}CHCOCH_3$

67%

In each case the product is a mixture of the *cis* and *trans* isomers, with the *cis* being present in larger amounts than the *trans*. The ratio of the isomers appears to vary with the conditions of the protonolysis stage and awaits further study and interpretation.

This oxygen-induced reaction of organoboranes to acetylacetylene produces α,β-unsaturated ketones. These have previously been shown to undergo the 1,4-addition reaction with organoboranes. Consequently, the reaction appears capable of being controlled to introduce two different groups from two different organoboranes, making available a wide variety of ketones (19.44).

$$HC{\equiv}CCOR \xrightarrow[H_2O]{R_3'B} \underset{R'}{\overset{H}{\diagdown}}C{=}CHCOR \xrightarrow[H_2O]{R_3''B}$$

$$\underset{R'}{\overset{R''}{\diagdown}}CHCH_2COR \quad 19.44$$

In the same way acrylonitrile fails to react with trialkylboranes in the absence of oxygen or other free-radical initiators. In the presence of oxygen, reaction readily occurs, with the triethylborane and acrylonitrile disappearing from the reaction mixture in a $1:1$ ratio[37] (19.45).

$$Et_3B + CH_2{=}CHCN \longrightarrow EtCH_2CH{=}C{=}NBEt_2 \quad 19.45$$
$$\mathbf{19.25}$$

However, the intermediate, **19.25,** evidently undergoes some other reaction prior to hydrolysis, since valeronitrile is not present in the reaction mixture.

[37] Research in progress with Dr. George W. Holland.

On the other hand, the reaction does proceed in the normal manner with a secondary alkyl derivative, tricyclohexylborane (19.46). In this case oxygen and other initiators induce a reaction resulting in the synthesis of 3-cyclohexylpropionitrile (19.46).

$$CH_2CH=C=NB(C_6H_{11})_2$$

$$\downarrow \quad H_2O \qquad 19.46$$

$$CH_2CH_2CN$$

This interesting difference in the behavior of triethylborane and tricyclohexylborane is under examination.[37] It will doubtless lead to an understanding of new interesting chemistry. Irrespective of the final interpretation, it is clear that organoboranes and acrylonitrile fail to react in the absence of free-radical initiators. The presence of such initiators evidently brings about chain reaction involving the usual 1,4-addition to the conjugated system.

No reaction takes place between 1,3-butadiene monoxide and triethylborane in the absence of oxygen. However, the reaction proceeds nicely with the introduction of small quantities of air[38] (19.47, 19.48).

$$Et_3B + CH_2=CHCH \overset{\displaystyle }{\underset{O}{\diagdown \diagup}} CH_2 \xrightarrow[25°, C_6H_6]{cat.\ O_2}$$

$$EtCH_2CH=CHCH_2OBEt_2 \quad 19.47$$

$$EtCH_2CH=CHCH_2OBEt_2 + H_2O \longrightarrow$$
$$EtCH_2CH=CHCH_2OH + Et_2BOH \quad 19.48$$
$$75\%$$

[38] A. Suzuki, N. Miyaura, M. Itoh, H. C. Brown, G. W. Holland, and E. Negishi, *J. Amer. Chem. Soc.*, **93**, 2792 (1971).

The yield was only 44 percent with equimolar amounts of the two reactants. However, use of an excess of 1,3-butadiene monoxide led to improved yields. The 2-hexen-1-ol product was predominantly the *trans* isomer, 89 percent *trans*, 11 percent *cis*.

Typical free-radical initiators, such as di-*t*-butylperoxide and azobisisobutyronitrile, as well as oxygen, are also effective in promoting the reaction. Moreover, a typical free-radical scavenger, galvinoxyl, inhibits the reaction. Thus, this reaction must involve a free-radical chain mechanism.

As was discussed earlier (Section XIX-1), trialkylboranes react with oxygen to generate alkyl radicals (19.49).

$$R_3B + O_2 \longrightarrow R\cdot + R_2BO_2\cdot \qquad 19.49$$

In the presence of 1,3-butadiene monoxide, these alkyl radicals must add to the double bond of the epoxide to yield an intermediate radical (19.50).

$$R\cdot + CH_2{=}CHCH\underset{\diagdown \; O \; \diagup}{}CH_2 \longrightarrow$$

$$RCH_2\overset{\cdot}{C}HCH\underset{\diagdown \; O \; \diagup}{}CH_2 \qquad 19.50$$

This free radical then rearranges with opening of the epoxide ring to give an alkoxy free radical (19.51).

$$RCH_2\overset{\cdot}{C}HCH\underset{\diagdown \; O \; \diagup}{}CH_2 \longrightarrow RCH_2CH{=}CHCH_2O\cdot \qquad 19.51$$

The alkoxy radical reacts with the trialkylborane to form a borinate, displacing an alkyl radical, which continues the chain (19.52).

$$RCH_2CH{=}CHCH_2O\cdot + BR_3 \longrightarrow$$
$$RCH_2CH{=}CHCH_2OBR_2 + R\cdot \qquad 19.52$$

Hydrolysis of the intermediate borinate produces the 4-alkyl-2-buten-1-ol (19.53).

$$\text{RCH}_2\text{CH}{=}\text{CHCH}_2\text{OBR}_2 + \text{H}_2\text{O} \longrightarrow$$
$$\text{RCH}_2\text{CH}{=}\text{CHCH}_2\text{OH} + \text{R}_2\text{BOH} \quad 19.53$$

The reaction gives reasonable yields with a variety of alkyl groups (19.54–19.56).

$$\text{CH}_3\text{CH}_2\text{CH}{=}\text{CH}_2 \xrightarrow[\text{H}_2\text{O, C}_6\text{H}_6]{\text{cat. O}_2}$$
$$\text{CH}_3(\text{CH}_2)_3\text{CH}_2\text{CH}{=}\text{CHCH}_2\text{OH} \quad 19.54$$
$$73\%$$

CH$_2$CH$=$CHCH$_2$OH

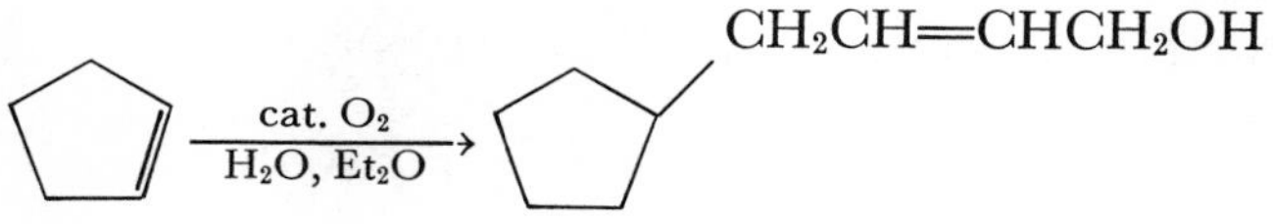

19.55

56%

CH$_2$CH$=$CHCH$_2$OH

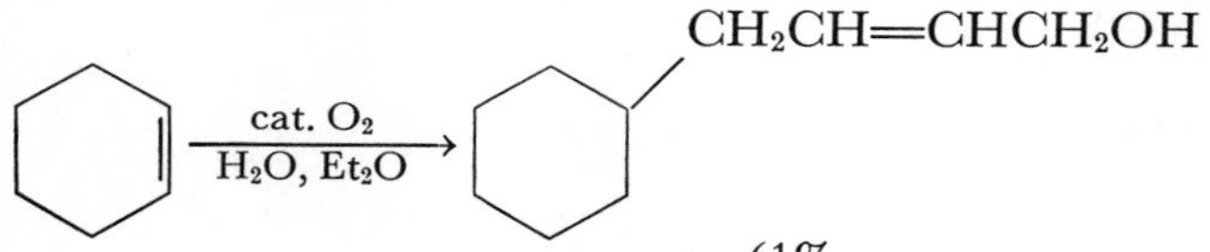

19.56

61%

Since trialkylboranes are generally prepared in tetrahydrofuran solutions, this has usually been the solvent of choice for the reactions of organoboranes. However, in the present case we observed the formation of considerable quantities of 4-(2'-tetrahydrofurfuryl)-2-buten-1-ol (**19.26**).

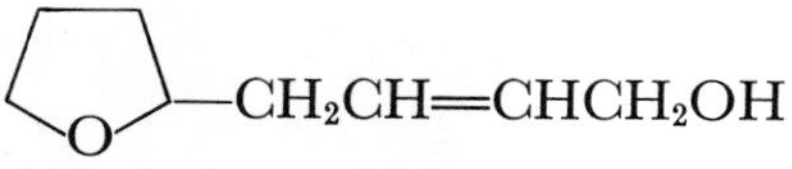

—CH$_2$CH$=$CHCH$_2$OH

19.26

Apparently the free-radical intermediates can attack the tetrahydrofuran molecule, producing a new free-radical intermediate, and this can add to the 1,3-butadiene monoxide. This side reaction can be avoided in benzene, and this solvent was preferred for the reaction of organoboranes with primary alkyl groups. For some reason, organoboranes with secondary alkyl groups (19.55, 19.56) gave slightly higher yields in ethyl ether as the solvent, in spite of a competitive reaction with solvent analogous to that observed in tetrahydrofuran (**19.26**).

Despite present limitations, which will doubtless yield to further study and understanding, this new reaction appears to be one of wide generality and provides a convenient, one-step four-carbon-atom homologation, leading to a variety of 4-alkyl-2-buten-1-ols. It nicely complements the base-induced reaction of organoboranes with ethyl 4-bromocrotonate (Section XVIII-6) that provides the 4-alkyl-3-butenoates, and thus provides a route to the isomeric 4-alkyl-3-buten-1-ols.

These initial results indicate that the free-radical addition reaction of organoboranes must be quite general—far more general than our initial explorations indicated. Thus, our studies directed to understanding why crotonaldehyde and 3-penten-2-one failed to undergo a reaction readily participated in by the very similar structures, acrolein and methyl vinyl ketone, opened the door to a wide extension of the original reaction.

6. The Initiation Step in the Autoxidation Reaction

The discovery that the autoxidation of certain organoboranes can be inhibited by galvinoxyl and N,N-dibutyldithiocarbamate led recently to the recognition that this reaction must be a chain reaction involving free radicals[1-3] (Section XIX-1). Almost concurrently we encountered the oxygen catalysis of the 1,4-addition reaction (Sections XIX-4 and XIX-5). Clearly the reaction of oxygen with organoboranes had many interesting features whose understanding might facilitate our explorations of the chemistry of organoboranes, so we undertook a systematic study of such reactions.

As reported earlier (Section XVI-7), this study was greatly facilitated by our adapting the small automatic hydrogenator[39] to the automatic generation of oxygen. We observed that 0.5 M THF solutions of various trialkylboranes undergo very rapid uptake of oxygen and that such oxidations can be readily utilized to provide essentially quantitative yields of the corresponding alcohols, provided the re-

[39] C. A. Brown and H. C. Brown, *J. Amer. Chem. Soc.*, **84**, 2829 (1962).

action was controlled to admit no more than the theoretical amount of oxygen required stoichiometrically to produce the alcohol[40] (19.57).

$$2R_3B + 3O_2 \xrightarrow{0°} R_2BO_2R + RB(O_2R)_2$$

$$\downarrow \text{NaOH, H}_2\text{O}$$

$$6ROH + 2NaB(OH)_4$$

19.57

Since the rate of oxygen absorption could be conveniently followed with the automatic gas generator, we decided to explore the effect of various added materials on the rate of uptake. To our surprise, added iodine proved to be an exceptionally effective inhibitor.[11,41,42] Indeed, it proved effective even for tri-*n*-butylborane, which had previously proven resistant to the operation of galvinoxyl.[3]

For example, exposure of a 0.5 M solution of tri-*n*-butylborane in tetrahydrofuran at 0° to oxygen at one atmosphere pressure in the automatic generator results in a rapid uptake of one mole of oxygen per mole of borane in a matter of one to two minutes. The presence of 5 mole percent of iodine in solution effectively halts any uptake of oxygen for 12.5 minutes. At that time the iodine color vanishes and the rate of oxygen absorption is approximately that observed in the absence of the inhibitor. Similar results are realized with *n*-hexane as a solvent.

The presence of a methyl substituent in the 2-position, as in tris-(2-methyl-1-pentyl)borane, results in a much longer inhibition period. Thus, as little as 1 mole percent of iodine produces an inhibition period of 32 minutes. However, the inhibition of the oxidation of *sec*-alkylboranes can be even more effective. Thus, 1 mole percent of iodine causes an inhibition period of 43 minutes in the

[40] H. C. Brown, M. M. Midland, and G. W. Kabalka, *J. Amer. Chem. Soc.*, **93,** 1024 (1971).

[41] Iodine had been previously tested in an attempt to inhibit the autoxidation of triethylborane, but was reported to be ineffective: R. L. Hansen and R. R. Hamann, *J. Phys. Chem.*, **67,** 2868 (1963).

[42] Alkyl iodides did not inhibit the rate of oxygen uptake, but caused the reaction to take a new, interesting course (Section XIX–8).

oxidation of tri-*sec*-butylborane under these conditions. The experimental results[11] are summarized in Table XIX-1.

The effectiveness of iodine as an inhibitor increases considerably with increasing initial concentrations of the halogen. Indeed, 0.5 M solutions of tri-*n*-butylborane, tri-*sec*-butylborane, and triisobutylborane in tetrahydrofuran containing 0.2 M iodine failed to reveal any significant oxygen uptake over several days.

Table XIX-1. Inhibition of the Autoxidation of Trialkylboranes by Iodine

Organoborane[a]	Iodine, mole %	Inhibition period, min[b]
Tri-*n*-butylborane[c]	5	12.5
Tris(2-methyl-1-pentyl)borane	1	32 .
Tri-*sec*-butylborane	0.5	12
	1	43
	2	150
Tricyclohexylborane	1	34
Tri-*exo*-norbornylborane	1	17

[a] All reactions were carried out in the automatic generator using 10 mmole of organoborane in 20 ml THF at 0°.

[b] Time for disappearance of the iodine color, followed by rapid uptake of oxygen.

[c] Reactions run in *n*-hexane behaved similarly.

The length of the induction periods produced by iodine suggests that the oxidation of organoboranes must involve a relatively slow initiation stage (19.58), followed by competitive reactions of the alkyl free radical with oxygen (19.59) or with iodine (19.60).

$$R_3B + O\!-\!O \longrightarrow R_2BO_2\cdot + R\cdot \qquad 19.58$$
$$R\cdot + O_2 \longrightarrow RO_2\cdot \qquad 19.59$$
$$R\cdot + I_2 \longrightarrow RI + I\cdot \qquad 19.60$$

Oxygen is a diradical. Consequently, it has been suggested that the initiation may involve displacement of an alkyl free radical from boron analogous to the known reaction of free alkoxy radicals.[12,13] Unfortunately, nothing is known at present about the properties and fate of the dialkylboraperoxy radicals, $R_2BO_2\cdot$, postulated to be produced in this initiation stage.

The peroxy free radicals, formed in the reaction of the alkyl free radicals produced in the initial stage with oxygen (19.59), react rapidly with the organoborane, displacing an alkyl free radical[12,13] (19.61) to continue the chain.

$$RO_2\cdot + BR_3 \longrightarrow RO_2BR_2 + R\cdot \qquad 19.61$$

However, the analogous reaction of iodine atoms with the organoborane must be very slow or insignificant. (This great difference in reactivity is readily accounted for in terms of the large difference in the strengths of bonds formed by iodine or by oxygen with boron.) Consequently, the chain is effectively stopped by iodine present in adequate concentrations.

At low iodine concentrations there must be a competition between the reaction of the alkyl radical with oxygen, favoring the chain pathway, and the reaction of the radical with iodine, inhibiting the chain.

At relatively high iodine concentrations, such as 40 mole percent, it is observed that no oxygen is absorbed over periods as long as several days, even under one atmosphere of 100 percent oxygen. At the same time alkyl iodide is produced at a slow, steady rate that varies considerably for different organoboranes (Figure XIX-1). The number of moles of alkyl iodide produced is roughly equal to that of molecular iodine that disappears from the reaction mixture. Even though oxygen is not absorbed during the reaction, its presence is necessary. In the absence of oxygen, the formation of alkyl iodides from the organoboranes and iodine is negligible.

These observations can be accounted for in terms of the following mechanism. Oxygen, a diradical, reacts with the organoborane to form $R_2BO_2\cdot$ and $R\cdot$ (19.58). At the high iodine concentration used, the alkyl radical $R\cdot$ fails to react with oxygen (19.59) but reacts preferentially with iodine to form alkyl iodide. Since there is no absorption of oxygen, we must postulate that iodine reacts with the dialkylboraperoxide radical to give the dialkyliodoborane and oxygen (19.62).

$$R_2BO_2\cdot + I_2 \longrightarrow R_2BI + O_2 + I\cdot \qquad 19.62$$

It follows that under conditions where no oxygen is absorbed, the rate of production of alkyl iodide should be the same as the rate of reaction of oxygen with the organoborane to displace free radicals. From the data shown in Figure XIX-1, it is possible to calculate the rates of initiation of the three tributylboranes. These give the values ($k \times 10^6$, 1 mole^{-1} min^{-1}): n-Bu$_3$B, 24.4; i-Bu$_3$B, 1.6; s-Bu$_3$B, 3.5.[43]

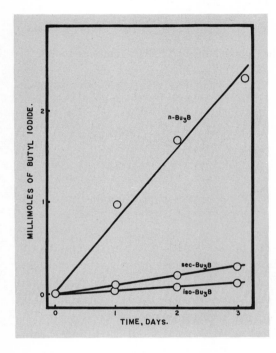

Figure XIX-1. Iodine-inhibited oxidation of the tributylboranes in tetrahydrofuran solution at 0° (10 mmole of a tributylborane and 4 mmole of iodine).

The rate of reaction of tri-*sec*-butylborane with oxygen is much faster than those of tri-*n*-butylborane and triisobutylborane.[11] Clearly, this must be the result of exceptionally favorable propagation and termination steps, rather than a favorable initiation stage.

The vast difference in the rates of initiation of the two primary derivatives, tri-*n*-butylborane and triisobutylborane, is also noteworthy. It suggests that the β-methyl substituent must play a major role in decreasing the rate of initiation. This suggests a powerful

---

[43] Using galvinoxyl as an inhibitor, the rates of initiation have been previously calculated[3] to be ($k \times 10^6$, 1 mole^{-1} min^{-1}): i-Bu$_3$B, 7; s-Bu$_3$B, 3.

437

steric influence on the ease of attack of the oxygen molecule on the organoborane as a major factor in the observed rates of initiation.[44]

This hypothesis was tested by subjecting two organoboranes, one with relatively small steric crowding about the boron atom, B-*n*-butyl-9-borabicyclo[3.3.1]nonane, and one with relatively large steric crowding, tris(3-methyl-2-butyl)borane (trisiamylborane), to autoxidation with added iodine. The results are summarized in Table XIX-2.

Table XIX–2. Inhibition of the Autoxidation of Trialkylboranes by Iodine

Organoborane[a]	Iodine, mole %	Inhibition period,[b] min
Tri-*n*-butylborane	5	12.5
B-*n*-butyl-9-BBN	5	0.4
Tri-*sec*-butylborane	0.5	12
Trisiamylborane	0.5	164

[a] 10 mmole R_3B in 20 ml THF at 0^0, 100 percent oxygen, 1 atm.
[b] Time for iodine to disappear, followed by rapid oxygen absorption.

Clearly, the bicyclic moiety of the 9-BBN structure opens up the steric environment of the boron atom. Consequently, the steric approach to the boron atom in B-*n*-butyl-9-BBN must be more open than the steric approach to the boron atom in tri-*n*-butylborane. The decrease by a factor of 30 in the inhibition period for B-*n*-butyl-9-BBN is therefore in agreement with the proposed steric interpretation. Similarly, the introduction of methyl groups in the 3-position, in going from tri-*sec*-butylborane to trisiamylborane, must also increase greatly the steric crowding in the latter structure. The increase in the length of the inhibition period by a factor of 14 is again consistent with the steric interpretation.

These results reveal that iodine is an exceptionally powerful inhibitor of the free-radical chain reactions of organoboranes. Utilization of this property of iodine has revealed that the fast reaction of iodine with organoboranes is initiated by a slow attack of oxygen

[44] H. C. Brown and M. M. Midland, *Chem. Commun.*, 699 (1971).

on the organoborane. Moreover, the rate of initiation depends strongly upon the structure of the organoborane, proceeding slower with increasing crowding about the boron atom. This suggests that the initiation step involves a direct attack of the oxygen molecule, as a diradical, upon the boron atom of the organoborane, displacing an alkyl radical.

7. Other Types of Free-Radical Chain Reactions

An essential feature of the chemistry discussed in this chapter so far has been the newly discovered facile reaction of alkoxy free radicals, RO·, with organoboranes to displace alkyl free radicals.[12,13] As pointed out earlier (Section XIX-1), this reaction apparently also occurs with ease with the related free radicals, $R_2N·$ and RS·, in which the odd electron is located on nitrogen[14] or sulfur.[15] On the other hand, bromine atoms (Section XVIII-8) and iodine atoms (Section XIX-6) do not exhibit the property of a facile displacement of alkyl free radicals from boron. Presumably, the difference in the behavior of the two groups of atoms and free radicals lies in the great difference in the strengths of the bonds formed by the individual atoms with boron. It is known that oxygen, nitrogen, and sulfur form bonds to boron of very high stability, presumably because of the great facility with which the lone pairs on oxygen, nitrogen, and sulfur can engage in resonance with the boron atom and its vacant orbital. Such resonance appears to be much less important with bromine and iodine.

These considerations suggest that there should exist a large number of chain reactions of organoboranes, involving $R_2N·$ and RS· as intermediates, analogous to the many chain reactions we have been reviewing in this chapter involving RO· and related species. Indeed, recognition of this possibility made it possible to resolve a puzzle of long standing.

It was originally observed that chloramine reacts with organoboranes in the presence of base to produce the corresponding amines[45] (19.63).

[45] H. C. Brown, W. R. Heydkamp, E. Breuer, and W. S. Murphy, *J. Amer. Chem. Soc.*, **86,** 3565 (1964).

$$R_3B + 2NH_2Cl + 3NaOH \longrightarrow$$
$$2RNH_2 + RB(OH)_3^-Na^+ + 2NaCl \quad 19.63$$

Although this reaction uses only two of the three groups of the organoborane, it provides a highly convenient stereospecific route to certain amines (Section XVI-10).

Sharefkin and Banks attempted to extend the reaction to the synthesis of the tertiary alkyldimethylamines[46] (19.64).

$$R_3B + ClNMe_2 \longrightarrow RNMe_2 + R_2BCl \quad\quad 19.64$$

However, the reaction took another course, the formation of alkyl chloride (19.65).

$$R_3B + ClNMe_2 \longrightarrow RCl + R_2BNMe_2 \quad\quad 19.65$$

It has now been shown that this last reaction, 19.65, is a free-radical chain reaction, proceeding through a free-radical intermediate, $Me_2N\cdot$[47] (19.66, 19.67).

$$Me_2N\cdot + n\text{-}Bu_3B \longrightarrow n\text{-}Bu_2BNMe_2 + n\text{-}Bu\cdot \quad 19.66$$
$$n\text{-}Bu\cdot + ClNMe_2 \longrightarrow n\text{-}BuCl + Me_2N\cdot \quad\quad 19.67$$

The addition of galvinoxyl inhibits the free-radical reaction, 19.65, and the reaction then follows the polar pathway, 19.64, leading to the formation of the tertiary amine, n-butyldimethylamine. Unfortunately, in its present state of development, it does yet not appear applicable for synthetic purposes.

Trialkylboranes either do not react, or react very slowly with disulfides, such as diphenyldisulfide and dimethyldisulfide. However, under the influence of light a rapid reaction takes place to yield the corresponding sulfides[48] (19.68, 19.69).

$$R_3B + PhSSPh \xrightarrow[65°]{h\nu} RSPh + R_2BSPh \quad\quad 19.68$$

$$R_3B + 2MeSSMe \xrightarrow[65°]{h\nu} 2RSMe + RB(SMe)_2 \quad\quad 19.69$$

[46] J. G. Sharefkin and H. D. Banks, *J. Org. Chem.*, **30**, 4313 (1965).

[47] A. G. Davies, S. C. W. Hook, and B. P. Roberts, *J. Organometal. Chem.*, **23**, C11 (1970).

[48] H. C. Brown and M. M. Midland, *J. Amer. Chem. Soc.*, **93**, 3291 (1971).

Passage of a slow stream of air into a mixture of the organoborane and dimethyldisulfide induces a facile reaction that appears to be broadly applicable, providing the corresponding sulfide in excellent yield (19.70–19.72).

$$CH_3(CH_2)_5CH{=}CH_2 \longrightarrow CH_3(CH_2)_7SCH_3 \qquad 19.70$$
$$84\%$$

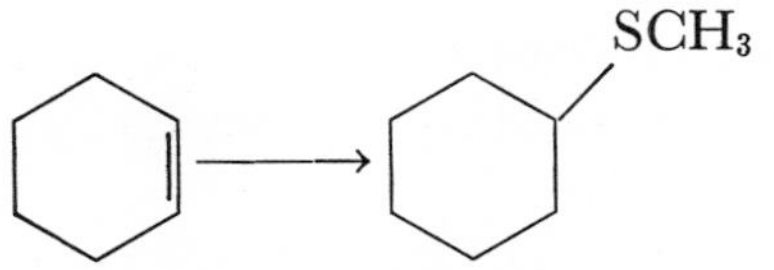

19.71

88%

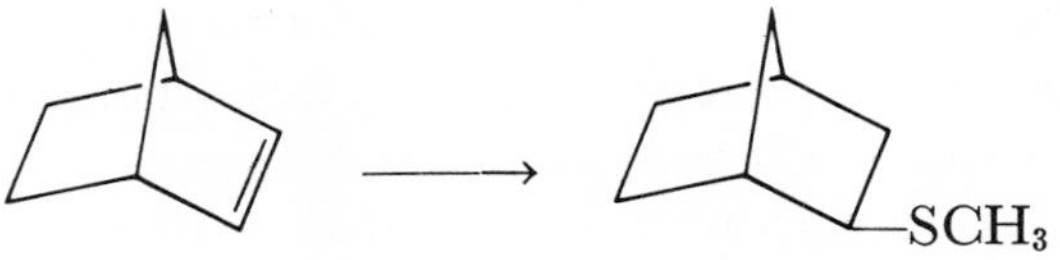

19.72

78%

The reaction must involve a free-radical chain reaction proceeding through methylthio radicals (19.73).

$$R_3B + O_2 \longrightarrow R_2BO_2{\cdot} + R{\cdot}$$
$$R{\cdot} + CH_3SSCH_3 \longrightarrow RSCH_3 + CH_3S{\cdot}$$
$$CH_3S{\cdot} + R_3B \longrightarrow CH_3SBR_2 + R{\cdot} \qquad 19.73$$
$$CH_3S{\cdot} + CH_3SBR_2 \longrightarrow (CH_3S)_2BR + R{\cdot}$$

An interesting feature is the conclusion that free alkyl radicals must attack one of the sulfur atoms in the disulfide linkages to form the corresponding sulfide, liberating an organothio radical. Such reactions of disulfides have long been accepted.[49]

We are presently only in the initial stages of the exploration of such chain reactions. However, the understanding of these systems already attained indicates that we should anticipate a large increase in the number and kinds of free-radical reactions involving organoboranes.

[49] W. A. Pryor and T. L. Pickering, *J. Amer. Chem. Soc.*, **84,** 2705 (1962).

8. The Oxygen-Induced Coupling Reaction of Alkyl Halides under the Influence of Organoboranes

As pointed out earlier (Section XIX-6), the rate of oxygen absorption by solutions of organoboranes could be conveniently followed with the automatic gas generator. Consequently, we established the rate of oxygen uptake by representative organoboranes and then began to explore the effect of various added materials on the rate of uptake of oxygen. Added iodine caused a complete inhibition of oxygen absorption, and this phenomenon was subjected to a detailed study[11] (Section XIX-6).

An examination of the effect of typical organic iodides on the rate of reaction of trialkylboranes with oxygen revealed that the rate is not significantly altered (unless significant amounts of iodine formed), but the products are very different. Thus, when 2 mmoles of typical organic iodides, such as benzyl, *p*-nitrobenzyl, or allyl iodide, iodoform, or 1,2-diiodoethane, were added to 10 mmoles of tri-*n*-butylborane in 20 ml of tetrahydrofuran and the resulting solution treated with oxygen in the automatic gas generator, *n*-butyl iodide was formed equivalent to the amount of organic iodide added. A small amount of bibenzyl was detected in the experiments with benzyl iodide.[50]

The products suggest that the *n*-Bu· radicals produced in the initiation step (19.58) can react with the organic iodide present (19.74) in competition with the usual reaction with oxygen (19.59).

$$n\text{-Bu·} + \text{RI} \longrightarrow n\text{-BuI} + \text{R·} \qquad 19.74$$

The new radicals, R·, can either react with oxygen or, under conditions of low oxygen concentration, undergo other reactions, such as coupling.

These considerations suggested that by controlling the rate and quantity of oxygen introduced, it might be possible to control the course of the reaction to produce alkyl iodide preferentially (19.74). Indeed, we discovered that by introducing air into a reaction mixture

---

[50] A. Suzuki, S. Nozawa, M. Harada, M. Itoh, H. C. Brown, and M. M. Midland, *J. Amer. Chem. Soc.*, **93**, 1508 (1971).

containing equimolar amounts of the organoborane, R_3B, and allyl iodide, it was possible to realize an essentially quantitative yield of the alkyl iodide, RI. The reaction was applied to a series of trialkylboranes with allyl iodide and to tri-n-hexylborane with a series of organic halides. The results are summarized in Table XIX-3.

Table XIX–3. The Air-Induced Reaction of Trialkylboranes with Alkyl Halides

Organoborane[a]	Alkyl halide[b]	Product	Yield,[c] %
Triethylborane	Allyl iodide	Ethyl iodide	100
Tri-n-butylborane	Allyl iodide	Butyl iodide	97 (10)
Tri-n-pentylborane	Allyl iodide	Pentyl iodide	98 (9)
Tri-n-hexylborane	Allyl iodide	Hexyl iodide	100 (11)
Tricyclohexylborane	Allyl iodide	Cyclohexyl iodide	75
Tri-n-hexylborane	Methyl iodide	Hexyl iodide	40 (20)
Tri-n-hexylborane	Ethyl iodide	Hexyl iodide	54 (13)
Tri-n-hexylborane	Isopropyl iodide	Hexyl iodide	64 (13)
Tri-n-hexylborane	t-Butyl iodide	Hexyl iodide	85 (10)[d]
Tri-n-hexylborane	Allyl chloride	Hexyl chloride	trace
Tri-n-hexylborane	Allyl bromide	Hexyl bromide	12
Tri-n-hexylborane	Benzyl bromide	Hexyl bromide	24

[a] All reactions used 10 mmole of borane in 20 ml THF with air introduced at 10 ml/min.

[b] 10 mmole.

[c] Based on alkyl halide reactant by glpc. The percent of secondary isomer of total iodide is in parentheses. The starting trialkylborane prepared from the olefin by hydroboration contained 6 percent secondary isomer.

[d] A trace of hexamethylethane was detected.

Even methyl iodide undergoes the chain transfer reaction and produces 40 percent of hexyl iodide. Allyl iodide gave quantitative results. Consequently, the reaction may be used for the preparation of alkyl iodides as an alternative to the base-induced iodination of trialkylboranes (Section XVI-6). It should be especially valuable for alkyl groups that are sensitive to the alkaline conditions of the latter synthesis.

In the reactions that have been discussed earlier in this chapter (other than oxidation), only a catalytic amount of oxygen has been required to induce the desired reaction. However, in the present reaction involving organic halides, the results reveal that approximately one mole of oxygen is required per mole of alkyl iodide formed. Thus, the reaction is not a chain reaction, but oxygen is used in roughly stoichiometric quantities to produce free radicals, which react with the organic halide introduced. The stable radicals produced after halogen abstraction, such as allyl and benzyl, do not continue the chain, but undergo coupling. The rate of reaction depends upon the rate of addition of air. The introduction of too much oxygen, however, can divert the reaction to the usual chain oxidations, so for high yields it is necessary to introduce the oxygen at such a rate that it can be utilized completely by the trialkylborane present.

With allyl iodide as the iodine donor, biallyl is produced. Initially, the molar quantity of biallyl formed was one-half that of the alkyl iodide produced in the reaction. However, in the latter stages of the reaction the amount of biallyl formed failed to increase significantly, even though additional alkyl iodide was being produced. A possible explanation might be that in the latter stages of the reaction the autoxidation of the residual organoborane (proceeding via alkyl radicals) fails to compete effectively against the allyl radicals for the oxygen dissolving in the solution.

It is clear that to obtain high yields of the coupled product, the reaction of the allyl radical (and related radicals) with oxygen must be minimized. This objective is readily achieved by increasing the amount of the organoborane over that required by the stoichiometry of the reaction. We finally adopted the readily available triethylborane as the reagent and utilized it in 100 percent excess to explore the efficacy of this new approach to biallyl and related coupled products.

The reaction was applied to a series of alkyl halides. The results are summarized in Table XIX-4.

The reaction was very clean. No detectable amount of coupling with ethyl radicals was observed. Evidently the ethyl radicals abstract iodine or react with oxygen rapidly and do not couple. On the other hand, mixed coupling products could be obtained from mixed allylic and benzylic iodides. The coupling gave a nearly sta-

Table XIX–4. Air-Induced Coupling of Benzylic and Allylic Iodides via Triethylborane

Alkyl iodide	Product	Yield,[a] %
Benzyl[b]	Bibenzyl	90 (87)
p-Nitrobenzyl[b]	4,4'-Dinitrobibenzyl	(43)
Allyl[b,c]	Biallyl	97
2-Methylallyl[b]	2,5-Dimethyl-1,5-hexadiene	90
Benzyl and allyl, 1:1[d]	4-Phenyl-1-butene	41
Benzyl and allyl, 1:2[d]	4-Phenyl-1-butene	60
Benzyl and allyl, 1:3[d]	4-Phenyl-1-butene	64
Benzyl and allyl, 1:4[d]	4-Phenyl-1-butene	72

[a] By glpc (isolated yields in parentheses).

[b] The reactions used 10 mmole alkyl iodide, 20 mmole triethylborane in 20 ml THF. Air was introduced at a rate of 50 ml/min, the complete reaction requiring approximately 45 min.

[c] The solvent was diethyl ether.

[d] Benzyl iodide, 5 mmole, was used in 20 ml THF. Triethylborane was used in a 2:1 molar ratio to total alkyl iodide.

tistical distribution of products. If one wanted to couple an expensive or rare iodide with a more available iodide, an excess of the latter could be used to increase the conversion of the more valuable material. Thus a 72 percent yield of 4-phenyl-1-butene was obtained when allyl iodide was used in a four-fold excess.

The free-radical coupling of olefins may be readily achieved by hydroboration followed by treatment of the organoborane with alkaline silver nitrate (Section XVI-13). The air-induced coupling of allylic and benzylic iodides now provides a mild method of coupling alkyl groups not available via hydroboration. Both methods may accommodate a wide variety of functional groups.

It is evident that the organoboranes constitute a versatile new source of free radicals and that these reactions can be readily controlled in many instances to give very clean, synthetically useful procedures. In many of these reactions oxygen is required in minor amounts to induce the desired free-radical chain reaction. Perhaps the most important aspect of the present development, the oxygen-induced reaction involving organic halides, is the recognition that in other cases it may be desirable to utilize oxygen in approximately equimolar amounts with organoboranes to induce a coupling reaction involving free radicals as intermediates in a nonchain process.

9. Conclusion

The reaction of oxygen with organoboranes was first studied in 1938.[5] Yet not until 1969 was it finally recognized that the reaction must involve a facile chain process proceeding through free radicals.[2,3] The 1,4-addition reaction of organoboranes with methyl vinyl ketone[16] and acrolein[17] was first reported in 1967. Recognition that these reactions also proceed through free-radical chain reactions was reported in 1970.[29,33,34]

At the time of this writing (February 1971) less than two years have passed since the effectiveness of the organoboranes as a source of free radicals was recognized. Much progress has been made in this short period, and doubtless much further progress will be made. Clearly, the results refute the pessimists who argue that little in the way of new exciting chemistry awaits discovery.

EPILOGUE

XX. In Retrospect

As I told at the start, I received the B.S. degree in 1936 from the University of Chicago and immediately undertook graduate work there for the M.S. and Ph.D. degrees. It may make the background more vivid for the younger reader to indicate some of the great changes that have taken place in American chemistry since that time. In 1936 the Department of Chemistry at Chicago had just one essential requirement for the M.S. degree for majors in organic chemistry —a course in macro combustions! This I took, but never had occasion to use, since microanalytical methods were already becoming generally accepted and utilized.

There have been other changes in chemical education. Unfortunately, in my opinion, not all have served the student and the progress of chemistry. For example, our courses at that time were much less hurried. At every level an effort was made to teach understanding, and the subject matter was not presented faster than a majority of the students could thoroughly assimilate it. Today, in many courses the material is presented too fast for understanding; instead, the instructor is content if the students learn the vocabulary and can regurgitate set replies. Chemistry is often taught now as a language rather than as a science.

The pace was leisurely even in our graduate courses. There was time for our lecturers to tell us stories of their experiences as graduate students and postdoctoral fellows. In this way, I learned how others had encountered experimental difficulties and had surmounted them. Thus, when I came to Wayne University and found no equipment, I did not despair. Instead, I recalled the experiences of Professor M. S. Kharasch, who had come to the University of Maryland in 1920 and found neither equipment nor available laboratory space. He used the time to institute detailed correlations based on heats of combustion and the electronegativities of organic radicals.

In recent years the pace has quickened enormously. We give fewer and fewer lecture courses, but direct the students to read more and more current literature. It is desirable, certainly, that students become acquainted with the current literature. At the same time, however, we have effectively divorced them from a genuine acquaintance with the past. They no longer gain a background appreciation of how previous workers solved experimental or other difficulties facing them, or how individual research programs were initiated.

Consequently, I decided to adopt a new approach for this book as an experiment. In describing the individual research programs that I and my associates have followed in the past thirty-five years, I would not restrict myself solely to describing the scientific data and the final conclusions. Instead, I would attempt to present the background, the influences and forces which led to the initiation of a particular research program and the considerations which caused it to take the course it did.

It is my hope that this book will give the reader some historical perspective, and some vicarious experience of the past, of the sort that helped the previous generation so much in facing their own problems.

Before closing, I would like to make one additional point. Pessimists among us preach that we have reached the summit in science and are now on the downgrade. They claim that the major developments have all been made and little remains to be discovered. I firmly disagree. Organoborane chemistry, only a few years old, is clearly a new continent awaiting further exploration, development, and exploitation. My associates and I have been engaged in the initial explorations. There has been little systematic application, as yet, to the problems of synthesis of complex structures, for pharmaceuticals or natural products. There has been no work, as yet, in exploring the possible utilization of this chemistry for industrial applications. (Indeed, industrial organizations appear to be largely unaware of what has been happening in this area.) Almost no work, as yet, has been done on the physical organic aspects of these new reactions.

Moreover, there are doubtless additional new continents around us awaiting discovery. They will not be discovered by pessimists, but by optimists, exploring with enthusiasm and hope. As I have said before:

"Tall oaks from little acorns grow."[1]

[1] H. C. Brown, *Hydroboration* (W. A. Benjamin, Inc., New York, 1962), p. 279.

INDEX

Index

Index

**Boranes in
Organic Chemistry**

Designed by R. E. Rosenbaum.
Composed by Kingsport Press, Inc.,
in 11 point Monotype Baskerville, 2 points leaded,
with display lines in Optima Semi-Bold.
Printed offset by Vail-Ballou Press
on P&S Smooth Offset, 60 pound basis.
Bound by Vail-Ballou Press
in Columbia Riverside Linen
and stamped in All Purpose foils.

Library of Congress Cataloging in Publication Data
(For library cataloging purposes only)

Brown, Herbert Charles, date.
  Boranes in organic chemistry.

  (The George Fisher Baker non-resident lectureship in
chemistry at Cornell University)
    1. Borane.   2. Chemistry, Organic—Synthesis.
I. Title.   II. Series.
QD181.B1B73   547'.05   79–165516
ISBN 0–8014–0681–1

**Boranes in
Organic Chemistry**

Designed by R. E. Rosenbaum.
Composed by Kingsport Press, Inc.,
in 11 point Monotype Baskerville, 2 points leaded,
with display lines in Optima Semi-Bold.
Printed offset by Vail-Ballou Press
on P&S Smooth Offset, 60 pound basis.
Bound by Vail-Ballou Press
in Columbia Riverside Linen
and stamped in All Purpose foils.

Library of Congress Cataloging in Publication Data
(For library cataloging purposes only)

Brown, Herbert Charles, date.
  Boranes in organic chemistry.

  (The George Fisher Baker non-resident lectureship in
chemistry at Cornell University)
  1. Borane. 2. Chemistry, Organic—Synthesis.
I. Title. II. Series.
QD181.B1B73   547'.05   79-165516
ISBN 0-8014-0681-1